IMPACT
MATHEMATICS

Algebra and More

Course 3

Developed by
Education Development Center, Inc.

Principal Investigator: Faye Nisonoff Ruopp

Senior Project Director: Cynthia J. Orrell

Senior Curriculum Developers: Michelle Manes, Susan Janssen, Sydney Foster, Daniel Lynn Watt, Nina Arshavsky, Ricky Carter, Joan Lukas

Curriculum Developers: Phil Lewis, Debbie Winkler

Special Contributors: E. Paul Goldenberg, Charles Lovitt

New York, New York   Columbus, Ohio   Chicago, Illinois   Peoria, Illinois   Woodland Hills, California

The McGraw-Hill Companies

The algebra content for *Impact Mathematics* was adapted from the series, *Access to Algebra*, by Neville Grace, Jayne Johnston, Barry Kissane, Ian Lowe, and Sue Willis. Permission to adapt this material was obtained from the publisher, Curriculum Corporation of Level 5, 2 Lonsdale Street, Melbourne, Australia.

Send all inquiries to:
Glencoe/McGraw-Hill
8787 Orion Place
Columbus, OH 43240-4027

ISBN 0-07-860929-1

3 4 5 6 7 8 9 10   079/055   14 13 12 11 10 09 08 07 06 05

Impact Mathematics Project Reviewers

Education Development Center appreciates all the feedback from the curriculum specialists and teachers who participated in review and testing.

Special thanks to:

Peter Braunfeld
Professor of Mathematics Emeritus
University of Illinois

Sherry L. Meier
Assistant Professor of Mathematics
Illinois State University

Judith Roitman
Professor of Mathematics
University of Kansas

....

Marcie Abramson
Thurston Middle School
Boston, Massachusetts

Alan Dallman
Amherst Middle School
Amherst, Massachusetts

Steven J. Fox
Bendle Middle School
Burton, Michigan

Denise Airola
Fayetteville Public Schools
Fayetteville, Arizona

Sharon DeCarlo
Sudbury Public Schools
Sudbury, Massachusetts

Kenneth L. Goodwin Jr.
Middletown Middle School
Middletown, Delaware

Chadley Anderson
Syracuse Junior High School
Syracuse, Utah

David P. DeLeon
Preston Area School
Lakewood, Pennsylvania

Fred E. Gross
Sudbury Public Schools
Sudbury, Massachusetts

Jeanne A. Arnold
Mead Junior High
Elk Grove Village, Illinois

Jacob J. Dick
Cedar Grove School
Cedar Grove, Wisconsin

Penny Hauben
Murray Avenue School
Huntingdon, Pennsylvania

Joanne J. Astin
Lincoln Middle School
Forrest City, Arkansas

Sharon Ann Dudek
Holabird Middle School
Baltimore, Maryland

Jean Hawkins
James River Day School
Lynchburg, Virginia

Jack Beard
Urbana Junior High
Urbana, Ohio

Cheryl Elisara
Centennial Middle School
Spokane, Washington

Robert Kalac
Butler Junior High
Frombell, Pennsylvania

Chad Cluver
Maroa-Forsyth Junior High
Maroa, Illinois

Patricia Elsroth
Wayne Highlands Middle School
Honesdale, Pennsylvania

Robin S. Kalder
Somers High School
Somers, New York

Robert C. Bieringer
Patchogue-Medford School Dist.
Center Moriches, New York

Dianne Fink
Bell Junior High
San Diego, California

Darrin Kamps
Lucille Umbarge Elementary
Burlington, Washington

Susan Coppleman
Nathaniel H. Wixon Middle School
South Dennis, Massachusetts

Terry Fleenore
E.B. Stanley Middle School
Abingdon, Virginia

Sandra Keller
Middletown Middle School
Middletown, Delaware

Sandi Curtiss
Gateway Middle School
Everett, Washington

Kathleen Forgac
Waring School
Massachusetts

Pat King
Holmes Junior High
Davis, California

Kim Lazarus
San Diego Jewish Academy
La Jolla, California

Ophria Levant
Webber Academy
Calgary, Alberta
Canada

Mary Lundquist
Farmington High School
Farmington, Connecticut

Ellen McDonald-Knight
San Diego Unified School District
San Diego, California

Ann Miller
Castle Rock Middle School
Castle Rock, Colorado

Julie Mootz
Ecker Hill Middle School
Park City, Utah

Jeanne Nelson
New Lisbon Junior High
New Lisbon, Wisconsin

DeAnne Oakley-Wimbush
Pulaski Middle School
Chester, Pennsylvania

Tom Patterson
Ponderosa Jr. High School
Klamath Falls, Oregon

Maria Peterson
Chenery Middle School
Belmont, Massachusetts

Lonnie Pilar
Tri-County Middle School
Howard City, Michigan

Karen Pizarek
Northern Hills Middle School
Grand Rapids, Michigan

Debbie Ryan
Overbrook Cluster
Philadelphia, Pennsylvania

Sue Saunders
Abell Jr. High School
Midland, Texas

Ivy Schram
Massachusetts Department of Youth
Services
Massachusetts

Robert Segall
Windham Public Schools
Willimantic, Connecticut

Kassandra Segars
Hubert Middle School
Savannah, Georgia

Laurie Shappee
Larson Middle School
Troy, Michigan

Sandra Silver
Windham Public Schools
Willimantic, Connecticut

Karen Smith
East Middle School
Braintree, Massachusetts

Kim Spillane
Oxford Central School
Oxford, New Jersey

Carol Struchtemeyer
Lexington R-5 Schools
Lexington, Missouri

Kathy L. Terwelp
Summit Public Schools
Summit, New Jersey

Laura Sosnoski Tracey
Somerville, Massachusetts

Marcia Uhls
Truesdale Middle School
Wichita, Kansas

Vendula Vogel
Westridge School for Girls
Pasadena, California

Judith A. Webber
Grand Blanc Middle School
Grand Blanc, Michigan

Sandy Weishaar
Woodland Junior High
Fayetteville, Arkansas

Tamara L. Weiss
Forest Hills Middle School
Forest Hills, Michigan

Kerrin Wertz
Haverford Middle School
Havertown, Pennsylvania

Anthony Williams
Jackie Robinson Middle School
Brooklyn, New York

Deborah Winkler
The Baker School
Brookline, Massachusetts

Lucy Zizka
Best Middle School
Ferndale, Michigan

CONTENTS

Chapter Three

Exponents and Exponential Variation ... 144

Chapter Four

Solving Equations 212

Chapter Seven

Solving Quadratic Equations 430

Chapter Eight

Functions and Their Graphs 486

Chapter Nine

Chapter Ten

CHAPTER 1

Linear Relationships

Real-Life Math

It's Only Natural! Linear relationships can be found in a variety of situations in nature. One of the foremost Renaissance artists, Leonardo da Vinci, believed that in the perfect body, the parts should be related by certain ratios. For instance, the length of the arm should be 3 times the length of the hand, and the length of the foot should be 6 times the length of the big toe. These relationships can be expressed by the linear equations $a = 3h$, where a is arm length and h is hand length, and $f = 6t$, where f is foot length and t is the length of the big toe.

Think About It Use a tape measure to find the lengths of your arm and hand. What is the ratio r of these two lengths? Write an equation that expresses the relationship between your hand length h and arm length a.

Family Letter

Dear Student and Family Members,

Our class is about to begin an exciting year of mathematics. Our first chapter is about *linear relationships*. These are relationships in which one amount, or variable, changes at a constant rate as another variable changes.

For example, Lara earns $8 per hour. This is a linear relationship between the variable *hours worked (H)* and the variable *dollars earned (D)*: for every hour Lara works, her earnings increase by $8. Using algebra, this relationship can be expressed as $D = 8 \times H$, or $D = 8H$.

We will learn how to recognize and work with linear relationships represented as tables, algebraic rules, and graphs. The graph of any linear relationship is a straight line, and just looking at the graph can tell you a lot about the relationship.

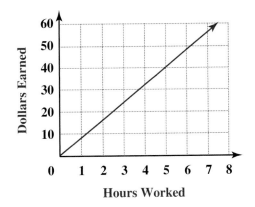

Hours	Dollars Earned
0	0
1	8
2	16
3	24
4	32

Many commonplace situations are linear relationships: the total cost for some number of CDs when the price is $15 per CD, the number of miles traveled if one drives at 40 miles per hour. Sometimes a relationship may not be exactly linear mathematically, but close enough to use a linear model to make predictions and estimates.

Vocabulary Along the way, we'll be learning about these new vocabulary terms:

coefficient   slope-intercept form   linear relationship

slope   direct variation

directly proportional   *y*-intercept

What can you do at home?

You and your student might enjoy looking for such relationships day to day and making some estimates or predictions based on them. If you can type 40 words per minute, about how long will it take to input a book report? How long will you have to baby-sit to earn $100 if you are paid $4 an hour? You might even find linear relationships a big help to your student in planning and saving!

1.1 Direct Variation

Lara earns $8 per hour in her after school job. She drew this graph to show the relationship between the number of hours she works and the number of dollars she earns.

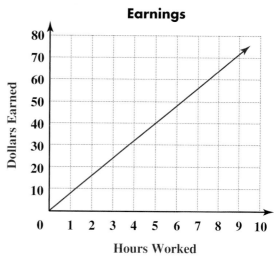

Earnings

Algebra is a useful tool for investigating relationships among *variables,* or quantities that vary. Lara's graph shows that there is a relationship between the variable *hours worked* and the variable *dollars earned.*

The graph of this relationship is a line. A straight-line graph indicates a *constant* rate of change. In the graph, each time the number of hours increases by 1, the number of dollars earned increases by 8. Relationships with straight-line graphs are called **linear relationships.**

VOCABULARY
linear relationship

In this chapter, you will explore graphs, tables, and equations for linear relationships. You will start by making a "human graph."

Explore

Select a team of nine students to make the first graph. The team should follow these rules:

• Line up along the *x*-axis. One student should stand on ⁻4, another on ⁻3, and so on, up to 4.

• Multiply the number you are standing on by 2.

- When your teacher says "Go!" walk forward or backward to the y value equal to the result you found in the previous step.

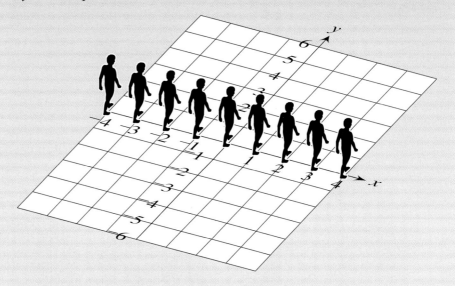

Describe the resulting "graph."

With the students on the first team staying where they are, select another team of nine students. The second team should follow these rules:

- Line up along the x-axis. One student should stand on ⁻4, another on ⁻3, and so on, up to 4.

- Multiply the value you are standing on by 2, and then add 3.

- When your teacher says "Go!" walk forward or backward to the y value equal to the result you found in the previous step. You may have to go around someone from the first team.

Are both graphs linear?

Does either graph pass through the origin? If so, which one?

Write an equation for each graph.

Explain why the two graphs will never intersect.

Investigation 1 ▶ Direct Linear Variation

The two human graphs you created illustrate two types of linear relationships, which you will investigate in the following problem sets.

Problem Set A

One weekend, Mikayla delivered pamphlets explaining her town's new recycling program.

On Saturday I delivered 350 pamphlets.

On Sunday I delivered even more—150 pamphlets an hour.

1. Copy and complete the table to show the number of pamphlets Mikayla delivered as she worked through the day on Sunday.

Hours Worked on Sunday, h	0	1	2	3	4	5	6
Sunday Deliveries, s	0	150					
Total Deliveries, t	350						

2. Look at your completed table.

    a. As the number of hours Mikayla worked doubled from 1 to 2, did the number of Sunday deliveries also double? As the number of hours worked doubled from 2 to 4, did the number of Sunday deliveries also double?

    b. As the number of hours worked doubled from 1 to 2, did the total number of deliveries also double? As the number of hours worked doubled from 2 to 4, did the total number of deliveries also double?

    c. As the number of hours worked tripled from 1 to 3, did the number of Sunday deliveries also triple? As the number of hours worked tripled from 2 to 6, did the number of Sunday deliveries also triple?

    d. As the number of hours worked tripled from 1 to 3, did the total number of deliveries also triple? As the number of hours worked tripled from 2 to 6, did the total number of deliveries also triple?

3. Look at the first two rows of your table. Write an equation describing the relationship between the number of Sunday deliveries, *s*, and the number of hours worked on Sunday, *h*.

4. Look at the first and third rows of your table. Write an equation describing the relationship between the total number of deliveries, *t*, and the number of hours worked on Sunday, *h*.

VOCABULARY
**directly
proportional**

Think about how the number of hours Mikayla worked and the number of Sunday deliveries are related. When you multiply the value of one variable by a quantity such as 2, 30, or 150, the value of the other variable is multiplied by the same quantity. That means the number of pamphlets delivered is **directly proportional** to the number of hours worked.

Another way to say this is that the ratio of Sunday deliveries to hours worked is constant:

$$\frac{\text{Sunday deliveries}}{\text{hours}} = \frac{s}{h} = \frac{150}{1} = \frac{300}{2} = \frac{450}{3} = 150$$

VOCABULARY
direct variation

A linear relationship in which two variables are directly proportional is a **direct variation.** The equation for any direct variation can be written in the form $y = mx$, where *x* and *y* are variables and *m* is a constant.

Not all linear relationships are direct variations. For example, although the relationship between the number of hours Mikayla worked and the total number of deliveries is linear, it is *not* a direct variation.

You will now examine graphs of the relationships related to Mikayla's pamphlet deliveries.

Problem Set B

1. On a grid like the one below, graph the equation you wrote in Problem Set A showing the relationship between the number of hours Mikayla worked and the number of Sunday deliveries. Label the graph with its equation.

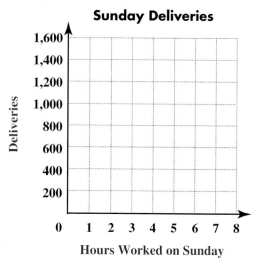

Sunday Deliveries

2. On the same grid, graph the equation you wrote showing the relationship between the number of hours Mikayla worked and the total number of deliveries.

3. How are the graphs similar? How are they different?

4. Think about the situation each graph represents.

   a. Explain why the graph for Sunday deliveries passes through the origin.

   b. Why doesn't the graph for the total number of deliveries pass through the origin?

5. Explain why the graphs will never intersect.

6. The equation for the number of pamphlets delivered on Sunday is $s = 150h$.

   a. What does 150 represent in the situation?

   b. How would changing 150 to 100 affect the graph?

   c. How would changing 150 to 200 affect the graph?

7. Must the graph of a direct variation pass through the origin? Explain.

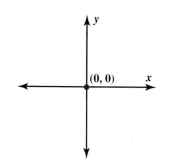

Remember
The *origin* is the point (0, 0).

Problem Set C

Remember

A relationship is a *direct variation* if when the value of one variable is multiplied by a quantity, the value of the other variable is multiplied by the same quantity.

These graphs show the relationship between the number of pamphlets delivered and the number of hours worked for five students.

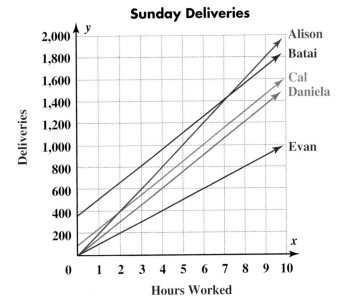

Sunday Deliveries

1. Whose graphs show the same delivery rate? Explain how you know.

2. Whose graphs show a direct variation? Explain how you know.

3. Whose graphs show a relationship that is not a direct variation? Explain how you know.

Share & Summarize

1. Describe in words two more linear situations, one that is a direct variation and one that is not.

2. How will the graphs for the two relationships differ?

3. How will the equations for the two relationships differ?

Investigation 2 ▶ Decreasing Linear Relationships

In Investigation 1, you looked at how the number of pamphlets Mikayla delivered increased with each hour she worked. You will now examine this situation by thinking about it in another way:

Mikayla started with a stack of pamphlets to deliver. For every hour she worked, the number of pamphlets in her stack decreased.

MATERIALS

graph paper

Problem Set D

Suppose Mikayla started with 1,000 pamphlets. On Saturday she delivered 350 of them. On Sunday she delivered the remaining pamphlets, at a constant rate of 150 an hour.

1. How many pamphlets did Mikayla have left to deliver when she began work on Sunday?

2. Copy and complete the table to show the number of pamphlets Mikayla has left to deliver after each hour of work on Sunday.

Hours Worked on Sunday, h	0	1	2	3	4
Pamphlets Remaining, r					

3. Write an equation to describe the relationship between the number of pamphlets remaining, r, and the number of hours worked on Sunday, h.

4. After how many hours did Mikayla run out of pamphlets? Explain how you found your answer.

5. Draw a graph of your equation. Is the relationship linear?

6. How is your graph different from the graphs you made in Investigation 1? What about the situation causes the difference?

7. Is the number of pamphlets remaining directly proportional to the number of hours Mikayla worked on Sunday? That is, is the relationship a direct variation? Explain how you know.

8. Consider the equation you wrote in Problem 3.

   a. In your equation, you should have added ⁻150*h* or subtracted 150*h*. What does the negative symbol or the minus sign before 150*h* indicate about the situation? How does it affect the graph?

   b. Your equation should also have the number 650 in it. What does 650 indicate about the situation? How does it affect the graph?

9. Lucita delivered pamphlets more slowly than Mikayla. She started with 1,000 pamphlets, delivered 200 on Saturday, and then delivered 100 pamphlets per hour on Sunday. Write an equation for the relationship between the number of hours Lucita worked on Sunday and the number of pamphlets she had left.

You will use what you have learned about decreasing linear relationships as you work on the next problem set.

Problem Set E

1. Invent a situation involving a decreasing linear relationship. Your situation should not be a direct variation.

   a. Describe your situation in words.

   b. Describe your situation with a table.

   c. Describe your situation with an equation.

   d. Describe your situation with a graph.

2. Challenge Invent a situation involving a decreasing linear relationship that *is* a direct variation.

Share & Summarize

Consider these six graphs.

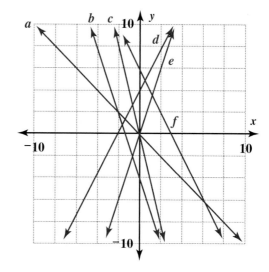

Sort the graphs into groups in at least two ways. Explain the criteria for each of your groups.

Investigation 3 ▶ Recognizing Direct Variation

In this investigation, you will practice identifying direct variations by examining written descriptions, graphs, tables, and equations.

Think & Discuss

How do you know a relationship is linear

• from its equation?

• from a table of values?

• from a description in words?

Problem Set F

Problems 1–4 each describe how the amount of money in a bank account changes over time. For each problem, do Parts a–c.

    a. Determine which of these descriptions fits the relationship:

- a direct variation
- linear but not a direct variation
- nonlinear (in other words, not linear)

    b. Explain how you decided which kind of relationship is described.

    c. If the relationship is linear, write an equation for it.

1. At the beginning of school vacation, Evan had nothing in the bank. He then started a part-time job and deposited $25 a week.

2. At the beginning of her vacation, Tamika had $150 in the bank. Each week she deposited another $25.

3. At the beginning of school, Ben had $150 in the bank that he had earned over the summer. During the first week of school he withdrew one-fifth of his savings, or $30. During the second week of school he withdrew one-fifth of the remaining $120. He continued to withdraw one-fifth of what was left in the account each week.

4. At the beginning of school, Diego had $150 in the bank. Each week he withdrew $25.

Problem Set G

For each graph, determine which of these descriptions fits the relationship, and explain how you decided:

- a direct variation
- linear but not a direct variation
- nonlinear

1.

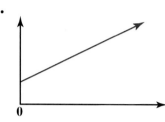

2.

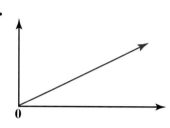

3.

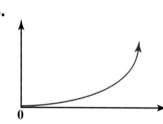

4.

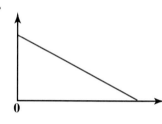

5.

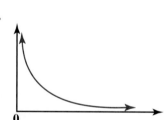

6.

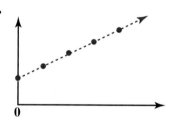

Problem Set H

For each equation, determine which of these descriptions fits the relationship, and explain how you decided:

- a direct variation
- linear but not a direct variation
- nonlinear

1. $y = 6p$

2. $y = 6p + 1$

3. $s = 5.3r$

4. $s = 5.3r - 2$

5. $t = 4s^2$

6. $y = 12 - 5p$

7. $y = \frac{5p}{2}$

8. $y = \frac{5}{2p}$

Problem Set I

For Problems 1–4, do Parts a and b.

a. Determine whether the table could describe a linear relationship. Explain how you decided.

b. If the relationship could be linear, determine whether it is a direct variation. Explain how you decided.

1.

x	1	2	3	4	5
y	3	7	11	15	19

2.

u	1	2	3	4	5
w	1	3	9	27	81

3.

p	2	4	6	8	10
q	26	46	66	86	106

4.

t	2	6	9	10	25
r	6	18	27	30	75

Share & Summarize

Copy and complete the table by telling how you can identify each type of relationship for each type of representation.

Relationship	Words	Graph	Equation	Table
nonlinear	doesn't have a constant rate of change			
direct variation		a line that passes through the origin		
linear but not a direct variation			can be represented in the form $y = mx + b$ with $b \neq 0$	

Practice & Apply

Remember

When graphing an ordered pair, put the first number on the horizontal axis and the second number on the vertical axis.

1. Carlos and Shondra were designing posters for the school play. During the first two days, they created 40 posters. By the third day, they had established a routine, and they calculated that together they would produce 20 posters an hour.

a. Make a table like the following that shows how many posters Carlos and Shondra make as they work through the third day.

Hours Worked, h								
Posters Made, p								

b. Draw a graph to represent the number of posters Carlos and Shondra will make as they work through the third day.

c. Write an equation to represent the number of posters they will make as they work through the third day.

d. Make a table to show the *total number* of posters they will have as they work through the third day.

Hours Worked, h							
Total Posters, t							

e. Draw a graph to show the total number of posters Carlos and Shondra will have as they work through the third day.

f. Write an equation that will allow you to calculate the total number of posters they will have based on the number of hours they work.

g. Explain how describing just the number of posters created the third day is different from describing the total number of posters created. Is direct variation involved? How are these differences represented in the tables, the graphs, and the equations?

2. Economics The Glitz mail order company charges $1.75 per pound for shipping and handling on customer orders.

The Lusterless mail order company charges $1.50 per pound for shipping and handling, plus a flat fee of $1.25 for all orders.

a. For each company, make a table showing the costs of shipping items of different whole-number weights from 1 to 10 pounds.

b. Write an equation for each company to help calculate how much you would pay for shipping, *C*, on an order of any weight, *W*.

c. Draw graphs of your equations, and label each with the corresponding company's name.

d. Which company offers the better deal on shipping?

e. Describe how the graphs you drew could help you answer Part d.

f. How would the Lusterless company have to change their rates to make them vary directly with the weight of a customer's order?

3. Marcus handed out advertising flyers last weekend. He distributed 400 flyers on Saturday and 200 per hour on Sunday.

a. Write an equation for the relationship between the number of hours Marcus worked on Sunday, *h*, and number of flyers he handed out on Sunday, *s*.

b. Write an equation for the relationship between the number of hours Marcus worked on Sunday, *h*, and total number of deliveries, *t*.

4. Which of these graphs represent decreasing relationships? Explain how you know.

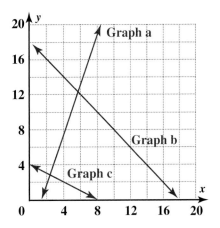

5. Which of these tables represent decreasing relationships? Explain how you know.

Table A

x	1	2	3	4	5	6	7	8	9
y	19	18	17	16	15	14	13	12	11

Table B

x	1	2	3	4	5	6	7	8	9
y	$^-2$	1	4	7	10	13	16	19	22

Table C

x	1	2	3	4	5	6	7	8	9
y	1.5	1	0.5	0	$^-0.5$	$^-1$	$^-1.5$	$^-2$	$^-2.5$

6. Which of these equations represent decreasing relationships? Explain how you know.

a. $y = {}^-x + 20$ **b.** $y = 3x - 5$ **c.** $y = {}^-\frac{1}{2}x + 2$

7. Five linear relationships for five businesses are described here in words, equations, tables, and graphs. Determine which equation, table, and graph match each description, and then tell whether or not the relationship is a direct variation. Record your answers in a table like this one.

Business	Equation Number	Table Number	Graph Number	Type of Relationship
Rent You Wrecks				
Get You There				
Internet Cafe				
Talk-a-Lot				
Walk 'em All				

Descriptions of the Businesses

- The *Rent You Wrecks* car rental agency charges $0.25 per mile plus $3.00 for a one day rental.

- The *Get You There* taxi company charges a rider $2.00 plus $0.10 per mile.

- The *Internet Cafe* charges $5.00 plus $0.30 per minute for internet access.

- The *Talk-a-Lot* phone company charges $0.75 plus $0.10 per minute for one call.

- The *Walk 'em All* pet-walking service charges $0.10 per minute to care for your dog.

Equations

i. $y = 0.3x + 5$

ii. $y = 0.25x + 3$

iii. $y = 0.1x + 2$

iv. $y = 0.1x$

v. $y = 0.1x + 0.75$

Tables

i.

x	1	2	3	4	5	6
y	0.85	0.95	1.05	1.15	1.25	1.35

ii.

x	1	2	3	4	5	6
y	2.10	2.20	2.30	2.40	2.50	2.60

iii.

x	1	2	3	4	5	6
y	3.25	3.50	3.75	4.00	4.25	4.50

iv.

x	1	2	3	4	5	6
y	5.30	5.60	5.90	6.20	6.50	6.80

v.

x	1	2	3	4	5	6
y	0.1	0.2	0.3	0.4	0.5	0.6

Graphs

i.

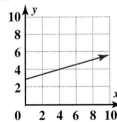

ii.

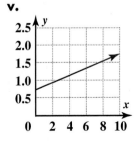

iii.

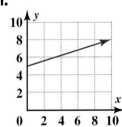

iv.

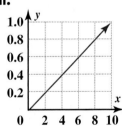

v.

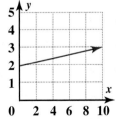

8. Kai is hiking in Haleakala Crater on the island of Maui. For the first hour he walks at a steady pace of 4 kph (kilometers per hour). He then reaches a steeper part of the trail and slows to 2 kph for the next 2 hours. Finally he reaches the top of the long climb and for the next 2 hours hikes downhill at a rate of 6 kph.

a. Make a graph of Kai's trip showing distance traveled d and hours walked h. Put distance traveled on the vertical axis.

b. Does the graph represent a linear relationship?

9. A blue plane flies across the country at a constant rate of 400 miles per hour.

a. Is the relationship between hours of flight and distance traveled linear?

b. Write an equation and sketch a graph to show the relationship between distance and hours traveled for the blue plane.

c. A smaller red plane starts off flying as fast as it can, at 400 miles per hour. As it travels it burns fuel and gets lighter. The more fuel it burns, the faster it flies. Will the relationship between hours of flight and distance traveled for the red plane be linear? Why or why not?

d. On the axes from Part b, sketch a graph of what you think the relationship between distance and hours traveled for the red plane might look like.

Haleakala Crater on the island of Maui, like all the Hawaiian islands, was created entirely from lava.

10. Three navy divers are trapped in an experimental submarine at a remote location. They radio their position to their base commander, calling for assistance and more oxygen. They can use the radio to broadcast a signal to help others find them, but their battery is running low. The base commander dispatches these three vehicles to help:

- a helicopter that can travel 45 miles per hour and is 300 miles from the sub

- an all-terrain vehicle that can travel 15 miles per hour and is 130 miles from the sub

- a boat that can travel 8 miles per hour and is 100 miles from the sub

Each vehicle is approaching from a different direction. The commander needs to keep track of which vehicle will reach the submarine next, so he can tell the sub to turn its radio antenna toward that vehicle.

a. To assist the base commander, create three graphs on one set of axes that show the distance each vehicle is from the sub over time. Put time on the horizontal axis, and label each graph with the vehicle's name.

b. Use your graphs to determine when the commander should direct the submarine to turn its antenna towards each of the following: the helicopter, the all-terrain vehicle, and the boat.

c. Write an equation for each graph that the commander could use to determine the exact distance d each vehicle is from the submarine at time h.

11. One day Lydia walked from Allentown to Brassville at a constant rate of 4 kilometers per hour. The towns are 30 kilometers apart.

a. Write an equation for the relationship between the distance Lydia traveled, d, and the hours she walked, h.

b. Graph your equation to show the relationship between hours walked and distance traveled. Put distance traveled on the vertical axis.

c. How many hours did it take Lydia to reach Brassville?

d. Now write an equation for the relationship between the hours walked, h, and the distance remaining to complete the trip, r.

e. Graph the equation you wrote for Part d on the same set of axes you used for Part b. Label the vertical axis for both d and r.

f. How can you use your graph from Part e to determine how many hours it took Lydia to reach Brassville?

12. Three cellular telephone companies have different fee plans for local calls.

    i. Talk-It-Up offers a flat rate of $50 per month. You can talk as much as you want for no extra charge.

    ii. One Thin Dime charges $0.10 for each half minute, with no flat rate.

    iii. CellBell charges $30 per month and then $0.10 per minute for all calls made.

a. For each company, write an equation that relates the cost of the phone service, *c,* to the number of minutes a customer talks during a month, *t.*

b. Any linear equation can be written in the form $y = mx + b$. Give the value of *m* and *b* for each equation you wrote in Part a.

c. For each company, make a graph that relates the cost of the phone service to the number of minutes a customer talks during a month.

d. Where do the values of *m* and *b* appear in the graph for each phone company?

13. Geometry You have studied formulas to calculate the area and perimeter of various shapes. Some of these formulas are linear, and some aren't. Tell whether the formula for each measurement below is linear or not, and explain your answer.

a. area of a circle

b. circumference of a circle

c. area of a square

d. perimeter of a square

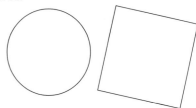

Remember

The area of a circle is the number pi, represented by π, multiplied by the radius of the circle squared.

Mixed Review

Use the distributive property to rewrite each expression without using parentheses.

14. $2a(0.5z + z^2)$

15. $ab(b^2 - 0.4a^2b)$

16. $-2c\left(\frac{2}{c^2} + c^2\right)$

17. $pq\left(\frac{1}{p^2} - \frac{q}{p}\right)$

Fill in the blanks to make true statements.

18. $3y = {}^{-}17y - \underline{\hspace{1cm}}$

19. $mn^2 - 0.2mn^2 = \underline{\hspace{1cm}}$

20. $\frac{8}{b} = \underline{\hspace{1cm}} + \frac{9}{b}$

21. $\underline{\hspace{1cm}} - \frac{a^2}{n^2} = 4$

Geometry Find the value of the variable in each drawing.

22.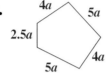

Perimeter = 61.5

23.

Area = 56.25π

24. Tell how many units in the *x* direction and how many units in the
y direction you must travel to take the shortest path from one point
to the other.

  a. Point *A* to Point *B*

  b. Point *B* to Point *C*

  c. Point *C* to Point *D*

  d. Point *D* to Point *A*

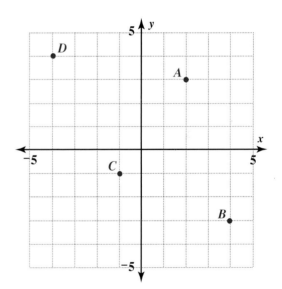

25. Graph the following points on a grid similar to the one shown above,
and connect them in the order they are given.

$(4, {}^{-}1)$ $(3, {}^{-}4)$ $({}^{-}3, {}^{-}4)$ $({}^{-}5, {}^{-}1)$ $({}^{-}2, {}^{-}1)$ $({}^{-}2, 4)$

$(1, 2.5)$ $({}^{-}1.5, 1.5)$ $({}^{-}1.5, {}^{-}1)$ $(4, {}^{-}1)$

1.2 Slope

People in many professions work with the concept of steepness. Highway engineers may need to measure the steepness of hills for a proposed highway. Architects may need to describe the steepness of a roof or a set of stairs. Manufacturers of ladders may need to test the stability of a ladder as it relates to its steepness as it leans against a wall.

Think & Discuss

Think about one of the situations described above: the steepness of the roof of a house, which is called the roof's *pitch*. Suppose you want to precisely describe the steepness of each of these three roofs.

- Would it help to just measure the length of the roof from the peak down to one edge of the roof? Explain.

- Can you think of another way you might measure steepness?

Investigation Describing Slope

In this investigation, you will explore a common way to describe steepness.

MATERIALS

metric ruler

Problem Set A

1. A ladder leans against a wall. The top of the ladder is 10 feet up the wall, and the base is 4 feet from the wall. In this scale drawing, 10 mm represents 1 foot of actual distance.

   Notice that the vertical distance (called the *rise*) between Point *O* on the ground and Point *A* on the ladder is 20 mm, and that the horizontal distance (called the *run*) between these two points is 8 mm.

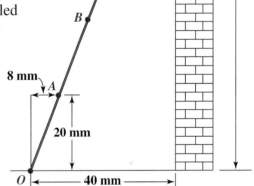

   a. What is the vertical distance, or rise, in the drawing from Point *O* to Point *E*? What is the horizontal distance, or run, from Point *O* to Point *E*?

   b. Copy and complete the table by measuring the rise and run between the given points on the scale drawing.

Points	A to B	A to C	B to C	A to D	B to D	D to E	O to E
Rise	20						
Run	8						

   c. The steepness of the ladder—or of any line between two points—can be described by the ratio $\frac{\text{rise}}{\text{run}}$. Add another row to your table, label it $\frac{\text{rise}}{\text{run}}$, and compute the ratio for each pair of points in the table.

   d. Choose any two unlabeled points on the ladder, and find the ratio $\frac{\text{rise}}{\text{run}}$ for your points. How does this ratio compare to the ratios for the points in the table?

2. Here is a scale drawing of a second ladder positioned with the top of the ladder 8 feet up the wall and the base 4 feet from the wall.

Select at least three pairs of points on this ladder, and calculate the ratio $\frac{\text{rise}}{\text{run}}$ for each pair. What do you find?

3. How does the $\frac{\text{rise}}{\text{run}}$ ratio for the first ladder compare to the $\frac{\text{rise}}{\text{run}}$ ratio for the second ladder? Which ladder appears to be steeper?

4. Imagine a third ladder positioned higher, 11 feet up the wall and 4 feet from the wall at its base. How would its $\frac{\text{rise}}{\text{run}}$ ratio compare to the ratios for the first two ladders?

5. Do you think using $\frac{\text{rise}}{\text{run}}$ is a good way to describe steepness? Explain.

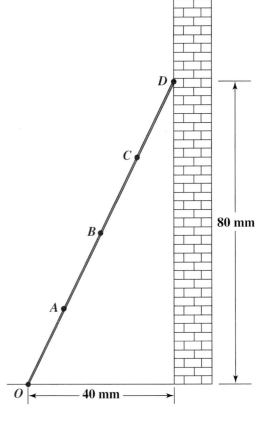

80 mm

40 mm

MATERIALS

metric ruler

Problem Set B

What happens if you try to find the ratio $\frac{\text{rise}}{\text{run}}$ for a curved object? The drawing below shows a cable attached to a wall.

1. Calculate the ratio $\frac{\text{rise}}{\text{run}}$ for each pair of points: Points P and Q, Points Q and R, and Points P and R. What do you find?

2. Describe the difference between the steepness of a ladder and the steepness of a curved cable. Be sure to discuss the ratio $\frac{\text{rise}}{\text{run}}$ for the two situations.

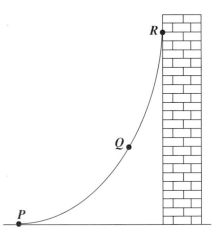

Using the ratio $\frac{\text{rise}}{\text{run}}$ is a good way to describe steepness for a ladder, but not for a curved cable. Since a ladder is straight, you can calculate $\frac{\text{rise}}{\text{run}}$ between any two points—the ratio will be the same regardless of which points you choose.

V O C A B U L A R Y

slope

The ratio $\frac{\text{rise}}{\text{run}}$ is also used to describe the steepness of a line. The ratio $\frac{\text{rise}}{\text{run}}$ for a line is called the line's **slope.**

Problem Set C

M A T E R I A L S

graph paper

Consider this line.

1. What are the coordinates of Points *R* and *S*?

2. Find the slope of the line through Points *R* and *S*.

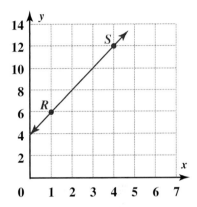

You might have found the slope of the line by subtracting coordinates. The rise from Point *R* to Point *S* is the difference in the *y*-coordinates for those points, and the run is the difference in the *x*-coordinates. However, you might not have thought about whether the *order* in which you subtract the coordinates affects the value of the slope.

3. In Parts a–c, find the slope of the line again by subtracting the coordinates for Point *R* from the coordinates for Point *S*.

    a. Find the rise by subtracting the *y*-coordinate of Point *R* from the *y*-coordinate of Point *S*.

    b. Find the run by subtracting the *x*-coordinate of Point *R* from the *x*-coordinate of Point *S*.

    c. Use your answers to calculate the slope of the line through Points *R* and *S*.

4. Would you find the same slope if you subtracted the coordinates of Point *S* from the coordinates of Point *R*? Try it and see.

5. Would you find a different value for $\frac{\text{rise}}{\text{run}}$ if you used a different pair of points on the line? Explain.

6. Ben calculated the slope of the line as $^-2$. Here is his calculation:

$$\frac{\text{rise}}{\text{run}} = \frac{12 - 6}{1 - 4} = \frac{6}{^-3} = {}^-2$$

    What was Ben's mistake?

7. Review your answers to Problems 4–6. Is the order in which you subtract the coordinates important? Explain.

Graph the line through the given pair of points, and find its slope.

8. (⁻3, 4) and (⁻7, 2) **9.** (2, 4) and (3, 3)

10. (3, 5) and (4, 5) **11.** (⁻3, 4) and (⁻4, 6)

12. Look back at your work in Problems 9–12. Two of the lines have a negative slope. What do you notice about these lines?

13. One of the lines in Problems 9–12 has a slope of 0. What do you notice about that line?

14. A line has slope $-\frac{2}{3}$. One point on the line is (4, 5). Find two more points on the line, and explain how you found them.

15. Consider the line passing through the points (2, 4) and (2, 7).

  a. Graph the line. What does it look like?

  b. Try to find the slope of the line. What happens?

  c. What is the *x*-coordinate of every point on the line?

  d. What is an equation of the line?

Surveyors can use instruments to determine the slope of a certain section of land.

Share & Summarize

1. Two lines with positive slope are graphed on one set of axes. Explain why a greater slope for one line means that line will be steeper than the other.

2. What does a negative slope tell you about a line?

3. What does a slope of 0 tell you about a line?

4. Give the coordinates of two points so that the line connecting them has a positive slope.

Investigation 2 ▶ Slope and Scale

Slope is a good measure of the steepness of objects such as ladders. However, when you are using a graph to find or to show the slopes of lines, you need to be careful.

MATERIALS
graph paper

Explore

Copy the grids below onto graph paper, and graph the equation $y = 2x + 1$ on each grid.

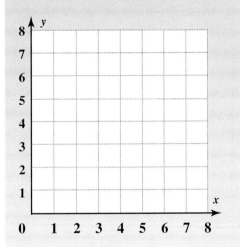 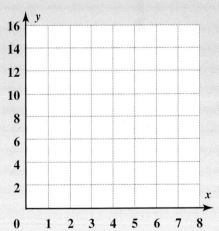

Describe the difference between the graphs. What do you think causes this difference?

In addition to slope, what other factor affects how steep the graph of a line looks?

Problem Set **D**

1. Gabriela's mother thinks Gabriela is spending too much money on CDs. Gabriela says that since CDs cost only $10 at Deep Discount Sounds, the amount she spends doesn't increase very quickly.

   a. Gabriela decides to make a graph showing how the total amount she spends changes as she buys more CDs. She thinks that if she chooses her scales carefully, she will convince her mother that the amount increases at a slow rate. Draw a graph Gabriela might use. (Hint: Use your observations from the Explore on page 29 to create a graph that doesn't look very steep.)

   b. Gabriela's mother knows a thing or two about graphing, too. She wants to make a graph to convince Gabriela that the total cost increases quickly as she adds to her CD collection. Draw a graph Gabriela's mother might use.

2. Imagine that you are using a graph to keep track of the amount of money remaining in a bank account. You start with $200 and withdraw $5 a week. Suppose you graph the time in weeks on the horizontal axis and the bank balance in dollars on the vertical axis.

   a. What will be the slope of the line?

   b. Draw the graph so it appears that the balance is decreasing very rapidly.

   c. Draw another graph of the same relationship so it appears that the balance is decreasing very slowly.

You can graph an equation on a graphing calculator and then adjust the window settings to change the appearance of the graph.

Problem Set **E**

Set the window of your graphing calculator to the standard window settings (x and y values from $^-10$ to 10).

1. Graph the equation $y = x$, and sketch the graph.

2. Now change the window using the square window setting on your calculator. This adjusts the scales so the screen shows 1 unit as the same length on both axes. Graph $y = x$ using the new setting. Sketch the graph.

3. Compare the graphs you made in Problems 1 and 2.

4. Adjust the window settings to make the line appear steeper than both graphs. Record the settings you use.

5. Adjust the window settings to make the line appear less steep than the other graphs. Record the window settings you used.

Share & Summarize

Work with a partner. One of you should use your calculator to graph the equation $y = 3x + 2$ so that it looks very steep, while the other graphs the same equation so that it does not look very steep.

1. Together, write a description of what you each did to make your graphs look as they do.

2. Try to explain why your method works.

Investigation 3 > Using Points and Slopes to Write Equations

You will now learn how to find equations for lines when you know two points on the line, or when you know the slope of the line and one point.

MATERIALS

graph paper

Explore

The table describes a linear relationship.

x	-2	-1	0	1	2	3
y	-3	-1	1	3	5	7

What is an equation of the line described by the table? Explain how you found the equation.

Use two data pairs (x, y) to find the slope of this line. How is the slope used in the equation you wrote?

Graph the equation.

What is the y value of the point at which the graph crosses the y-axis? How is this value used in the equation?

VOCABULARY
coefficient
y-intercept

You have seen that linear equations can be written in the form $y = mx + b$. The multiplier of a variable such as x is called its **coefficient.** In a linear equation of the form $y = mx + b$, the value of m is the slope of the line. The constant term, b, is the **y-intercept** of the line. That is, b is the y-coordinate of the point at which the line crosses (or *intercepts*) the y-axis.

In the Explore on page 31, you were probably able to find the values for *m* and *b* fairly easily. But now look at this table, which also shows data pairs for a linear relationship:

x	$^-6$	$^-4$	$^-1$	$1\frac{1}{2}$	3	7
y	$^-3\frac{3}{4}$	$^-2\frac{1}{4}$	0	$1\frac{7}{8}$	3	6

Finding an equation of the line for these data is a slightly more complex task. You could calculate the slope, but the *y*-intercept is not given—and you can't be certain what it is by graphing the data pairs and looking at the graph.

However, you *can* determine an equation of a line if you know the *slope* and *one point* on the line. The fact that linear equations take the form $y = mx + b$ helps you do this.

EXAMPLE

What is an equation of the line that has slope 3 and passes through the point (2, 5)?

Start with the fact that the equation of a line can be written in the form $y = mx + b$. The slope is 3, so $m = 3$. This gives the equation $y = 3x + b$.

Because the point (2, 5) is on this line, substituting 2 for *x* and 5 for *y* will make the equation a true statement. We say that the point (2, 5) *satisfies* the equation $y = 3x + b$.

$$y = 3x + b$$
$$5 = 3(2) + b$$
$$5 = 6 + b$$
$$^-1 = b$$

Now you know that the value of the *y*-intercept, *b,* is $^-1$, and you can write the final equation:

$$y = 3x - 1$$

Problem Set F

1. What is an equation of the line that has slope 4 and passes through the point (1, 5)?

2. What is an equation of the line that has slope 3 and passes through the point (2, 4)?

3. What is an equation of the line that has slope ⁻2 and passes through the point (8, ⁻12)?

4. What is an equation of the line that has slope 0 and passes through the point (3, 5)?

Problem Set G

Suppose you know only two points on a line but not the slope—how can you find an equation for the line? For example, suppose you want to write an equation of the line that contains the points (1, 3) and (3, 11).

1. What is the slope of the line connecting these points? Show how you find it.

2. If the equation of this line is in the form $y = mx + b$, what is the value of m?

3. Now find the value of b (the y-intercept) without drawing a graph. Show your work. Hint: Look back at your work in Problem Set F if you need to.

4. Write an equation of the line.

5. Check that the point (3, 11) satisfies your equation by substituting 3 for x and 11 for y, and then evaluating. Also check that the point (1, 3) satisfies your equation. If either does not, was your error in determining the value of m or b? Write down what you find out, and adjust the equation if necessary.

Remember

slope $= \frac{\text{rise}}{\text{run}}$

Problem Set H

Find an equation of the line through each pair of points. Plot the points and draw the line to check that the equation is correct. If you get stuck, review the process you followed in Problem Set G.

1. (3, 7) and (8, 12)

2. (6, 11) and (18, 17)

3. (0, 0) and (100, 100)

4. (3, 5) and ($^-$1, 5)

MATERIALS

graph paper

Problem Set I

Researchers have discovered that people shake salt over their food for about the same amount of time regardless of how many holes there are in the saltshaker or how large the holes are.

When you use a saltshaker with large holes, you will probably use more salt than when you use one with small holes. In fact, there appears to be a linear relationship between the average amount of salt people sprinkle on their food and the total area of the holes in their saltshakers. The following data were collected:

Total Area of Holes (mm^2), a	4.5	8
Average Amount of Salt Applied (g), s	0.45	0.73

Assume the researchers are correct and that the amount of salt is linearly related to the total area of the holes. You can use the small amount of data in the table to estimate how much salt would be shaken onto food from shakers with different-sized holes.

1. Plot the two points from the table, with total area on the horizontal axis, and use them to graph a linear relationship. Think carefully about your axes, and size the scales so they can be read easily.

2. Use your graph to estimate the amount of salt delivered from a shaker with a total hole area of 6 mm^2.

3. If each hole has a radius of 1.1 mm, what is the area of one hole?

4. Use your graph and your answer to Problem 3 to estimate the amount of salt delivered by a shaker having 10 such holes.

5. If you want to limit your salt intake at each meal to 0.5 g, what should be the total area of the holes in your saltshaker?

6. Find an equation of the line that passes through the points (4.5, 0.45) and (8, 0.73).

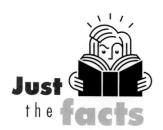

Just the facts

The average adult human's body contains about 250 grams of salt.

7. Use the equation to find the amount of salt delivered by the given total hole areas.

    a. 2.7 mm^2, approximately the smallest total hole area used in commercial saltshakers

    b. 44.7 mm^2, approximately the largest total hole area used

Share & Summarize

1. Without using numbers, describe a general method for writing an equation for a line if all you know is the slope of the line and one point on the line.

2. Without using numbers, describe a general method for writing an equation for a line if all you know is two points on the line.

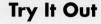

Lab Investigation ▶ Making Linear Designs

You can make some very interesting designs with linear equations on your graphing calculator.

MATERIALS

graphing calculator

Remember

The four quadrants of a graph are numbered like this:

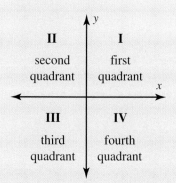

Try It Out

See if you can produce a starburst design, like the one below, by graphing four linear equations in one window. Your design doesn't have to look exactly like this one, but it should include two lines that pass through Quadrants I and III and two lines that pass through Quadrants II and IV. Make a sketch of your design.

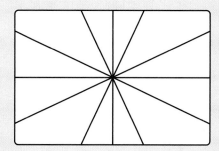

1. Record the equations you used.

2. What do you notice about the *y*-intercepts in your equations?

3. What do you notice about the slopes in your equations?

Make It Rain

On weather maps, a set of parallel lines is often used to symbolize sheets of rain. On your calculator, try to create four parallel lines that are evenly spaced, similar to the design below. Make a sketch of your design.

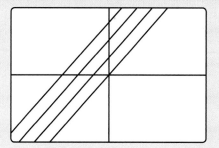

4. Record the equations you used.

5. What do you notice about the slopes in your equations? How is this reflected in your design?

6. What do you notice about the *y*-intercepts in your equations? How is this reflected in your design?

Making Diamonds

Try to make your own diamond shape, like this one, on your calculator. Make a sketch of your design.

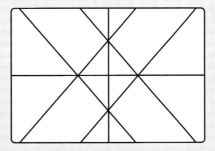

7. Record the equations you used.

8. What do you notice about the slopes in your equations?

9. What do you notice about the *y*-intercepts in your equations?

What Have You Learned?

10. Think about the equations you wrote to make different kinds of lines and shapes: parallel lines, lines radiating from a central point as in a starburst, and intersecting lines that form squares or diamonds. Make a design of your own with at least four lines.

Share what you have learned by preparing a written report about making linear designs on a graphing calculator. Include sketches of your designs and the equations that can reproduce them.

Practice & Apply

1. Consider these tables of data for two linear relationships.

Relationship 1

x	y
1	4.5
2	6
3	7.5
4	9

Relationship 2

x	y
−3	1
−1	3
1	5
3	7

a. Use the (x, y) pairs in the tables to draw each line on graph paper.

b. Find the slope of each line by finding the ratio $\frac{\text{rise}}{\text{run}}$ between two points.

c. Did you use the tables or the graphs in Part b? Does it matter which you use? Explain.

d. Check your results by finding the slope of each line again, using points different from those you used before.

2. Look at the roofs on these three barns. All measurements are in feet.

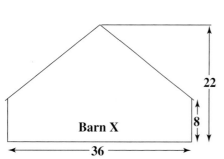
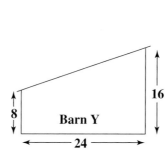
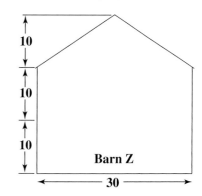

a. Which roof looks the steepest?

b. Find the *pitch*, or slope, of the roof of each barn. Was your prediction correct?

3. Make up an equation for a linear relationship.

a. Construct a table of (x, y) values for your equation. Include at least five pairs of coordinates.

b. Draw a graph using the values in your table. If your graph is not a line, check that the values in your table are correct.

c. Choose two points from the table, and use them to determine the slope of the line. Check the slope using two other (x, y) pairs.

impactmath.com/self_check_quiz

In Exercises 4–7, find the slope of the line by identifying two points on the line and using them to determine the slope.

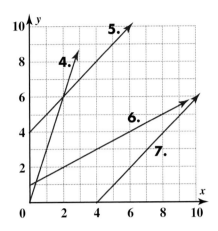

In Exercises 8–11, you are given a slope and a point on a line. Find another point on the same line. Then draw the line on graph paper.

8. slope: $\frac{1}{2}$; point: (3, 4) **9.** slope: $^-1$; point: (2, 5)

10. slope: 3; point: (2, 8) **11.** slope: $\frac{1}{4}$; point (4, 5)

12. Here are two linear relationships: $y = 3x + 2$ and $y = 2x + 3$. Graph both relationships on copies of the grids shown below. Each grid will have two graphs.

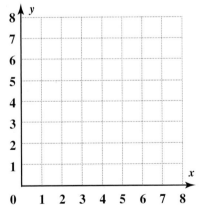 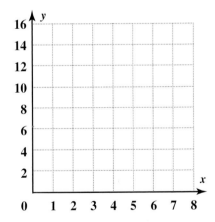

a. What aspects of the graphs depend on the grid you drew them on?

b. What parts of the graphs are not affected by the scale of the grid? For example, do the points of intersections with the x- and y-axes change from one grid to the other?

13. Study these graphs.

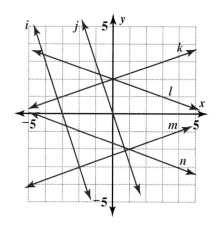

 a. Which pairs of lines have the same slope?

 b. Find the slope of each line.

Find an equation of each line.

 14. a line with slope $^-1$ and passing through the point $(1, 4)$

 15. a line with slope $\frac{1}{3}$ and passing through the point $(3, 3)$

 16. a line with slope $^-2$ and passing through the point $(3, 6)$

Find an equation of the line passing through the given points.

 17. $(3, 4)$ and $(7, 8)$

 18. $(2, 7)$ and $(6, 6)$

 19. $(3, 5)$ and $(9, 9)$

Connect & Extend

Remember

Lines have a *constant difference* in their y values: when the x values change by a certain amount, the y values also change by a certain amount.

For Exercises 20–28, answer Parts a and b.

 a. What is the constant difference between the y values as the x values increase by 1?

 b. What is the constant difference between the y values as the x values decrease by 2?

 20. $y = x$ **21.** $y = x + 2$ **22.** $y = 3x - 3$

 23. $y = {}^-2x + 12$ **24.** $y = 5x$ **25.** $y = \frac{1}{2}x$

 26. $y = 23x - 18$ **27.** $y = {}^-x$ **28.** $y = {}^-2x + 6$

29. Architecture An architect is designing several staircases for a home in which the distance between floors is 10 feet. In designing a staircase, she considers these two ratios:

$$\frac{\text{total rise}}{\text{total run}} \qquad \frac{\text{step rise}}{\text{step run}}$$

The diagram shows how these quantities are measured.

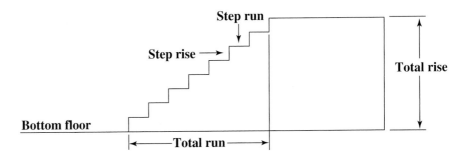

    a. One staircase is to have 18 steps with a total run of 14 ft. What is the ratio $\frac{\text{total rise}}{\text{total run}}$ for the staircase?

    b. Find the step rise and step run, in inches, for each stair in this staircase. What is the ratio $\frac{\text{step rise}}{\text{step run}}$?

    c. Compare your results for Parts a and b. Explain what you find.

30. Design a staircase with a total rise of 14 ft. The step rise should be between 6 in. and 8 in. All steps must have the same rise and the same run. (See the diagram above.) The sum of the step rise and the step run should be between 17 in. and 18 in. You must determine these things:

    a. the number of steps

    b. the height of each step (in inches)

    c. the run of each step (in inches)

    d. the ratio $\frac{\text{step rise}}{\text{step run}}$

    e. the total run (in feet)

    f. the ratio $\frac{\text{total rise}}{\text{total run}}$ for the staircase

31. Mr. Arthur has two bank accounts, Account A and Account B. Each graph shows how much money is in his two accounts over time.

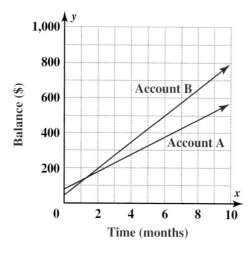

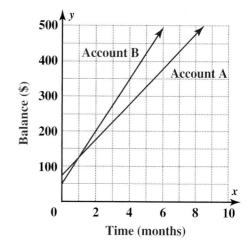

a. Which account is growing faster?

b. Which account started out with more money?

c. Does it matter which grid you use if you are comparing the two accounts' growth rates? Explain.

32. Alejandro looked at the equations $y = \frac{3}{2}x - 1$ and $y = -\frac{2}{3}x + 2$ and said, "These lines form a right angle."

a. Graph both lines on *two different* grids, with the axes labeled as shown here.

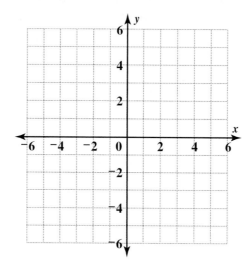

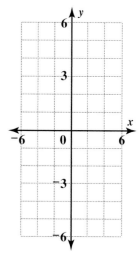

b. Compare the lines on each grid. Do both pairs of lines form a right angle?

c. What kind of assumption must Alejandro have made when he said the lines form a right angle?

Remember

A right angle has a measure of 90°.

For each equation, identify the slope and the y-intercept. Graph the line to check your answer.

33. $y = 2x + 0.25$

34. $y = {}^-x + 5$

35. $y = x - 3$

36. $y = {}^-2x$

37. $y = \frac{3}{4} + \frac{1}{2}x$

38. $y = 3x$

39. The table shows x and y values for a particular relationship.

x	6	3	1	2.5
y	7	1	$^-3$	0

a. Graph the ordered pairs (x, y). Make each axis scale from $^-10$ to 10.

b. Could the points represent a linear relationship? If so, write an equation for the line.

c. From your graph, predict the y value for an x value of $^-2$. Check your answer by substituting it into the equation.

d. From your graph, find the x value for a y value of $^-2$. Check your answer by substituting it into the equation.

e. Use your equation to find the y value for each of these x values: 0, $^-1$, $^-1.5$, $^-2.5$. Check that the corresponding points all lie on the line.

In your
own
words

How can you determine the slope of a line from a graph? If you are given the slope of a line, what else do you need to know before you can graph the line?

40. Consider these four equations.

    i. $y = 2x - 3$ **ii.** $y = -\frac{1}{2}x - 6$

    iii. $y = \frac{2}{5}x + 4$ **iv.** $y = -\frac{5}{2}x$

  a. Graph the four equations on one set of axes. Use the same scale for each axis. Label the lines with the appropriate roman numerals.

  b. What is the slope of each line?

  c. What do you notice about the angle of intersection between Lines i and ii? Between Lines iii and iv?

  d. What is the relationship between the slopes of Lines i and ii? Between the slopes of Lines iii and iv?

  e. Make a conjecture about the slopes of perpendicular lines.

  f. Create two more lines with slopes that fit your conjecture. Are they perpendicular?

  g. Write an equation for the line that passes through the point $(-1, 4)$ and is perpendicular to $y = \frac{1}{3}x + 4$. Check your answer by graphing both lines on one set of axes. Use the same scale for each axis.

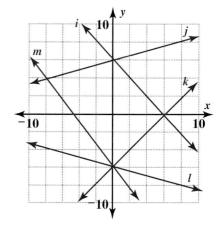

Remember

A conjecture is a statement someone believes to be true but has not yet been proven true.

Mixed Review

Evaluate each expression for $a = 2$ and $b = 3$.

41. $a^b + b^a$ **42.** $\left(\frac{a}{b} + a\right)^a$ **43.** $b^a \cdot b^a$

44. Match each equation to a line.

  a. $y = -\frac{5}{4}x - 6$

  b. $y = 0.25x + 6$

  c. $y = -0.25x - 6$

  d. $y = -x + 6$

  e. $y = x - 6$

45. Refer to the graphs in Problem 44. Tell which line or lines each given point is on.

  a. $(0, -6)$ **b.** $(6, 0)$ **c.** $(-4, -1)$

  d. $(-2, 8)$ **e.** $(-8, -4)$ **f.** $(-2, -2)$

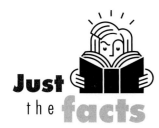

Just the facts

Because a total solar eclipse is observable in only a limited geographic area, most people rarely experience one.

Remember

The *median* is the middle value of a set of values arranged in ascending or descending order. If there are an even number of data values, it is the average of the two middle values.

46. Earth Science The table gives the duration and width of the first 10 total solar eclipses after the year 2000.

Solar Eclipses

Date	Duration (minutes:seconds)	Width (miles)
June 21, 2001	4:56	125
Dec 4, 2002	2:04	54
Nov 23, 2003	1:57	338
Apr 8, 2005	0:42	17
Mar 29, 2006	4:07	118
Aug 1, 2008	2:27	157
July 22, 2009	6:39	160
July 11, 2010	5:20	164
Nov 13, 2012	4:02	112
Nov 3, 2013	1:40	36

Source: *World Almanac and Book of Facts 1999.* Copyright © 1998 Primedia Reference Inc.

a. On what given date is the longest total solar eclipse?

b. What is the median duration time for the eclipses listed? What is the median width?

c. Plot the data from the table on a grid like the one below.

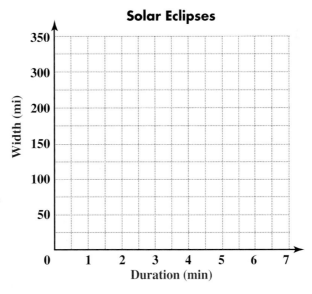

d. Notice that one of the points seems very far from the general location of all of the other points. What is the date associated with that point?

More Explorations with Lines

You have worked with equations in the form $y = mx + b$, where m is the slope of the line and b is the y-intercept. Both m and b are constants, while x and y are variables.

The values of m and b affect what the graph will look like. In this lesson, you will study these effects to help you analyze patterns in graphs and equations.

MATERIALS

graphing calculator

Explore

Each group contains equations in the form $y = mx + b$.

Group I	Group II
$y = x + 2$	$y = {}^-2x - 1$
$y = 2x + 2$	$y = {}^-2x$
$y = {}^-2x + 2$	$y = {}^-2x + 1$
$y = \frac{1}{2}x + 2$	$y = {}^-2x + 2$

• Graph the four equations in Group I in a single viewing window of your calculator. Make a sketch of the graphs. Label the minimum and maximum values on each axis.

What do the four equations in Group I have in common? Give another equation that belongs in this group.

• Now graph the four equations in Group II in a single viewing window. Sketch and label the graphs.

What do the four equations in Group II have in common? Give another equation that belongs in this group.

• In one group, the equations have different values for m but the same value for b. If you start with a specific equation and change the value of m, how will the graph of the new equation be different?

• In the other group, the equations have different values for b but the same value for m. If you start with a specific equation and change the value of b, how will the graph of the new equation be different?

The lines in Group I are a *family of lines* that all pass through the point $(0, 2)$. The lines in Group II are a family of lines with slope $^-2$.

Investigation Parallel Lines and Collinear Points

As you saw in Lesson 1.2, you can use two points to find the slope of the line through the points. In this investigation, you will work more with the slopes of lines. First you will use the connection between parallel lines and their slopes.

MATERIALS

graph paper

Problem Set A

1. Look back at the graphs of the equations in Group II of the Explore. What do you notice about them?

2. If two equations in $y = mx + b$ form have the same m value, what do you know about their graphs? If you're not sure, write a few equations that have the same m value and graph them. Explain why your observation makes sense.

3. Without graphing, decide which of these equations represent parallel lines. Explain.

   a. $y = 2x + 3$ **b.** $y = 2x^2 + 3$

   c. $y = 2x - 7$ **d.** $y = 5x + 3$

4. Consider the line $y = 5x + 4$.

   a. A second line is parallel to this line. What do you know about the equation of the second line?

   b. Write an equation for the line parallel to $y = 5x + 4$ that passes through the origin.

   c. Write an equation for the line parallel to $y = 5x + 4$ that crosses the y-axis at the point $(0, 3)$.

   d. Write an equation for the line parallel to $y = 5x + 4$ that passes through the point $(2, 11)$. Hint: The data pair $(2, 11)$ must satisfy the equation.

5. A line has the equation $y = 3x - 1$. Find an equation for the line parallel to it and passing through the point $(3, 4)$.

6. A line has the equation $y = 4$.

   a. Graph the line.

   b. Find an equation for the line parallel to this line and passing through the point $(3, 6)$.

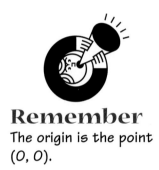

Remember

The origin is the point $(0, 0)$.

Points *A, B,* and *C* below are *collinear.* In other words, they all lie on the same line. Points *D, E,* and *F* are not collinear.

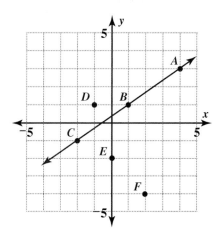

If you are given three points, how can you tell whether they are collinear? You could plot the points to see whether they look like they are on the same line, but you wouldn't know for sure. In Problem Set B, you will develop a method for determining whether three points are on the same line without making a graph.

Problem Set **B**

1. Find a way to determine—*without graphing*—whether the three points below are collinear. Explain your method.

$$(3, 5) \qquad (10, 26) \qquad (8, 20)$$

2. The points in one of the two sets below are collinear; the points in the other set are not. Which set is which? Test your method on both sets to be sure it works.

Set A	Set B
(⁻4, ⁻3)	(⁻1, 1)
(1, 2)	(0, ⁻1)
(7, 7)	(2, ⁻5)

3. Determine whether the three points in each set are collinear.

   a. (⁻3, ⁻2), (0, 4), (1.5, 7)

   b. (1.25, 1.37), (1.28, 1.48), (1.36, 1.70)

1. How would you find an equation for the line through Point C and parallel to the line through Points A and B? Assume you know the coordinates of each point and that Point C is not on the line through Points A and B.

2. Describe a method for determining whether Point D is on the line through Points A and B, if you know the coordinates of each point.

Investigation 2 ▶ Rearranging and Simplifying Linear Equations

Equations can be complicated, and sometimes it isn't obvious whether a relationship is linear. A graph can tell you whether a relationship *looks* linear, but it still might not be.

Think Discuss

You know by its form that the equation $y = 3x + 2$ represents a line, but these equations aren't in that form:

$$4x + 3y = 12 \qquad y = \frac{4x + 2(3 + x) - 2}{2}$$

How can you determine whether these equations are linear? Explain your reasoning.

VOCABULARY
slope-intercept form

If you can write an equation in the form $y = mx + b$, you know it is linear and you can easily identify the slope m and the y-intercept b. In fact, $y = mx + b$ is often called the **slope-intercept form** of a linear equation.

Problem Set C

Determine whether each equation is linear. If an equation is linear, identify the values of m and b. If an equation is nonlinear, explain how you know.

1. $3y = \frac{x}{2} - 8$ **2.** $y = 3$

3. $y = \frac{2}{x}$ **4.** $5y - 7x = 10$

5. $2y = 10 - 2(x + 3)$ **6.** $y = x(x - 1) - 2(1 - x)$

7. $y = 2x + \frac{1}{2}(3x + 1) + \frac{1}{4}(2x + 8)$

1. Which of these equations describe the same relationship?

   a. $p = 2q + 4$ **b.** $p - 2q = 4$

   c. $p - 2q + 4 = 0$ **d.** $0.5p = q + 2$

   e. $p - 4 = 2q$ **f.** $2p = 8 + 4q$

Write each equation in slope-intercept form.

2. $y - 1 = 2x$

3. $2y - 4x = 3$

4. $2x + 4y = 3$

5. $6y - 12x = 0$

6. $x = 2y - 3$

7. $y + 4 = {}^-2$

8. Choose the equations in Problems 2–7 that would have graphs that are parallel lines, and sketch them on the same coordinate axes.

9. Group these equations into sets of parallel lines.

   a. $y = (2x - 7) - 3$ **b.** $y + 5 + 2x = 5$

   c. $y = 30 + 4(x - 7)$ **d.** $4y - 5x = 3x - 2$

   e. $y + 3(10 - x) = x$ **f.** $y = 5x + 3(10 - x)$

   g. $y = 8x - \frac{1}{3}(12x - 30)$ **h.** $y = 1 + \frac{1}{2}(2 - 4x)$

   i. $y - 3 = {}^-2x$ **j.** $2y = {}^-4(3 - x)$

Share & Summarize

What are some of the strategies you used to simplify the equations in this investigation?

Investigation 3 ▶ Fitting Lines to Data

Sometimes when you graph data, the points will lie close to—but not exactly on—a line.

Think & Discuss

Students in a science class were measuring how far a cart traveled each time the wheels rotated once. They measured the distance for one rotation, two rotations, three rotations, and so on. The wheels of the cart are 2.5 feet in circumference.

Would you expect the students' data to lie on a line? If so, explain why, and give an equation of that line.

The graph shows a plot of the students' data. Notice that the points do not appear to lie exactly on a line. Why might the data they collected not fit precisely on a line?

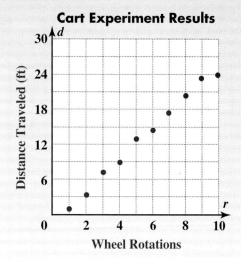

Cart Experiment Results

Using mathematics to describe something, such as a set of data from an experiment, is called *modeling*. Modeling is important in many professions, especially in the fields of science and statistics.

People who gather data are often uncertain what kind of relationship the data will show. When data are graphed, they sometimes are close to, but not exactly on, a line. It may be that the variables are linearly related but that there were inaccuracies in the measurements. Or the relationship between the variables may not be *exactly* linear, but close enough to use a line as a reasonable model.

In cases like these, you can use the data to help find a *line of best fit*—a line that fits all the data points as closely as possible. You can then use a line of best fit to make predictions or to solve problems.

There are several ways to find such a line. Some of the techniques are sophisticated, but you can make reasonably good estimates by using more simple techniques.

Problem Set E

Breaths per minute and heartbeats per minute were measured for 16 people after they had each walked for 20 minutes. The data are in the table.

Breaths per Minute	16	16	19	20	20	23	24	26	27	28	28	30	34	36	41	44
Heartbeats per Minute	57	59	66	68	71	70	72	84	82	80	83	91	94	105	116	120

MATERIALS

- graph paper
- transparent ruler or piece of dry spaghetti

Here is a graph of the data.

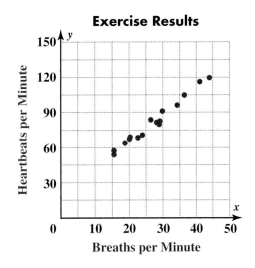

It seems that the more rapid the breathing, the more rapid the heart rate. The relationship is not exactly linear since it is not possible to draw a single line that passes through every point on the graph. However, if you can find a line that fits the points reasonably well, you can use it to predict the value of one variable—heartbeats per minute or breaths per minute—from the other.

1. With your partner, plot the points in the table on a grid.

   a. Draw a straight line that fits the points as well as possible. Try to draw the line so that about the same number of points lie on either side of it. A transparent ruler or a piece of dry spaghetti will help with this.

   b. Write an equation of the line you drew. To do this, find the slope and the *y*-intercept of the line, or the slope and any point on the line, or two points on the line—and use that information to write an equation.

2. One technique for improving the "fit" of your line is to use the means of the data.

   a. Find the mean number of breaths per minute and the mean number of heartbeats per minute from the data in the table.

   b. Plot the point that has these two means as its coordinates, and adjust the line you made in Problem 1 to go through this point.

3. Use your graph to write an equation for your new line. Why might your line be different from someone else's line?

4. Use your equation for Problem 3 to predict the heart rate for a person who takes 35 breaths per minute after walking for 20 minutes. Compare your prediction with other students' predictions.

5. Use your equation to predict the heart rate for a person who takes 100 breaths per minute after walking for 20 minutes. Do you think your prediction is reasonable? Explain.

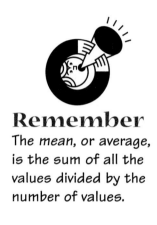

Remember

The *mean*, or average, is the sum of all the values divided by the number of values.

Just the **facts**

The winning time for the 400-m relay has decreased steadily since 1928. In the 2000 Olympics, the winner of the gold medal, on the Bahamas team, completed the relay in 41.95 s.

Problem Set F

The table shows winning times for the women's 400-meter relay in the Olympic Games from 1928 to 1980. No games were held in 1940 or 1944 because of World War II.

Year	Country	Time (s)
1928	Canada	48.4
1932	United States	46.9
1936	United States	46.9
1948	Netherlands	47.5
1952	United States	45.9
1956	Australia	44.5

Year	Country	Time (s)
1960	United States	44.5
1964	Poland	43.6
1968	United States	42.8
1972	West Germany	42.81
1976	East Germany	42.55
1980	East Germany	41.60

The data are plotted below. The points are quite close to falling on a line, with one exception. When a data point seems very different from the others, it is often called an *outlier*. Outliers are typically given less emphasis when analyzing general trends, or patterns.

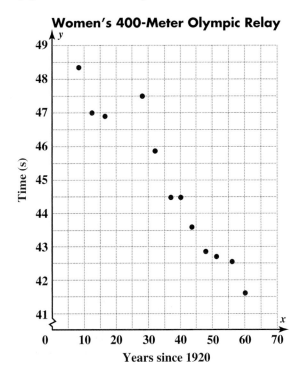

Women's 400-Meter Olympic Relay

1. Which point is the outlier?

2. Can you suggest a reason why that point is so far from the trend of the other values?

3. Copy the data points on a graph of your own. Using the techniques you learned in Problem Set E, find an equation for a line that fits the data.

Problem Set G

Graphing calculators use a sophisticated mathematical technique to find lines to fit a set of data. You will now use your graphing calculator to find a line of best fit.

1. Enter the 400-meter-relay data from Problem Set F into your calculator in two lists.

    a. Use your calculator to determine an equation of a line of best fit.

    b. How close is the calculator's equation to the equation you found in Problem Set F?

2. The winning times in the table on page 54 are decreasing fairly steadily over the years. Do you think they will always continue to do so? Why or why not?

3. Use your equation from Problem 1 to predict the winning times for 1984, 1988, 1992, and 1996.

4. The real winning times for these years are given below. How do your predictions compare?

Year	1984	1988	1992	1996
Time (s)	41.65	41.98	42.11	41.95

5. Assume that the winning times actually continue to decrease after 1980 at the same rate as your linear model from Problem 1 predicts they will.

    a. Write an equation you can use to find when the winning time will be 0 seconds. Solve your equation.

    b. According to your model, in which year do you predict this impossible winning time of 0 seconds will occur?

Share & Summarize

You have just examined two situations that can be modeled with linear equations. Draw a graph containing 10 points for which a linear equation would *not* be a good model. Describe why you would have difficulty determining a line of best fit for your graph.

On Your Own Exercises

For each set of equations, tell what the graphs of all four relationships have in common *without* drawing the graphs. Explain your answers.

1. $y = {}^-1.1x + 1.5$

$y = {}^-1.1x - 4$

$y = {}^-1.1x + 7$

$y = {}^-1.1x$

2. $y = 2x$

$y = {}^-2x$

$y = 3x$

$y = {}^-3x$

3. $y = 2x$

$y - 1 = 2x$

$y = 2x + 4$

$y = 2x + 7$

4. $y = 1 - x$

$y = 1 - 2x$

$y = 1 - 3x$

$y = 1 - 4x$

5. In this exercise, you will apply what you have learned about writing equations for parallel lines.

a. Write three equations whose graphs are parallel lines with positive slopes. Write the equations so that the graphs are equally spaced.

b. Graph the lines, and verify that they are parallel.

c. Write three equations whose graphs are parallel lines with negative slopes and are equally spaced.

d. Graph the lines, and verify that they are parallel.

6. You can tell whether a particular point might be on a line by graphing it and seeing whether it seems to lie on the line. But to know for certain whether a particular point is on a line—and not just *close* to it—you must test whether its coordinates satisfy the equation for that line.

a. Graph the equation $y = \frac{13}{8}x - 3$.

b. Using the graph alone, decide which points below look like they might be on the line. You may want to plot the points.

$(0, {}^-3)$ $(3, 2)$ $(4, 4)$ $(5, 5)$ $(8, 10)$

c. For each point, substitute the coordinates into the equation and evaluate to determine whether the point satisfies the equation. Which points, if any, are on the line?

impactmath.com/self_check_quiz

If possible, write each equation in the form $y = mx + b$. Then identify the slope and the y-intercept.

7. $y = 5x + \frac{1}{3}(6x + 12)$

8. $y = \frac{1}{5}(10x + 5) - 5 + 7x$

9. $3x + 2(x + 1) = -\frac{1}{2}(4x + 6) + y$

10. $3x^2 - y = 3x + 5$

11. $y - 19 = {}^-2(x - 3)$

12. Within these equations are five pairs of parallel lines. Identify the parallel lines, and give the slope for each pair.

  a. $y = 3x - 5(x + 3)$ **b.** $\frac{x - 2y}{2} = 7$

  c. $y = 17 - 3(3 + x) + x$ **d.** $3x + 2y = 4$

  e. $y = {}^-\frac{x}{2} + 2\left(6 - \frac{x}{2}\right)$ **f.** $x - y + 3 = 0$

  g. $4x - 2y - 17 = 20$ **h.** $4\left(\frac{y}{2} - x\right) = 10$

  i. $y + x = 4x + 5 - 2(4 + x)$ **j.** $y = \frac{3x + 4}{2} - \frac{7 + 2x}{2}$

13. Give the slope-intercept form of each equation, and tell which of these eight equations describe the same relationship.

  a. $4x - 2y = 4$ **b.** $2y - 4x = 4$

  c. $2x - y = 2$ **d.** $y - 2x = 2$

  e. $y - 2x = {}^-2$ **f.** $y = 2x - 2$

  g. $y = 2x + 2$ **h.** $4x + 2y = 4$

How many sets of parallel lines can you find in this house, designed by the famous architect Frank Lloyd Wright?

14. Life Science The average length of gestation, or pregnancy, and the average life span for several animals are listed below.

Animal	Average Gestation (days)	Average Life Span (years)
Cat	63	11
Chicken	22	8
Cow	280	11
Dog	63	11
Duck	28	10
Elephant	624	35
Goat	151	12
Guinea pig	68	3
Hamster	16	2
Horse	336	23
Mule	365	19

a. Plot the points for each animal on one set of axes. Put gestation on the horizontal axis and life span on the vertical axis.

b. Draw a line that fits the data reasonably well. Write an equation for your line.

c. Use your equation to predict the life span of a hog, which has a gestation of about 114 days.

15. Fine Arts The genius Wolfgang Amadeus Mozart composed music for most of his very short life. His compositions were numbered in the order that he wrote them.

These data relate the total number of compositions, *K,* to Mozart's age when they were written, *a.*

Age (years), *a*	8	12	16	20	24	27	32	35
Total Number of Compositions, *K*	16	45	133	250	338	425	551	626

a. Plot the points from the table, with *a* on the horizontal axis and *K* on the vertical axis.

b. Do the data suggest a general rate at which Mozart wrote new compositions? What is that rate?

c. Draw a line that fits the data points as well as possible. Use the technique of finding a line that passes through the mean age and mean composition number for the given data points.

Just the **facts**

The letter K is used for the composition number variable to honor the Austrian scientist Ludwig von Köchel, who sorted the 626 compositions in the mid-nineteenth century.

d. Use your graph to find a linear equation for predicting the number of compositions Mozart wrote at a given age.

e. Mozart died at the young age of 35. Would it be reasonable to use your equation to predict the number of compositions Mozart would have produced if he had lived to age 70?

f. What is the value of K for $a = 0$? Does this data point make sense? What does this tell you about your linear model?

16. The lines for these three equations all pass through a common point.

$$y = \frac{x}{2} - 1 \qquad y = -\frac{2x}{3} + 6 \qquad y = -\frac{x}{6} + 3$$

a. Draw graphs for the three equations, and find the common point.

b. Verify that the point you found satisfies all three equations by substituting the x- and y-coordinates into each equation.

Each table describes a linear relationship. For each relationship, find the slope of the line and the y-intercept. Then write an equation for the relationship in the form $y = mx + b$.

17.

x	2	4	6	8	10
y	8	12	16	20	24

18.

x	$^-8$	$^-3$	3	5	10
y	26	11	$^-7$	$^-13$	$^-28$

19.

x	9	7	5	3	1
y	5	4	3	2	1

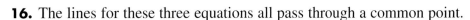

20. Hoshi drew graphs for $y = x$ and $y = {}^-x$ and noticed that the lines crossed at right angles at the point (0, 0). Then he drew graphs for $y = x + 4$ and $y = {}^-x + 4$ and noticed that the lines crossed at right angles again, this time at the point (0, 4). He tried one more pair, $y = x - 4$ and $y = {}^-x - 4$. Once again the lines crossed at right angles, at the point (${}^-4$, 0).

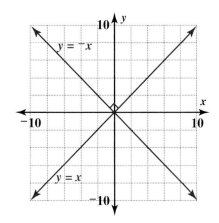 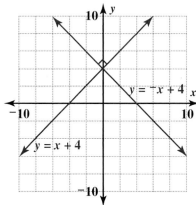 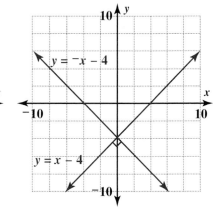

Hoshi made this conjecture: "When you graph two linear equations and one has a slope that is the negative of the other, you always get a right angle."

a. Do you agree with Hoshi's conjecture? Why or why not?

b. Draw several more pairs of lines that fit the conditions of Hoshi's conjecture, with different slope values. Do your drawings prove or disprove Hoshi's conjecture?

c. If you think Hoshi's conjecture is false, where do you think he made his mistake?

21. Review Problem Set C. Some of the equations given there are nonlinear. Write some guidelines for quickly identifying linear and nonlinear equations.

Just as the y-intercept of a line is the y value at which the line crosses the y-axis, the x-intercept is the x value at which the line crosses the x-axis. In Exercises 22–25, find an equation of a line with the given x-intercept and slope.

22. x-intercept 3, slope 2

23. x-intercept ${}^-2$, slope $-\frac{1}{2}$

24. x-intercept 1, slope ${}^-6$

25. Challenge x-intercept 3, no slope (Hint: If slope is $\frac{\text{rise}}{\text{run}}$, when would there be no slope?)

In your
own
words

Describe why determining a line of best fit can be useful in working with data.

26. You have been using the form $y = mx + b$ to represent linear equations. Linear equations are sometimes represented in the form $Ax + By = C$, where A, B, and C are constants.

   a. Rewrite the equation $Ax + By = C$ in the $y = mx + b$ form. To do this, you will need to express m and b in terms of A, B, and C.

   b. What is the slope of a line with an equation in the form $Ax + By = C$? What is the y-intercept?

27. Social Studies Below is world population data for the years 1950 through 1990.

World Population

Year	Population (billions)
1950	2.52
1960	3.02
1970	3.70
1980	4.45
1990	5.29

   a. Plot the points on a graph with "Years since 1900" on the horizontal axis and "Population (billions)" on the vertical axis. Try to fit a line to the data.

   b. Write an equation to fit your line.

   c. Use your equation to project the world population for the year 2010, which is 110 years after 1900.

   d. What does your equation tell you about world population in 1900? Does this make sense? Explain.

   e. According to United Nations figures, the world population in 1900 was 1.65 billion. The UN has predicted that world population in the year 2010 will be 6.79 billion. Are the 1900 data and the prediction for 2010 different from your predictions? How do you explain your answer?

28. Consider this data set.

x	0	2	4	6	8
y	2	20	6	8	10

a. Graph the data set.

b. One point is an outlier. Which point is it?

c. Find the mean of the x values and the mean of the y values.

d. Try to find a line that is a good fit for the data and goes through the point (mean of x values, mean of y values). Write an equation for your line.

e. Now find the means of the variables, *ignoring the outlier.* In other words, do not include the values for the outlier in your calculations.

f. Try to find a new line that is a good fit for the data, using the means you calculated in Part e for the (mean of x values, mean of y values) point. Write an equation for your line.

g. Do you think either line should be considered the best fit for the data? Explain.

Mixed Review

29. Each point below satisfies one of the equations. Match each equation with a point.

$(^-1, 1)$ $(7, 0)$ $(^-1, ^-1)$ $(10, 1)$ $(^-2, 0)$ $(^-1, 0.9)$

a. $y = 2x - 14$ **b.** $y = x^2 - 4$

c. $y = 0.1x + x^2$ **d.** $y = x^3$

e. $y = ^-x^3$ **f.** $y = x^2 - 99$

30. Geometry The formula for the volume of a cylinder is derived by multiplying the area of the cylinder's base, πr^2, by the cylinder's height, h:

$$\text{Volume} = \pi r^2 h$$

Use this formula to find the value of each variable.

a.

Volume $= 8\pi$

b.

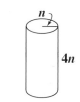

Volume $= 108\pi$

c.

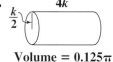

Volume $= 0.125\pi$

31. Geometry The Pythagorean Theorem states that if a and b are the lengths of the legs of a right triangle and c is the length of the hypotenuse (the longest side), then $a^2 + b^2 = c^2$.

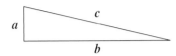

Use the Pythagorean Theorem to find the value of each variable.

a.

5
3
2k

b.

8
6
20m

c.

2.5p
6p
13

32. Science Civil engineers are designing a holding tank for the water system of a small community. The tank will be in the shape of a cylinder and will have a height of 15 meters. They are trying to determine the best size for the tank's radius.

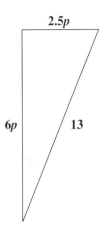

15 m

a. Fill in a table like the one below for the volume of the holding tank for each given radius.

Radius (m)	4	5	6	7	8	9
Volume (m³)						

b. Plot the points from your table on a grid like this one.

c. Use your graph to estimate what the radius of the tank should be if the tank is to hold 3,500 m³ of water.

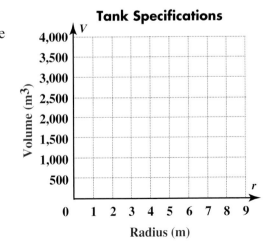

Tank Specifications

Chapter 1 Review & Self-Assessment

Chapter Summary

In this chapter, you looked at *linear relationships*—relationships with straight-line graphs. You investigated examples of increasing and decreasing linear relationships. In some cases, these relationships were *direct variations,* or *directly proportional* relationships.

You explored the connection between the $y = mx + b$ form of a linear equation and its graph. You found that the *coefficient, m,* represents the *slope* of the line. The constant, *b,* called the *y-intercept,* indicates where the line crosses the *y*-axis.

You learned to find an equation of a line from different kinds of given information: a written description, a table of values, a graph, a slope and a point on the line, or two points on the line. Finally, you saw that if plotted data show a linear trend, you can fit a line to the data and use the line or its equation to make predictions.

Strategies and Applications

The questions in this section will help you review and apply the important ideas and strategies developed in this chapter.

Recognizing linear relationships and writing linear equations

In Questions 1–6, tell which of the following descriptions fit the relationship, and explain how you decided:

- a direct variation

- linear but not a direct variation

- nonlinear

1. Aisha paid $2.50 admission to a carnival and $1.25 for each ride. Consider the relationship between the total he spent for admission and rides, and the number of rides he took.

2. The Scrooge Loan company charges a penalty of $10 the first time a borrower is late making a payment. The penalty is $20 for the second late payment, $40 for the third late payment, and so on, doubling with each late payment. Consider the relationship between the total penalty and the number of late payments.

3. $d = 65t$

4. $y = 13 - 12x$

5.

x	0	10	20	30	40	50	60
y	5	55	105	155	205	255	305

6.

x	40	30	20	15	10	5	0
y	184	138	92	69	46	23	0

Understanding the connection between a linear equation in the form $y = mx + b$ and its graph

7. Consider the equation $y = 300 - 25x$.

a. How would changing $^-25$ to $^-30$ affect the graph of this equation?

b. How would changing $^-25$ to $^-20$ affect the graph of this equation?

c. How would changing $^-25$ to 25 affect the graph of this equation?

d. How would changing 300 to $^-100$ affect the graph of this equation?

Understanding and applying the idea of slope

Determine whether the points in each set are collinear. Explain how you know.

8. $(^-2, 8), \left(\frac{1}{3}, 1\right), (5, ^-13)$

9. $\left(2, \frac{7}{2}\right), (6, 6), \left(^-3, \frac{3}{2}\right)$

10. $(^-6, ^-3), \left(8, \frac{5}{3}\right), (0, ^-1)$

11. Find an equation of the line that passes through the point $(^-4, 1)$ and is parallel to the line $y = ^-3x + 1$.

Using a linear graph to gather information or to make predictions

12. There is a linear relationship between temperature in degrees Fahrenheit and temperature in degrees Celsius. Two temperature equivalents are 0°C = 32°F and 30°C = 86°F.

a. Make a four-quadrant graph of this relationship, and plot the points (0, 32) and (30, 86). Connect the points with a line.

b. Use the line joining the two points to convert these temperatures.

i. 5°F   **ii.** 20°C   **iii.** ⁻30°C

Fitting a line to data

13. Every day for a week Lemeka practiced doing pull-ups. Each day she timed how many pull-ups she could do in a certain number of seconds. She added 10 seconds to the time each day.

Seconds	10	20	30	40	50	60	70
Pull-ups	15	25	44	35	42	50	55

a. Graph the data.

b. Do the data appear to be approximately linear?

c. Are there any outliers in the data—points that do not seem to fit the general trend of the data? If so, which point or points?

d. Draw a line that fits the data as well as possible, and find an equation of your line.

e. Use your equation or graph to predict how many pull-ups Lemeka will do when she reaches a time of 2 minutes.

Demonstrating Skills

In Questions 14–16, estimate the slope of the line.

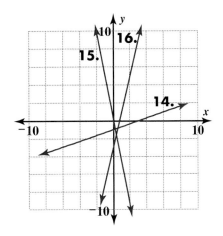

Find the slope of the line that passes through each given pair of points.

17. $(5, ^-3)$ and $(^-1, 9)$

18. $(3, 4)$ and $(^-1, ^-2)$

19. $(^-6, ^-4)$ and $(^-2, 5)$

20. Find an equation of the line that has slope $^-2$ and passes through the point $(^-1, ^-1)$.

21. Find an equation of the line that passes through the points $(4, 4)$ and $(8, ^-2)$.

22. Identify the equations that represent parallel lines.

a. $y = 3x + 1$ **b.** $y = \frac{1}{2}x - 1$ **c.** $\frac{1}{2}y = 3 + \frac{1}{2}x$

d. $y = ^-x + 2$ **e.** $y = 2x(1 - x)$ **f.** $y = 2x + (1 - x)$

g. $3y = 1 - 3x$ **h.** $^-4y = 2x$ **i.** $y = \frac{1}{2}x - x - 1$

Rewrite each equation in slope-intercept form.

23. $y - x - 1 = 2x + 1$

24. $2(y - 1) = 3x + 1$

25. $1 - y = x + 2(1 - x)$

CHAPTER 2

Quadratic and Inverse Relationships

Real-Life Math

Dropping the Ball Many applications in physical science involve quadratic and inverse relationships. In the late 1500's, the great Italian astronomer and physicist Galileo Galilei conjectured that two objects dropped at the same time from the same height would hit the ground at the same time, regardless of their masses.

According to legend, he proved his theory by dropping two cannonballs, one large and one small, from the top of the Leaning Tower of Pisa. The balls hit the ground at nearly the same time. His work led to the quadratic equation $d = 16t^2$, which gives the distance an object has fallen in feet t seconds after it has dropped.

Think About It Can you think of a situation in which it would be important for someone to know the relationship described above?

Family Letter

Dear Student and Family Members,

Our next chapter in mathematics is about two fundamental components of algebra: *quadratic relationships* and *inverse relationships*.

We will begin our study of quadratic relationships by analyzing and graphing equations in the form $y = ax^2$, where a represents a constant. For example, the formula for the area of a circle is $A = \pi r^2$. The constant is π, which is approximately 3.14. The distance that a dropped object has fallen is also quadratic: $d = 4.9t^2$, where d represents the distance in meters and t represents the time in seconds. Here the constant is 4.9.

The graph of any quadratic relationship is a symmetric, U-shaped curve called a *parabola*. For more complex quadratic relationships, like the motion of objects that are thrown or shot into the air, the equations take the form $y = ax^2 + bx + c$.

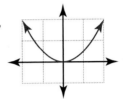

We will also look at inverse relationships, in which the product of two quantities is always the same. For example, suppose you have only $20 to pay for a baby-sitter: the number of hours of baby-sitting you can afford depends on the rate. If you pay $2 an hour, you can afford 10 hours; if you pay $8, you can afford only 2.5 hours.

We will close the chapter with making and proving conjectures. A *conjecture* is an educated guess that has not been proven correct. Knowing how to make and prove conjectures is a valuable skill students will use in mathematics and elsewhere throughout their future.

Vocabulary Along the way, we'll be learning about several new vocabulary terms:

conjecture	inverse variation	quadratic expression
cubic equation	parabola	reciprocal relationship
hyperbola	quadratic equation	vertex
inversely proportional		

What can you do at home?

Encourage your student to identify inverse relationships like the one described above in his or her everyday life. Use these relationships to solve problems relevant to your student's life.

Quadratic Relationships

The equation for the relationship between a circle's radius, r, and area, A, is $A = \pi r^2$. The equation for the relationship between the time in seconds an object has fallen, t, and the distance in meters it has fallen, d, is $d = 4.9t^2$. In this chapter, you will explore relationships like these.

Explore

Select a team of nine students to make a "human graph." The team should follow these rules:

- Line up along the x-axis. One student should stand on $^-4$, another on $^-3$, and so on, up to 4.

- Multiply the number you are standing on by itself. That is, if your number is x, calculate x^2. Remember the result.

- When your teacher says "Go!" walk forward or backward a number of paces equal to your result from the previous step.

- The number you began on is your x value, and the number of paces you took is your y value.

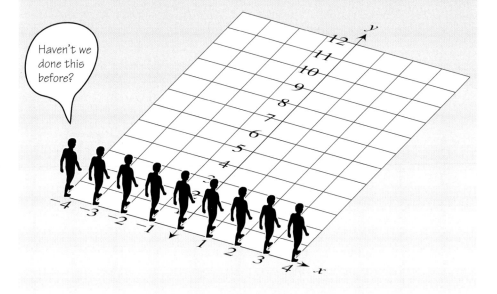

Haven't we done this before?

Each of the nine students on the graph should now report his or her coordinates. Record the information in a class table like this one.

x	$^-4$	$^-3$	$^-2$	$^-1$	0	1	2	3	4
y									

Make a class graph by plotting the points on the board or a large sheet of paper. Connect the points with a smooth curve.

Describe the graph.

When making the human graph, did any student walk backward? Why or why not?

If someone had started at 1.5 on the x-axis, how far forward or backward would that person have moved? What if someone had started at $^-1.5$?

Could this graph be extended to the left of $^-4$ or to the right of 4? Explain.

Is the graph *symmetrical*? That is, is there a line along which you could fold the graph so that the two halves match up exactly? If so, describe this *line of symmetry,* and tell whether any students were standing on it.

What equation describes the relationship shown in the graph?

Investigation Quadratic Equations and Graphs

The human graph your class created was a graph of $y = x^2$, which is a *quadratic equation.* The simplest quadratic equations can be written in the form $y = ax^2$. This form consists of a constant—represented by the letter a—multiplied by the square of a variable.

The formula for the area of a circle, $A = \pi r^2$, has this form. In the area formula, the constant is π. In the equation that gives the distance in meters fallen by an object after t seconds, $d = 4.9t^2$, the constant is 4.9. In the equation $y = x^2$, the constant is 1.

V O C A B U L A R Y
parabola

The graph of each of these relationships is a symmetric, U-shaped curve called a **parabola.** In this investigation, you will look at graphs, tables, and equations for more quadratic relationships.

Remember
The volume of a rectangular prism is the product of its length, width, and height. Volume is measured in cubic units such as cm^3.

Problem Set A

This box has height 9 cm and a length that is twice its width.

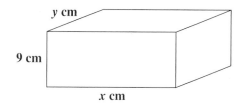

1. You can describe the box's volume with either of these quadratic equations.

$$V = 4.5x^2 \qquad V = 18y^2$$

How can two different equations describe the same volume?

2. On a grid like the one below, plot the equation $V = 4.5x^2$. Start by plotting points (x, V) for x values from 0 to 50. (You will look at x values less than 0 a little later.) Plot as many points as you need to, until you are confident of the shape of the graph. Then draw a smooth curve through the points.

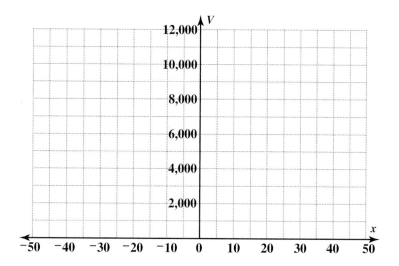

Just the facts
The path of any projectile—a golf ball hit into the air, a textbook tossed onto a couch, a flare launched from a boat—follows the shape of a parabola.

3. Suppose you want to construct a box with a particular volume and a length twice as long as its width. You can use your graph to find possible lengths for the box.

   a. Choose a value for the volume, V. (Do not use one of the values you plotted in Problem 2.) Use your graph to estimate the length x that corresponds to this volume.

   b. Check your estimate by substituting it for x in the equation $V = 4.5x^2$. How close is this result to the volume you chose in Part a?

4. Of course, boxes don't have negative lengths, but if you think of $V = 4.5x^2$ simply as a rule to generate ordered pairs, you can look at the graph for x values less than or equal to 0. Plot some (x, V) points for $x = 0$ and several negative values of x. Use the points to extend your curve into the second quadrant.

5. What is the line of symmetry for your graph?

6. Locate both points on the graph where $V = 9{,}000$.

   a. Use the graph to estimate the corresponding values of x.

   b. Use a calculator to guess-check-and-improve for a more accurate estimate of the x values for $V = 9{,}000$.

Share & Summarize

Tell whether each graph could represent a quadratic relationship. Explain how you decided.

1.

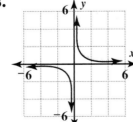

2.

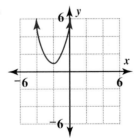

3.

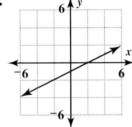

4.

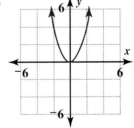

Investigation ▶2▶ Quadratic Patterns

In your study of linear equations, you saw them written in many forms, such as $y = mx + b$ and $Ay + Bx = C$. Quadratic equations can also be written in different forms.

In this investigation, you will explore some geometric patterns that can be represented by quadratic equations in forms other than $y = ax^2$.

MATERIALS

- square tiles (optional)
- graph paper

Problem Set B

Consider this pattern of square tiles.

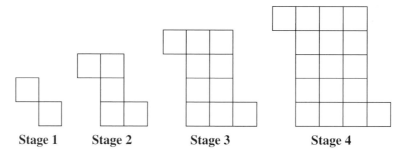

Stage 1   Stage 2   Stage 3   Stage 4

1. Describe the pattern in words.

2. Copy and complete the table to show the number of tiles used in each of the first four stages.

Stage, S	1	2	3	4
Tiles, T				

3. Describe the pattern in the way the number of tiles T increases as the stage number S increases.

4. Use your answer to Problem 3 to predict the number of tiles in Stages 5 through 8. Extend your table to include these values.

5. Check your answers to Problem 4 by building or drawing the next four stages of the pattern.

6. Evan found an equation for the number of tiles T in Stage S of the pattern. He reasoned this way: "Taking away the two tiles on the corners leaves a rectangle that is $S - 1$ tiles across the top and $S + 1$ tiles down the side. I multiply those numbers to find the number of tiles in the rectangle. Then I add the two corner tiles to find the total number of tiles." What equation do you think Evan found?

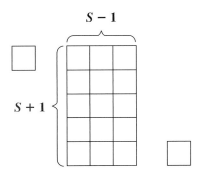

7. Tala found an equation for T by using her table. "First I square S. Then I see how S^2 compares to the number of tiles for that value of S." Try her idea with your table: square the numbers in the first row, and see what you could do to the squares to get the numbers in the second row. What equation do you think Tala found?

8. Mikayla found the same equation as Tala, but in a different way. "There are S tiles across the bottom of Stage S. If I remove this row, and turn it vertically, I can make a square with S tiles on each side, plus one extra tile." Explain how Mikayla's reasoning leads to the same equation Tala found.

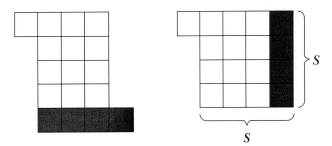

9. Consider the equation Tala and Mikayla found for the pattern.

  a. Graph the relationship between T and S. Even though it doesn't make sense to have a negative number of tiles, think of the equation as a rule to generate ordered pairs and consider both positive and negative values of S. Plot enough points so you can draw a smooth curve.

  b. Describe the graph.

  c. Does the graph have a line of symmetry? If so, describe its location.

  d. What is the smallest value of T shown on the graph?

Problem Set C

Here's a geometric problem that has a connection to a quadratic equation, although the connection is not obvious at first:

How many diagonals does a polygon with n sides have?

You probably can't answer this question yet. But if you start with a few simple polygons, you can look for a pattern to help you find the answer.

The drawings show the diagonals for polygons with 3, 4, and 5 sides.

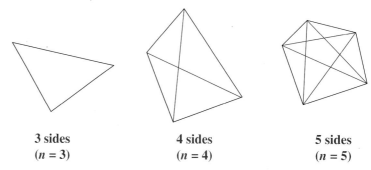

3 sides
(*n* = 3)

4 sides
(*n* = 4)

5 sides
(*n* = 5)

Remember

A *polygon* is a closed, two-dimensional figure whose sides are made of line segments.

1. Draw a hexagon (*n* = 6) with a complete set of diagonals. Make sure you join each vertex to every other vertex of the hexagon.

2. Copy and complete the table to show the number of diagonals connected to each vertex and the total number of diagonals for each polygon listed.

Sides, *n*	3	4	5	6
Diagonals Connected to Each Vertex	0	1		
Total Number of Diagonals, *d*	0	2		

3. Look at the second row of your table.

   a. Describe the pattern in the way the number of diagonals connected to each vertex changes as the number of sides increases.

   b. Is there a linear relationship between the number of sides and the number of diagonals connected to each vertex? Explain.

4. Look at the third row of your table.

   a. Describe the pattern in the way the total number of diagonals changes as the number of sides increases.

   b. Is there a linear relationship between the number of sides and the total number of diagonals? Explain.

5. Consider a heptagon, a seven-sided polygon.

   a. Use the pattern in your table to predict the number of diagonals connected to each vertex in a heptagon.

   b. Predict the total number of diagonals in a heptagon.

   c. Check your predictions from Parts a and b by drawing a heptagon and carefully drawing and counting its diagonals. Add the data for the heptagon to your table.

6. Without drawing any more polygons, extend your table to include data for 8-, 9-, and 10-sided polygons. Explain how you found your results.

7. How many diagonals does a 20-sided polygon have? Find the answer without extending your table, if you can.

8. Write an equation for the total number of diagonals in an *n*-sided polygon. Explain how you found your answer.

9. Use your equation to find the total number of diagonals in a 100-sided polygon.

10. What shape do you think the graph of your equation will have? Use your graphing calculator to check your answer.

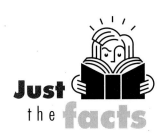

Just t h e **facts**

A polygon with 20 sides is called an *icosagon*. Mathematicians would also call it a *20-gon*.

Share & Summarize

Consider the quadratic relationships you have explored in this lesson.

   1. Write the equations of these quadratic relationships.

   2. Look back at the graphs of these equations. What do they all have in common?

   3. How can you tell by looking at the equations that their graphs will not be lines?

On Your Own Exercises

Practice
Apply

Remember

If you don't have a calculator with a button for π, you can use the approximation 3.14 in your calculations.

1. The area of a circle with radius r is given by the formula $A = \pi r^2$.

   a. Make a table of values for A as r increases from 0 to 10 units.

   b. Plot the values on a graph with r on the horizontal axis and A on the vertical axis. Connect the points with a smooth curve.

   c. The areas of three circles are given. Use your graph to find an approximate value for the radius of each circle.

   i. 25 square units

   ii. 100 square units

   iii. 300 square units

   d. Check your results by substituting each radius into the area formula.

2. Imagine several stacks of blocks like those shown below.

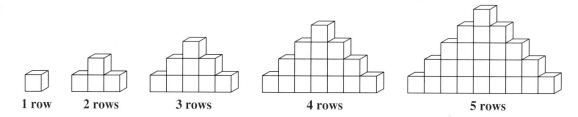

1 row   2 rows   3 rows   4 rows   5 rows

   a. Complete the table to show the number of blocks in the bottom row as the number of rows increases.

Number of Rows, n	1	2	3	4	5	6	7	8	9	10
Number of Blocks in Bottom Row, b										

   b. How many blocks would be in the bottom row of a stack with 25 rows? With n rows?

c. Think about the relationship between the number of rows and the number of blocks in the bottom row. What type of relationship is this?

d. What is the *total* number of blocks T needed for a stack with 5 rows?

e. Add a new row to your table for the total number of blocks T in each stack, and complete that row.

f. Write an equation that gives the total number of blocks T in terms of n.

g. Use this diagram to explain why your equation from Part f makes sense.

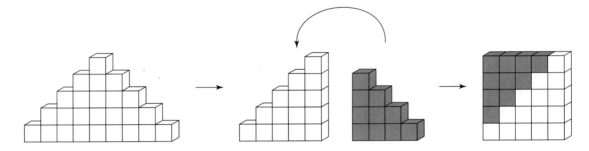

3. Look closely at this sequence of figures. Each square has an area of 1 square unit.

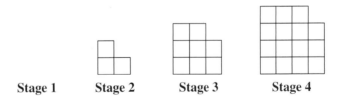

Stage 1   Stage 2   Stage 3   Stage 4

a. Find a formula for the area in square units, A, of Stage n.

b. Use your equation to explain why there are no squares in Stage 1.

c. Copy and complete the table.

n	1	2	3	4	5	6	7	8	9	10
A	0	3	8	15						

d. Extend your table to include negative integer values of n, from $^-10$ to 0. Plot enough points on a graph between $^-10$ and 10 to draw a smooth curve.

e. Describe the graph.

f. Find the line of symmetry for the graph. What value of n does the line of symmetry pass through?

Connect & Extend

4. Consider these three equations.

$$y = x^2 \qquad y = x^2 + 2 \qquad y = x^2 - 2$$

a. Make a table of values for each equation, using positive and negative values of x. Plot all three graphs on the same set of axes.

b. What similarities do you see in the three graphs?

c. What differences do you notice?

5. Consider these three equations.

$$y = {}^-x^2 \qquad y = {}^-x^2 + 2 \qquad y = {}^-x^2 - 2$$

a. Make a table of values for each equation, using positive and negative values of x. Plot all three graphs on the same set of axes.

b. What similarities do you see in the three graphs?

c. What differences do you notice?

6. Economics Rosalinda makes wooden birdhouses and sells them at arts and crafts fairs. She has found that she can sell more birdhouses when the price is lower. Looking back at past fairs, she estimates that for a price of p dollars, she can sell $200 - 2p$ birdhouses during a two-day fair. For example, if she sets the price at $20, she generally sells around 160 birdhouses.

a. When a person or business sells a product, the money from the sales is called *revenue*. The revenue for a product can be calculated by multiplying the number of items sold by the price. Find a formula for Rosalinda's revenue R at a two-day fair if she charges p dollars for each birdhouse.

b. Make a table of values for price p and revenue R. List at least 10 prices from $0 to $100.

c. Graph the values in your table, with p on the horizontal axis and R on the vertical axis. You may need to find additional points so you can draw a smooth curve.

d. For what price does Rosalinda earn the most revenue? What is that revenue?

7. Darlene made this table of ordered pairs.

x	1	2	3	4	5	6	7	8	9	10
y	0	3	8	15	24					

a. Describe the pattern of change in the y values as the x values increase by 1.

In your
own
words

How do graphs for quadratic relationships differ from those for linear relationships? How do you think the equations differ?

b. Copy and complete Darlene's table.

c. Use your table to plot a graph of these ordered pairs. Connect the points with a smooth curve or line.

d. Describe your graph.

e. Could Darlene's table represent a quadratic relationship? Explain.

8. Kai made this table of ordered pairs.

x	1	2	3	4	5	6	7	8	9	10
y	7	10	13	16	19					

a. Describe the pattern of change in the y values as the x values increase by 1.

b. Complete Kai's table.

c. Use your table to plot a graph of these ordered pairs. Connect the points with a smooth curve or line.

d. Describe your graph.

e. Could Kai's table represent a quadratic relationship? Explain.

9. Physical Science The distance it takes for a car to stop depends on the car's speed.

a. The *reaction distance* is the distance the car travels after the driver realizes he needs to stop and before he applies the brakes. It can be represented by the equation $d = 0.25s$, where s is speed in kilometers per second and d is distance in meters. The *braking distance* is the distance traveled after the driver applies the brakes until the car comes to a complete stop. It can be represented by $d = 0.006s^2$. Make a table showing the reaction distance and the braking distance for speeds from 0 km to 80 km.

b. Plot the points for reaction distance and braking distance on one set of axes. Draw a smooth line or curve through each set of points. Label each graph with its equation.

c. To find the total stopping distance, reaction distance and braking distance must be added. Add the values in your table for each speed to find the total stopping distance for that speed.

d. Plot the values in your table for total stopping distance on your graph, and connect them with a smooth line or curve. What is the equation of this relationship? What kind of relationship does this appear to be? (Hint: Think about what you did to generate the points for stopping distance.)

Mixed Review

Evaluate each expression for $a = {}^-2$ and $b = 3$.

10. a^a

11. $a^b \cdot a^a$

12. $(b - a)^{-a}$

13. a^{b-a}

14. $(ab)^{-a}$

15. $\dfrac{a^2}{b^2}$

Give the slope and y-intercept for the graph of each equation.

16. $y = {}^-3x + 5$ **17.** $2y = 2 - 3.2x$

18. $3y + 1.5 = x$ **19.** $x = y$

20. Daniela has a certain number of CDs. Let the variable d represent the number of CDs she has, and write an expression for each of the following.

   a. the number Andrés has, which is 3 times the number Daniela has

   b. the number Tyree has, which is one-tenth the number Andrés has

   c. the number Sandra has, which is 7 more than half the number Tyree has

   d. If Sandra has 13 CDs, how many do each of the other three people have?

Write an equation for a line that would pass through the given points.

21. Points A and C

22. Points A and B

23. Points D and E

24. Points A and F

25. Point F and the origin

26. Points G and H

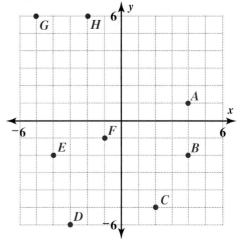

2.2 Families of Quadratics

In Lesson 2.1, you explored several quadratic relationships:

$$y = x^2 \qquad d = \frac{n^2 - 3n}{2} \qquad V = 4.5x^2 \qquad T = S^2 + 1$$

VOCABULARY
quadratic expression

Any expression that can be written in the form $ax^2 + bx + c$, where a, b, and c are constants, and $a \neq 0$, is called a **quadratic expression** in x.

The first term, ax^2, is called the x^2 *term*. The number multiplying x^2, represented by a, is called the *coefficient* of the x^2 term. In the same way, bx is the x *term* and b is its coefficient. The coefficients a and b, and the constant c, can stand for positive or negative numbers, but a cannot be 0. (Can you see why?)

VOCABULARY
quadratic equation

A **quadratic equation** can be written in the form $y = ax^2 + bx + c$, where $a \neq 0$. Like those in Lesson 2.1, all quadratic equations have symmetric, U-shaped graphs called *parabolas*.

Before you explore the connection among graphs, tables, and equations for quadratic relationships, it will help to review graphs, tables, and equations for linear relationships.

Just the facts

Mirrors in the shape of parabolas are used in flashlights and headlights to focus light into a narrow beam.

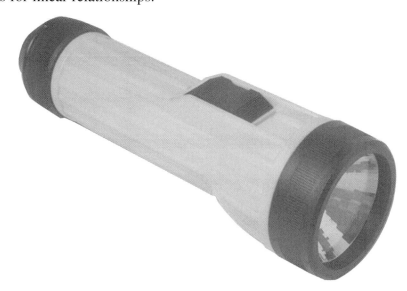

Think & Discuss

Equations, graphs, and tables of four linear relationships are shown here. Match each equation with its graph and table, and explain how you made each match.

Equation 1: $y = x$

Equation 2: $y = 2x$

Equation 3: $y = 2x + 3$

Equation 4: $y = ^-2x + 3$

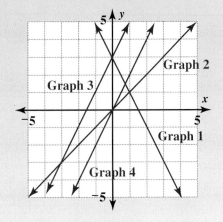

Table 1

x	$^-2$	$^-1$	0	1	2	3
y	7	5	3	1	$^-1$	$^-3$

Table 2

x	$^-2$	$^-1$	0	1	2	3
y	$^-4$	$^-2$	0	2	4	6

Table 3

x	$^-2$	$^-1$	0	1	2	3
y	$^-2$	$^-1$	0	1	2	3

Table 4

x	$^-2$	$^-1$	0	1	2	3
y	$^-1$	1	3	5	7	9

For linear equations of the form $y = mx + b$, how does the value of m affect the graph?

For linear equations of the form $y = mx + b$, how does the value of b affect the graph?

Investigation Quadratic Equations, Tables, and Graphs

Now you will explore the connection among equations, tables, and graphs for quadratic relationships.

Problem Set A

1. The table on the next page has columns for six quadratic equations. Copy and complete the table to give the y values for the given x values. (Hint: You can save time by completing Column A first, and then looking for connections between its equation, $y = x^2$, and the other equations. For example, if you know the value of x^2 in Column A, how can you easily determine $x^2 + 1$ in Column B?)

Remember

The expression $^-n^2$ means $^-(n^2)$, not $(^-n)^2$. To calculate $^-n^2$, square n and take the opposite of the result.

x	**A** $y = x^2$	**B** $y = x^2 + 1$	**C** $y = x^2 - 1$	**D** $y = {}^-x^2$	**E** $y = (x + 1)^2$	**F** $y = (x - 1)^2$
$^-4$	16					
$^-3.2$	10.24					
$^-2.2$						
$^-1$						
$^-0.5$						
0						
0.5						
1						
2.2						
3.2						
4						

2. Compare the values in each of Columns B–F with those in Column A. Explain how each comparison you make is related to the equations.

For the graphs in Problems 3–8, complete Parts a–c.

3.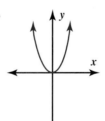

4.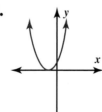

5.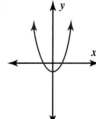

6.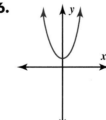

7.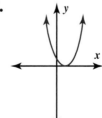

8.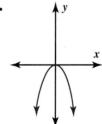

a. Match each graph with one of the quadratic equations in Columns A–F, and explain your reasoning.

b. Describe how the graph differs from the graph of $y = x^2$.

c. Describe the graph's line of symmetry.

Problem Set B

For the tables in Problems 1–7, find a quadratic equation. Hint: For Problems 2–7, compare the values with those in the tables from previous problems.

1.

x	y
$^-3$	9
$^-2$	4
$^-1$	1
0	0
1	1
2	4
3	9

2.

x	y
$^-3$	109
$^-2$	104
$^-1$	101
0	100
1	101
2	104
3	109

3.

x	y
$^-3$	5
$^-2$	0
$^-1$	$^-3$
0	$^-4$
1	$^-3$
2	0
3	5

4.

x	y
$^-3$	13
$^-2$	8
$^-1$	5
0	4
1	5
2	8
3	13

5.

x	y
$^-3$	90
$^-2$	40
$^-1$	10
0	0
1	10
2	40
3	90

6.

x	y
$^-3$	70
$^-2$	20
$^-1$	$^-10$
0	$^-20$
1	$^-10$
2	20
3	70

7.

x	y
$^-3$	36
$^-2$	16
$^-1$	4
0	0
1	4
2	16
3	36

8.

x	y
$^-3$	37
$^-2$	17
$^-1$	5
0	1
1	5
2	17
3	37

Share & Summarize

1. Graph A is the graph of $y = x^2$. Write equations for the other graphs, and explain how you know the equations fit the graphs.

Graph A

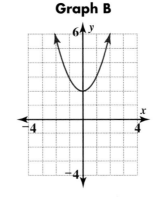

Graph B

Graph C

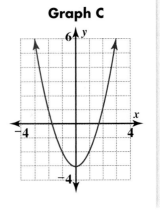

2. Table A represents the relationship $y = x^2$. Write the quadratic equations represented by Tables B and C. Explain how you found your equations.

<table>
<tr><th colspan="2">Table A</th></tr>
</table>

Table A			Table B			Table C	
x	*y*		*x*	*y*		*x*	*y*
⁻4	16		⁻4	⁻48		⁻4	36
⁻3	9		⁻3	⁻27		⁻3	25
⁻2	4		⁻2	⁻12		⁻2	16
⁻1	1		⁻1	⁻3		⁻1	9
0	0		0	0		0	4
1	1		1	⁻3		1	1
2	4		2	⁻12		2	0
3	9		3	⁻27		3	1
4	16		4	⁻48		4	4

Investigation ▶ 2 ◀ Quadratic Equations and Their Graphs

Just the facts

The cables of a suspension bridge form the shape of a parabola.

Quadratic equations can be written in the form $y = ax^2 + bx + c$, where a, b, and c are constants and a is not 0. Like the coefficient m and constant b in a linear equation $y = mx + b$, the coefficients and constant in a quadratic equation tell you something about the graph of the equation.

In Problem Set C, you will explore how the coefficient a and constant c affect the graphs of quadratic equations. In Problem Set D, you will look at the effect of b, which isn't as easy to see.

Problem Set **C**

1. Complete Parts a–c for each group of quadratic equations.

Group I	Group II	Group III	Group IV
$y = x^2$	$y = x^2$	$y = x^2$	$y = {}^-x^2$
$y = x^2 + 1$	$y = x^2 - 1$	$y = 2x^2$	$y = {}^-2x^2$
$y = x^2 + 3$	$y = x^2 - 3$	$y = \frac{1}{2}x^2$	$y = {}^-\frac{1}{2}x^2$

   a. Graph the three equations in the same window of your calculator. Choose a window that shows all three graphs clearly. Make a sketch of the graphs. Label the minimum and maximum values on each axis. Also label each graph with its equation.

   b. For each group of equations, write a sentence or two about how the graphs are similar and how they are different.

   c. For each group of equations, give one more quadratic equation that also belongs in that group.

2. Describe how the graphs in Group I are like the graphs in Group II and how they are different.

3. Describe how the graphs in Group III are like the graphs in Group IV and how they are different.

4. Use what you learned in Problems 1–3 to predict what the graph of each equation below will look like. Make a quick sketch of the graphs on the same set of axes. Be sure to label the axes, and label each graph with its equation. Check your predictions with your calculator.

   a. $y = x^2 + 2$

   b. $y = 3x^2 + 2$

   c. $y = {}^-3x^2 + 2$

   d. $y = \frac{1}{2}x^2 - 3$

5. All the equations you looked at in this problem set are in the form $y = ax^2 + bx + c$, but the coefficient b is equal to 0. Explain how the values of a and c affect the graph of an equation.

In Problem Set C, you probably saw that equations of the form $y = ax^2 + c$ have their highest or lowest point at the point $(0, c)$. The highest or lowest point of a parabola is called its **vertex.**

Not all parabolas have their vertices on the y-axis. In the next problem set, you will look at the properties of an equation that determine where the vertex of its graph will be.

Problem Set D

Consider these four groups of quadratic equations.

Group I	Group II	Group III	Group IV
$y = x^2 + 2$	$y = x^2 + 2$	$y = {}^-2x^2 + 2$	$y = {}^-2x^2 + 2$
$y = x^2 + 3x + 2$	$y = x^2 - 3x + 2$	$y = {}^-2x^2 + 3x + 2$	$y = {}^-2x^2 - 3x + 2$
$y = x^2 + 6x + 2$	$y = x^2 - 6x + 2$	$y = {}^-2x^2 + 6x + 2$	$y = {}^-2x^2 - 6x + 2$

1. In each group, all the equations have the same values of a and c. Use what you learned about the effects of a and c to make predictions about how the graphs in each group will be alike.

2. For each group of equations, complete Parts a–c.

   a. Graph the three equations in the same window of your calculator. Choose a window that shows all three graphs clearly. Make a sketch of the graphs. Remember to label the axes, and label the graphs with their equations.

   b. Were your predictions in Problem 1 correct?

   c. For each group of equations, write a sentence or two about how the graphs are similar and how they are different.

3. What patterns can you see in how the locations of the parabolas change as b increases or decreases from 0?

Share & Summarize

1. Imagine moving the graph of $y = {}^-2x^2 + 3$ up 2 units without changing its shape. What would the equation of the new parabola be?

2. Briefly describe or make a rough sketch of the graph of $y = {}^-\frac{1}{2}x^2 - 2$.

3. For each quadratic equation, tell whether the vertex is on the y-axis.

   a. $y = \frac{1}{2}x^2 - 3$ **b.** $y = x^2 - 3x + 1$

   c. $y = {}^-x^2$ **d.** $y = {}^-3x^2 + x + 13$

   e. $y = {}^-x^2 + 4$ **f.** $y = 7x^2 + 3x$

Investigation 3 ▶ Using Quadratic Relationships

Quadratic equations and graphs can help you understand the motion of objects that are thrown or shot into the air. If an object, like a ball, is launched at an angle between 0° and 90° (not straight up or down), its *trajectory*—the path it follows—approximates the shape of a parabola.

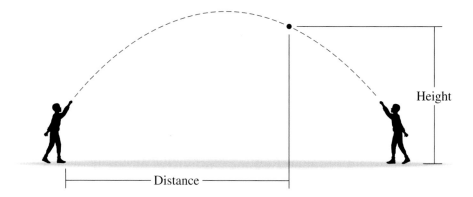

This means that to approximate the relationship between the object's height and the horizontal distance it travels, you can use a quadratic equation. As you will see in Problem Set A, there is also a quadratic relationship between the length of time the object is in the air and the object's height.

Problem Set E

A photographer set up a camera to take pictures of an arrow that had been shot into the air. The camera took a picture every half second. The points on the graph show the height of the arrow in meters each half second after the camera started taking photos. A parabola has been drawn through the points.

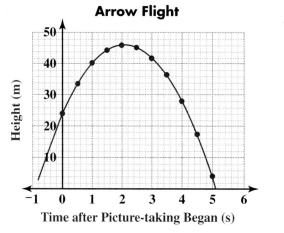

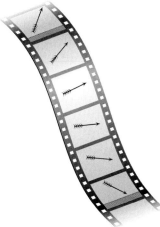

1. Consider the arrow's height when the camera took the first picture.

   a. How high was the arrow at that time? Explain how you know.

   b. How much time elapsed from that point until the arrow returned to that height? Explain how you know.

2. How long after the first photo was taken did the arrow hit the ground? How did you determine this?

3. The arrow was shot from a height of 1.5 m (just above shoulder height). About how long before the time of the first photo was the arrow shot? How did you determine this?

4. Approximately how long was the arrow in flight? Explain how you found your answer.

5. Now think about how high the arrow rose.

   a. What was the arrow's greatest height? Explain how you found your answer.

   b. How many seconds after the arrow was shot did it reach its maximum height? Explain how you found your answer.

MATERIALS

graphing calculator

Problem Set F

A quarterback threw a pass in such a way that the relationship between its height y and the horizontal distance it traveled x could be described by this equation:

$$y = 2 + 0.8x - 0.02x^2$$

Both y and x are measured in yards.

1. Will the graph of this quadratic equation open upward or downward? Explain.

2. Use your calculator to graph the equation. Use a window that gives a view of the graph for the entire time the ball was in the air. You can change the window settings by adjusting the minimum and maximum x and y values for the axes and the x and y scales. For this problem, try an x scale of 10 and a y scale of 2. Sketch the graph, being sure to label the axes.

For Problems 3–5, use your calculator graph or the equation to answer the question.

3. From what height was the football thrown? Explain how you found your answer.

4. What was the greatest height reached by the ball? Explain how you found your answer.

5. A receiver caught the ball in the end zone at the same height it was thrown. What distance did the pass cover before it was caught? Explain how you found your answer.

Share & Summarize

Marcus hit a tennis ball into the air. The graph shows the ball's height h in feet over time t in seconds. The point P shows the ball's height at $t = 1$.

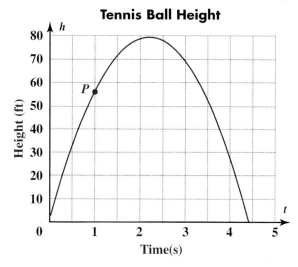

Tennis Ball Height

1. Explain how to use the graph to estimate when the ball reached its highest point.

2. Explain how to use the graph to find another time when the ball had the same height it had at $t = 1$.

Investigation ▶ 4 ▶ Comparing Quadratics and Other Relationships

These problems will help you distinguish between quadratic relationships and other types of relationships.

Problem Set G

One way to distinguish different types of relationships is to examine their equations.

Tell whether each equation is quadratic. If an equation is not quadratic, explain how you know.

1. $y = 3m^2 + 2m + 7$

2. $y = (x + 3)^2 + 7$

3. $y = (x - 2)^2$

4. $y = 10$

5. $y = \frac{2}{x^2}$

6. $y = b(b + 1)$

7. $y = 2^x$

8. $y = 3x^3 + 2x^2 + 3$

9. $y = {}^-2.5x^2$

10. $y = 4n^2 - 7n$

11. $y = 7p + 3$

12. $y = n(n^2 - 3)$

VOCABULARY
cubic equation

Some of the equations in Problem Set G are *cubic* equations. A **cubic equation** can be written in the form $y = ax^3 + bx^2 + cx + d$, where $a \neq 0$. These are all examples of cubic equations:

$$y = x^3 \qquad y = 2x^3 \qquad y = 0.5x^3 - x + 3 \qquad y = x^3 + 2x^2$$

The graphs of cubic equations have their own characteristics, different from those of linear and quadratic equations.

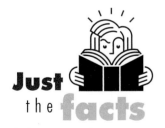

Just the facts

Cubic equations are used by computer design programs to help draw curved lines and surfaces.

Problem Set H

1. Consider these three equations.

$$y = x \qquad y = x^2 \qquad y = x^3$$

    a. Graph the three equations in the same window of your calculator. Choose a window that shows all three graphs clearly. Sketch the graphs, and remember to label the axes and the graphs.

    b. Write a sentence or two about how the graphs are the same and how they are different.

    c. Give the coordinates of the two points where all three graphs intersect.

You have seen how the values of m and b affect the graph of $y = mx + b$ and how the values of a, b, and c affect the graph of $y = ax^2 + bx + c$.

You will now consider simple cubic equations of the form $y = ax^3 + d$ to see how changing the coefficient a and the constant d affect the graphs.

2. Complete Parts a–c for each of these three groups of equations.

Group I	Group II	Group III
$y = x^3$	$y = 2x^3$	$y = 3x^3 + 1$
$y = x^3 + 3$	$y = \frac{1}{2}x^3$	$y = 3x^3 - 1$
$y = x^3 - 3$	$y = {}^-2x^3$	$y = {}^-3x^3 - 1$

    a. Graph the three equations in the same window of your calculator. Choose a window that shows all three graphs clearly. Sketch and label the graphs.

    b. Write a sentence or two about how the graphs are the same and how they are different.

    c. What does the coefficient of the x^3 term seem to tell you about the graph?

    d. What does the constant tell you about the graph?

Just the facts

The words *linear, quadratic,* and *cubic* come from the idea that lines, squares (also called quadrangles), and cubes have 1, 2, and 3 dimensions, respectively.

Problem Set I

Now consider what happens when a cubic equation has x^2 and x terms. Consider these three equations.

$$y = x^3 - x \qquad y = {}^{-}x^3 + 2x^2 + 5x - 6 \qquad y = 2x^3 - x^2 - 5x - 2$$

1. Graph the equations in the same window of your calculator, and sketch the graphs. Be sure to label the graphs and the axes.

2. Describe in words the general shapes of the graphs.

3. For each graph, find the points where the graph crosses the x-axis and y-axis.

4. Suppose you moved the graph of $y = x^3 - x$ up 1 unit.

   a. What is the equation of the new graph?

   b. Graph your new equation. How many x-intercepts does the graph have?

Share & Summarize

Look back at the graphs you made for this investigation. Describe what you observe about how graphs of linear, quadratic, and cubic relationships differ.

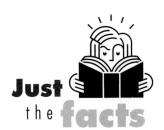

Just the facts

Cubic equations are often involved in the design of the curved sections of sailboats.

Lab Investigation ▶ Solving Graph Design Puzzles

MATERIALS

graphing calculator

Remember
The point where the x-axis and y-axis intersect, (0, 0), is called the *origin*.

These simple designs are made from the graphs of quadratic equations.

Design A

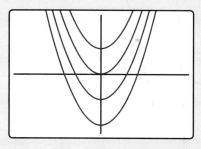

In this design, the parabolas are equally spaced and have their vertices on the *y*-axis.

Design B

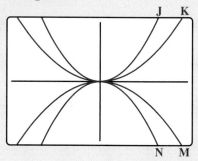

In this design, each parabola has a vertex at the origin. Parabola J has the same width as Parabola N. Parabola K has the same width as Parabola M.

Design C

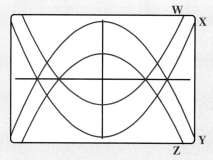

In this design, Parabola W has the same width as Parabola Z, and their vertices are the same distance from the origin. Parabola X has the same width as Parabola Y, and their vertices are the same distance from the origin.

Design D

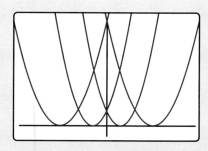

In this design, the vertices of the four parabolas are equally spaced along the *x*-axis. Two of those points are to the left of the origin, and two are to the right.

You will use what you have learned about quadratic relationships to try to re-create these designs on your calculator. Then you will create your own design.

Try It Out

With your group, choose one of Designs A, B, or C. Try to create the design on your calculator. Use equations in the form $y = ax^2 + bx + c$ and experiment with different values of a, b, and c. You may need to adjust the viewing window to make the design look the way you want. (Design D is the most difficult. You will have a chance to try it later.)

1. When you have created the design, make a sketch of the graph. Label each curve with its equation. Also label the axes, including the maximum and minimum values on each axis.

2. Different sets of equations and window settings can give the same design. Compare your results for Question 1 with the other members of your group or with other groups who chose your design. Did you record the same equations and window settings?

Try It Again

3. Now create each of the other three designs. For each design, make a sketch and record the equations and window settings you used.

Take It Further

4. Work with your group to create a new design from the graphs of four quadratic equations.

   a. Make a sketch of your design and—*on a separate sheet of paper*—record the equations and window settings you used.

   b. Exchange designs with another group and try to re-create their design.

What Did You Learn?

5. Write a report about the strategies you used to re-create the designs. For each design, discuss these points:

   a. How did you change the coefficients, a or b, or the constant, c, to create each design?

   b. Did any of the coefficients or constants have a value of 0? If so, why was having that value equal to 0 necessary to create the design?

   c. Did you change the window settings to make any of the designs? If so, explain how changing either the range of x values or the range of y values affects the design.

Just the facts

The design of satellite dishes is based on the parabola.

On Your Own Exercises

1. For each table of values, find an equation it may represent. Look for connections between the tables that may help you determine the equations. Explain how your found your solutions.

a.

x	y
-3	9
-2	4
-1	1
0	0
1	1
2	4
3	9

b.

x	y
-3	25
-2	16
-1	9
0	4
1	1
2	0
3	1

c.

x	y
-3	0
-2	1
-1	4
0	9
1	16
2	25
3	36

d.

x	y
-3	-18
-2	-8
-1	-2
0	0
1	-2
2	-8
3	-18

e.

x	y
-3	-6
-2	-4
-1	-2
0	0
1	2
2	4
3	6

f.

x	y
-3	-8
-2	2
-1	8
0	10
1	8
2	2
3	-8

In Exercises 2–5, match the equation with one of the graphs below. Explain your reasoning.

2. $y = x^2$

3. $y = (x - 2)^2$

4. $y = (x + 3)^2$

5. $y = {}^-2x^2$

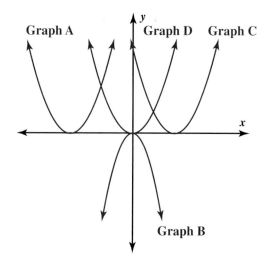

impactmath.com/self_check_quiz

In Exercises 6 and 7, match the equation with one of the graphs below. Explain your reasoning.

6. $y = {}^-2x^2 - 2x + 3$

7. $y = 2x^2 - 2x + 3$

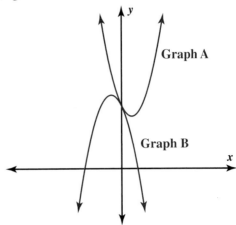

Graph A

Graph B

8. Could this be the graph of the equation $y = {}^-x^2 + 1$? Explain.

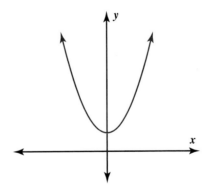

9. Could this be the graph of the equation $y = {}^-x^2 - 1$? Explain.

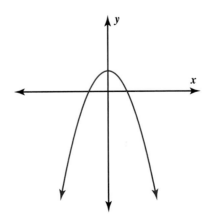

10. Could this be the graph of the equation $y = x^2 + 1$? Explain.

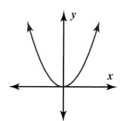

11. Graph A is the graph of $y = x^2$.

  a. Graph B is the graph of either $y = 2x^2$ or $y = \frac{x^2}{2}$. Which equation is correct? Explain how you know.

  b. Graph C is the graph of either $y = 3x^2$ or $y = \frac{x^2}{3}$. Which equation is correct? Explain how you know.

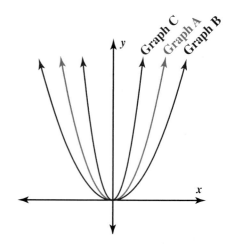

12. Consider these three graphs.

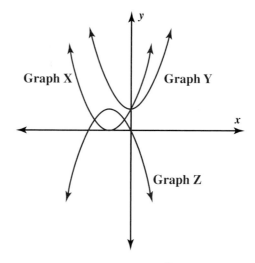

  a. Could Graph X be the graph of $y = x^2 + 1$? Explain.

  b. Could Graph Y be the graph of $y = (x + 1)^2$? Explain.

  c. Could Graph Z be the graph of $y = {}^-x^2 + 1$? Explain.

13. Sports A place kicker on a football team attempted three field goals during a game. All three were kicked from the opponent's 40-yard line, which is 50 yards from the goalpost. For a field goal to count, it must clear the crossbar, which is 10 feet high.

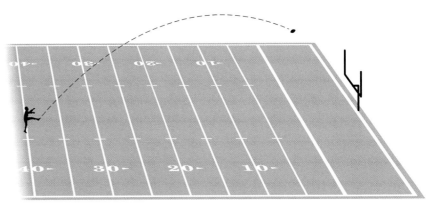

The football followed a different path through the air for each kick. These equations give the height of the kick in feet, h, for any distance from the kicker in yards, d. Each kick was aimed directly at the center of the goalpost.

Kick 1: $h = 3.56d - 0.079d^2$

Kick 2: $h = 1.4d - 0.0246d^2$

Kick 3: $h = 2d - 0.033d^2$

a. For each kick, plot enough points to draw a smooth curve. Plot all three graphs on the same axes and label them Kick 1, Kick 2, and Kick 3. Put distance from the kicker, from 0 to 70 yards, on the horizontal axis. Put height, from 0 to 50 feet, on the vertical axis.

b. Use your graphs to estimate the maximum height of each kick.

c. Use your graphs to estimate how many yards each kick traveled over the field before it struck the ground.

d. To make a field goal, the football must cross over the goalpost. That means it must be at least 10 feet high when it reaches the post, which is 50 yards from the kicker.

Use your graphs to estimate whether any of the kicks could have scored a field goal. Explain your reasoning.

Tell whether each relationship is quadratic.

14. $y = 3x + 5$

15. $y = {}^-(x + 1)^2$

16. $y = \frac{5}{x}$

17. $y = (x - 1)^2$

18. $y = {}^-(x + 1)$

19. $y = \frac{5}{x^2}$

20. $y = (3x + 5)^2$

21. $y = 7x^3 + 3x^2 + 2$

22. $y = x^3 + 7x + 5$

23. $y = 2^x$

24. Consider these three equations.

$$y = x + 3 \qquad y = x^2 + 3 \qquad y = x^3 + 3$$

a. For each equation, draw a rough sketch showing the general shape of the graph. Put all three graphs on one set of axes.

b. Write a sentence or two about how the graphs are the same and how they are different.

25. Graph A is the graph of $y = x^3$.

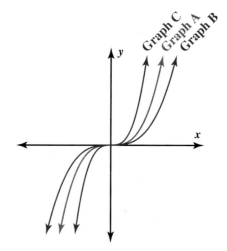

a. Graph B is the graph of either $y = \frac{x^3}{2}$ or $y = 2x^3$. Which equation is correct? Explain.

b. Graph C is the graph of either $y = \frac{x^3}{3}$ or $y = 3x^3$. Which equation is correct? Explain.

Connect & Extend

26. The quadratic equations below are more complicated than those you worked with in Investigation 1. Just as with the simpler equations, you can make a table of values and plot a graph of the equation.

$$y = 3m^2 + 2m + 7 \qquad p = n^2 + n - 6 \qquad s = 2t^2 - 3t + 1$$

Here is a table of values and a graph for $y = 3m^2 + 2m + 7$.

m	$^-3$	$^-2.5$	$^-2$	$^-1$	$^-0.5$	0	0.5	1	2	2.5
y	28	20.75	15	8	6.75	7	8.75	12	23	30.75

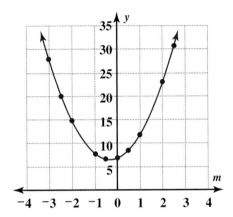

a. Prepare a table of ordered pairs for $p = n^2 + n - 6$. Plot the points on graph paper, and draw a smooth curve through them.

b. Prepare a table of ordered pairs for $s = 2t^2 - 3t + 1$. Plot the points on graph paper, and draw a smooth curve through them.

c. How are the graphs for the three equations alike?

d. How do the three graphs differ? In particular, where are their lowest points, and where are their lines of symmetry?

27. Each of these four graphs represents either $y = x^2 + 1$ or $y = 2x^2 + 1$.

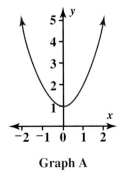

Graph A

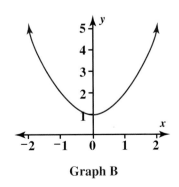

Graph B

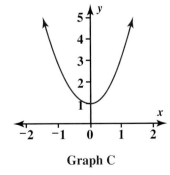

Graph C

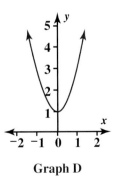

Graph D

a. Match each graph to its equation. Explain your reasoning.

b. How can graphs that look different have the same equation?

28. Challenge Consider the equation $y = x^2 - 4x$.

a. Identify the values of a, b, and c in this quadratic equation.

b. Where does the graph of $y = x^2 - 4x$ cross the y-axis? How does this point relate to the value of c in the equation?

c. Graph $y = x^2 - 4x$ by making a table and plotting points. Make sure your graph shows both halves of the parabola.

d. Give the coordinates of the points where the graph crosses the x-axis.

e. Use the distributive property to write $x^2 - 4x$ as a product of two factors.

f. How do the points where the graph crosses the x-axis relate to this factored form?

29. Passengers in a hot air balloon can see greater and greater distances as the balloon rises. The table shows data relating the height of a hot air balloon with the distance the passengers can see—the distance to the horizon.*

Study the table to observe what happens to d as h increases by equal amounts.

Just the **facts**

In May 1931, Auguste Piccard of Switzerland became the first person to reach the stratosphere when he ballooned to almost 52,000 feet. In October 1934, Jeanette Piccard became the first woman to reach the stratosphere when she and her husband (Auguste's twin brother Jean) ballooned to almost 58,000 feet.

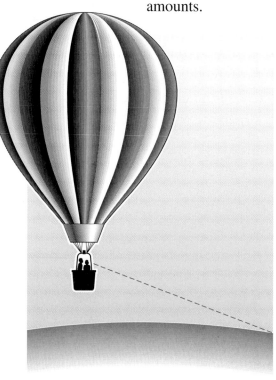

Height (meters), h	Distance to Horizon (kilometers), d
0	5
10	11
20	16
30	20
40	23
50	25

*Adapted with permission from the *Language of Functions and Graphs*, p. 110. Shell Centre for Mathematical Education, University of Nottingham. Published Dec. 1985 by the Joint Matriculation Board, Manchester.

a. Below are three graphs. Which is most likely to fit the data in the table? Explain.

In this graph, d increases by constant amounts as h increases.

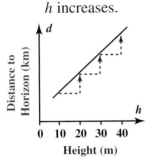

In this graph, d increases by greater amounts as h increases.

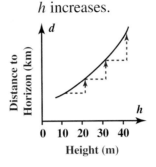

In this graph, d increases by smaller amounts as h increases.

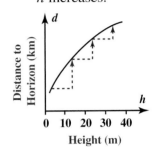

b. Check your answer to Part a by sketching a graph that represents the relationship between d and h.

c. Look at d in relation to h. Do you think these data could represent a quadratic relationship? Explain.

30. Consider this pattern of cube figures.

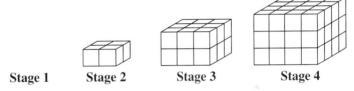

Stage 1   Stage 2   Stage 3   Stage 4

a. Copy and complete the table for the pattern.

Stage, n	1	2	3	4
Cubes, C				

b. How many cubes will be needed to build Stage 5?

c. Write an equation for the number of cubes in Stage n. Explain how you found your answer.

31. Sometimes people confuse the *quadratic* relationship $y = x^2$ with the *exponential* relationship $y = 2^x$.

a. Copy and complete the table of values for $y = x^2$ and $y = 2^x$.

x	-4	-3	-2	-1	0	1	2	3	4	5	6
$y = x^2$	16	9	4	1	0	1					
$y = 2^x$	$\frac{1}{16}$	$\frac{1}{8}$	$\frac{1}{4}$	$\frac{1}{2}$	1	2					

b. For which values of x shown are x^2 and 2^x equal?

c. For which values of x shown is x^2 greater than 2^x?

d. For which values of x shown is 2^x greater than x^2?

e. Compare the way the values of $y = x^2$ change as x increases by 1 to the way the values of $y = 2^x$ change as x increases by 1.

f. How do you think x^2 and 2^x compare for values of x greater than 6?

g. Use the table of values to graph $y = x^2$ and $y = 2^x$ on the same set of axes.

Mixed Review

Evaluate each expression for the given values.

32. $h(g - h^i)$ for $h = 2$, $g = -1$, and $i = 5$

33. $a^d + ab - b^c$ for $a = -3$, $b = -2$, $c = 3$, and $d = 2$

34. $(m - km)^k$ for $m = -4$ and $k = 3$

Find the value of b in each equation.

35. $3^b = 27$

36. $b^b = 256$

37. $(-b)^b = -27$

Find the slope of the line through the given points.

38. $(-2, 3)$ and $(5, 8)$

39. $(0, -6)$ and $(-8, 0)$

40. $(-3.5, 1.5)$ and $(0.5, 2)$

41. $(-7, -2)$ and $(-9, -2)$

42. Consider this rectangle.

f
9

a. Write an expression for the rectangle's area.

b. Use your expression to write an equation stating that the rectangle's area is 108 square units.

c. Solve your equation to find the value of f.

43. Consider this rectangle.

f
9

a. Write an expression for the rectangle's perimeter.

b. Use your expression to write an equation stating that the rectangle's perimeter is 39 units.

c. Solve your equation to find the value of f.

44. Rachel divided her fossil collection into three categories: plant fossils, insect fossils, and other animal fossils. She had eight more plant fossils than animal fossils. The number of insect fossils she has is three less than four times the number of plant fossils she has.

a. If a represents the number of animal fossils Rachel has, write an expression for the number of plant fossils she has.

b. Write an expression for the number of insect fossils Rachel has.

c. Rachel has 41 more insect fossils than animal fossils. Use this fact, and your expression from Part b, to write an equation for this situation.

d. Solve your equation to find how many fossils Rachel has in each category.

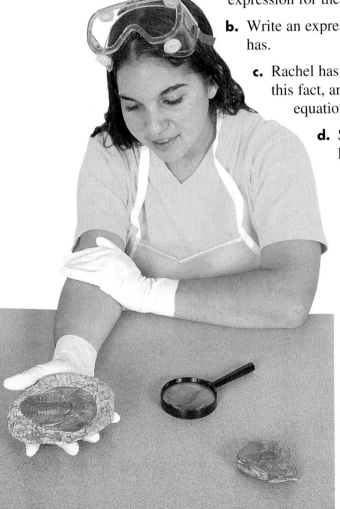

2.3 Inverse Variation

In this lesson, you will study a new type of relationship: relationships in which the product of two quantities is always the same. As with quadratics, you will discover that relationships of this type

- have graphs with a distinctive shape
- have tables that show a recognizable pattern
- have equations of a particular form

Explore

A group of students volunteered to paint a fence at a local park.

The more volunteers we get, the sooner we'll finish the job.

Assume all the volunteers work at the same rate. If you created a graph with the number of volunteers on the horizontal axis and the time needed to paint the fence on the vertical axis, what would the graph look like?

Discuss this question in your group and then sketch the graph. Just think about the general shape of the graph; don't worry about exact numbers.

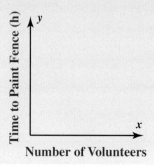

Now discuss these questions with your class:

- Will the graph increase or decrease from left to right? Why?

- Will the graph be a straight line? Why or why not?

- Will the graph intersect the axes? Why or why not?

- If one student who starts at 9:00 A.M. and doesn't take a break can finish the job by 5:00 P.M., it would make sense to put the point (1, 8) on the graph. The point (2, 4) would also be on the graph. Explain what this point represents and how you know it would be on the graph.

- What would the point (4, 2) represent?

- Which of the following points might reasonably be on the graph that includes (1, 8), (2, 4), and (4, 2):

$$(5, 1.6) \qquad (8, 1) \qquad (80, 0.1)$$

Explain your answers.

If one volunteer can complete the job in 8 hours, two can finish in 4 hours, 3 in $2\frac{2}{3}$ hours, and so on. In each case, the number of *person-hours*—the number of people multiplied by the number of hours—is 8. So, one way to express the relationship between the time it takes to finish the job T and the number of volunteers V is

$$TV = 8$$

In this lesson, you will explore several other situations in which the product of two variables is constant.

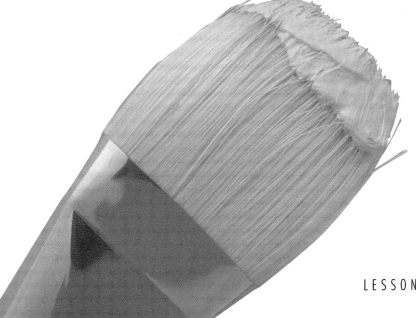

Investigation ▶1▶ When *xy* Is Constant

As you work on these problems, look for similarities in the equations, graphs, and tables.

MATERIALS

graph paper

Problem Set A

Ms. Anwar is considering renting a house that has a large rectangular backyard. She wants to figure out if there will be room for her children's play equipment. The owner told her, "The backyard has an area of 2,000 square feet." Ms. Anwar thought about what he said and tried to imagine what the actual dimensions of the yard might be.

1. Use x to represent length, in feet, of one side of the yard and y to represent length, in feet, of an adjacent side. Select some possible values for x and y and complete a table like this one.

x	50	25	100					
y	40							
Area, xy	2,000							

2. Write an equation that shows how x and y are related.

3. Make a graph of the relationship between x and y. Start by plotting the points in your table. Add more points, if you need to, until you can clearly see the shape of the graph. Draw a smooth curve through the points.

4. Think about how y changes as x changes for this situation.

   a. Copy and complete the table to show what happens.

x	20	30	40	50	60	70	80	90	100
y	100	66.7							
Decrease in y	—	33.3							

   b. As x increases by a fixed amount, say 10, will y decrease by a fixed amount, an increasing amount, or a decreasing amount?

   c. If x doubles, what happens to y? If x triples, what happens to y? What happens to y if x is multiplied by N?

   d. As x gets very small (nears 0), what happens to y? As x grows very large, what happens to y?

Problem Set B

The owner of the house Ms. Anwar is renting decided to have the over-grown yard cleaned up. He contacted Rob, a high school student, and after much bargaining agreed to pay him $240 for the job.

If Rob works very hard, the job will be done quickly and the hourly rate of pay will be high. If he works at a more leisurely pace, the job will take longer and the hourly rate of pay will be lower.

1. Complete the table.

Hours of Work	$\frac{1}{2}$		2		10	20	
Hourly Pay Rate		$240		$48			$8

2. Write an equation for this situation using h for number of hours to complete the job and d for number of dollars earned per hour.

3. Sketch a smooth graph by first plotting enough number pairs (h, d) to see a pattern.

In all the situations you have seen so far, the variables made sense for positive values only. The dimensions of Ms. Anwar's yard, Rob's hourly pay rate, and the number of hours Rob works must all be greater than 0.

In the next problem set, you will look at another relationship in which a product of variables is constant. This time you will consider both positive and negative values of the variables.

Problem Set C

Consider the relationship $xy = 3$, where x and y can be positive or negative.

1. If you start with values of x, it is often easier to find values of y if you first rewrite the equation so y is alone on one side. This is some-times referred to as *solving for y in terms of x*. Solve the equation $xy = 3$ for y in terms of x.

2. Find values of y for five negative values and five positive values of x between ⁻10 and 10. Record the (x, y) values in a table. Be adventurous! Try some values that aren't integers, including values between 0 and 1.

3. What is y when $x = 0$? What happens if you use your calculator to find y when $x = 0$?

4. Make a smooth graph of this relationship by first plotting the pairs in your table. Add more points if necessary to get a good sense of the graph's shape.

Think about what happens to the graph near the *y*-axis. Can the graph cross the *y*-axis?

5. How are this graph and equation like the yardwork graph and equation from Problem Set B? How are they different?

Share & Summarize

1. How are the equations in Problem Sets A, B, and C alike? How are they different?

2. How are the graphs alike? How are they different?

Investigation Inverse Proportion

You have been working with relationships in which multiplying two variables gives a constant product. That is,

$$xy = a$$

where *x* and *y* are the variables and *a* is a nonzero constant.

The graph of such a relationship is a curve like this:

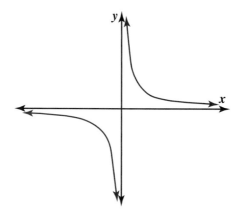

This type of curve is called a **hyperbola.**

Think & Discuss

Think about the three relationships you worked with in Investigation 1:

$$xy = 2{,}000 \qquad dh = 240 \qquad xy = 3$$

In each case, if you double the value of one variable, what happens to the other?

What happens if you triple the value of one variable? What happens if you halve the value of one variable?

VOCABULARY
**inversely
  proportional**
inverse variation

In Chapter 1, you learned that two variables are *directly proportional* if the values of the variables have a constant *ratio*. When two variables have a constant nonzero *product,* they are said to be **inversely proportional.** A relationship in which two variables are inversely proportional is called an **inverse variation.**

You will now explore the connection between inverse proportions and *reciprocals.*

MATERIALS
graph paper

Problem Set D

1. Find the reciprocal of each number. Express your answers as integers or fractions, not as mixed numbers or decimals.

   a. 2   **b.** 5

   c. $^{-}1$ **d.** $\frac{1}{3}$

   e. $\frac{3}{4}$ **f.** $-\frac{9}{7}$

   g. $\frac{7}{3}$ **h.** $^{-}10$

2. Look at your answers to Problem 1.

   a. Find the reciprocal of each answer.

   b. What do you notice?

3. If x is not 0, what is the reciprocal of x?

4. What is the reciprocal of $\frac{1}{x}$?

5. What is the reciprocal of the reciprocal of x?

Remember
The *reciprocal of a
number* is what the
number is multiplied by
to get 1. For example,
the reciprocal of 7 is $\frac{1}{7}$,
since $7 \cdot \frac{1}{7} = 1$.

Remember

x^{-1} is another way of writing $\frac{1}{x}$. Your calculator probably has a reciprocal key labeled $\boxed{1/x}$ or $\boxed{x^{-1}}$.

6. Find the reciprocal of each number. Express your answers as decimals.

  a. 8

  b. 100

  c. 0.2

  d. $^-0.25$

  e. 12

  f. 7.5

  g. $^-1$

  h. 0.0004

7. Look at your answers to Problem 6.

  a. Find the reciprocal of each answer.

  b. What do you notice?

8. Consider the relationship $y = \frac{1}{x}$.

  a. What happens to y if you double x? If you triple x? If you quadruple x? If you halve x?

  b. Is y inversely proportional to x? That is, is $y = \frac{1}{x}$ an inverse variation? Explain.

  c. Sketch a graph of $y = \frac{1}{x}$.

Problem Set E

Consider the relationship $y = \frac{5}{x}$.

  1. What happens to y if you double x? If you triple x? If you quadruple x? If you halve x?

  2. Is y inversely proportional to x? Explain.

  3. What is the value of $y = \frac{5}{x}$ when $x = 0$? What happens if you calculate $\frac{5}{0}$ on your calculator?

  4. Copy and complete the table. Write the entries as integers or fractions.

x	$^-5$	$^-4$	$^-3$	$^-2$	$^-1$	$^-\frac{1}{2}$	$^-\frac{1}{4}$	0	$\frac{1}{4}$	$\frac{1}{2}$	1	2	3	4	5
$y = \frac{1}{x}$															
$y = \frac{5}{x}$															

  5. Sketch a graph of $y = \frac{5}{x}$ on the same axes you used for the graph of $y = \frac{1}{x}$.

The relationships $y = \frac{1}{x}$ and $y = \frac{5}{x}$ are inverse variations. Notice that when the value of one variable is multiplied by a number, the value of the other variable is multiplied by the *reciprocal* of that number. For example, for $y = \frac{5}{x}$, when $x = 1$, $y = 5$. But when you multiply x by 3, y becomes $\frac{5}{3}$, or $5 \cdot \frac{1}{3}$. For this reason, inverse variations are sometimes called **reciprocal relationships.**

V O C A B U L A R Y
reciprocal relationship

Problem Set F

Tell whether each equation represents a reciprocal relationship. If the answer is no, tell what type of relationship the equation *does* represent.

1. $st = 6$

2. $s = 2t^2$

3. $s = 6t$

4. $y = \frac{x}{7}$

5. $y = \frac{7}{x}$

6. $x = \frac{2}{y}$

For Problems 7–9, complete Parts a and b.

a. Tell which kind of relationship could exist between the variables: linear, quadratic, or reciprocal.

b. Write an equation that relates the quantities.

7.

x	0.5	1	2	3	4	5	10
y	⁻2.5	⁻5	⁻10	⁻15	⁻20	⁻25	⁻50

8.

p	0.5	1	2	3	4	5	10
q	1.25	2	5	10	17	26	101

9.

t	0.5	1	2	3	4	5	10
r	3	1.5	0.75	0.5	0.375	0.3	0.15

Share & Summarize

1. If you are given a table of values for two variables, how can you tell whether the relationship between the variables could be an inverse variation?

2. If you are given an equation with two variables, how can you tell whether the relationship between the variables is an inverse variation?

Investigation Relationships with Constant Products

You have been studying relationships in which the product of two quantities is a positive constant. All these relationships can be written using equations in the form $xy = a$ or, equivalently, $y = \frac{a}{x}$, where a is a positive number. A group of equations of the same general form is sometimes called a *family* of equations.

Now you will explore other families of equations that describe relationships in which a product is constant. You will look at equations in the following forms:

- $xy = a$ or, equivalently, $y = \frac{a}{x}$, where a is a negative constant
- $(x + b)y = 1$ or, equivalently, $y = \frac{1}{x + b}$, where b is a constant
- $x(y - c) = 1$ or, equivalently, $y = \frac{1}{x} + c$, where c is a constant

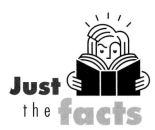

Just the facts

This type of relationship is used in the design of telescopic lenses, which collect and focus light in telescopes.

Allegheny Observatory, Pittsburgh, Pennsylvania

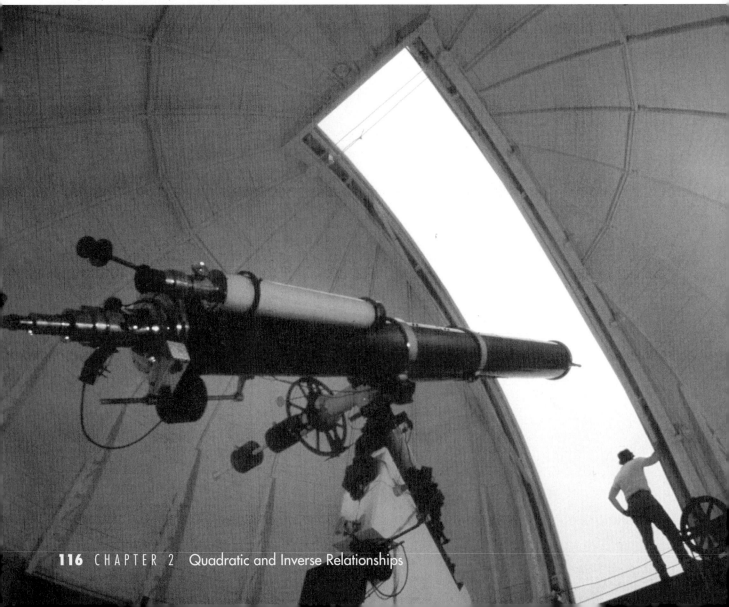

Think & Discuss

You have already studied the relationship $y = \frac{1}{x}$. Now think about $y = -\frac{1}{x}$. Here are a table and a graph for this relationship.

x	$^-10$	$^-0.2$	0.5	1	2	5	10
y	0.1	5	$^-2$	$^-1$	$^-0.5$	$^-0.2$	$^-0.1$

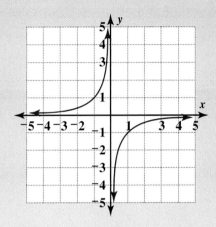

Think about parts of the graph that are not shown.

What is the y value for $x = 100$? For $x = 1,000$?

What happens as the x values grow even greater than 1,000? Will y ever equal 0?

What happens to the y values as positive x values get closer to 0? What happens when $x = 0$? What does this mean for the graph of $y = -\frac{1}{x}$?

What is the y value for $x = ^-100$? For $x = ^-1,000$?

What happens as the x values decrease beyond $^-1,000$? Will y ever equal 0?

What happens to the y values as negative x values grow closer to 0? What happens when $x = 0$?

As you probably just saw, for $y = \frac{1}{x}$ and $y = \frac{-1}{x}$, as the absolute value of x grows larger and larger, y approaches 0 without ever reaching it.

x	1	10	100	1,000	10,000	100,000	$\cdots$
$\frac{1}{x}$	1	0.1	0.01	0.001	0.0001	0.00001	$\cdots$

x	$^-1$	$^-10$	$^-100$	$^-1,000$	$^-10,000$	$^-100,000$	$\cdots$
$\frac{1}{x}$	$^-1$	$^-0.1$	$^-0.01$	$^-0.001$	$^-0.0001$	$^-0.00001$	$\cdots$

Remember

A number's *absolute value* is its distance from 0 on the number line. The absolute value of $^-4$ is 4.

MATERIALS

graphing calculator

In the same way, as the absolute value of y grows larger and larger, x approaches 0 without ever reaching it. This is a characteristic of all inverse variations: there is always some value—not always 0—that x approaches but does not reach as the absolute value of y becomes very large, and there is some value that y approaches but does not reach as the absolute value of x becomes very large.

Problem Set G

Work with your group to explore one of the families of equations described below.

Family A
Equations of the form $y = \frac{a}{x}$, such as $y = \frac{3}{x}$ and $y = \frac{-7}{x}$

Family B
Equations of the form $y = \frac{1}{x + b}$, such as $y = \frac{1}{x + 2}$ and $y = \frac{1}{x - 1}$

Family C
Equations of the form $y = \frac{1}{x} + c$, such as $y = \frac{1}{x} + 5$ and $y = \frac{1}{x} - 2$

Your goal is to understand what the graphs of relationships in the family you choose look like, and how the graphs change as the constant—a, b, or c—changes. Your group should prepare a report describing its findings, including the following:

- a series of sketches, carefully drawn and labeled to show how the graphs change as the constant changes

- a written explanation of your findings, including answers to Problems 1–6

1. How does the graph change as the constant changes?

2. How does changing the constant from positive to negative affect the graph?

3. What value does x approach as the absolute value of y grows very large? Does this change as the constant changes?

4. What value does y approach as the absolute value of x grows very large? Does this change as the constant changes?

5. For what value or values of x is there no value of y? Does this change as the constant changes?

6. How do the graphs compare to the graph of $y = \frac{1}{x}$?

Just the facts

Equations involving constant products are used to describe the path that some comets take through the sky.

Share & Summarize

1. Compare how the value of a affects graphs of equations of the form $y = ax^2$ to how it affects graphs of equations of the form $y = \frac{a}{x}$.

2. Compare how the value of b affects graphs of equations of the form $y = (x + b)^2$ to how it affects graphs of equations of the form $y = \frac{1}{x + b}$.

3. Compare how the value of c affects graphs of equations of the form $y = x^2 + c$ to how it affects graphs of equations of the form $y = \frac{1}{x} + c$.

Halley's Comet

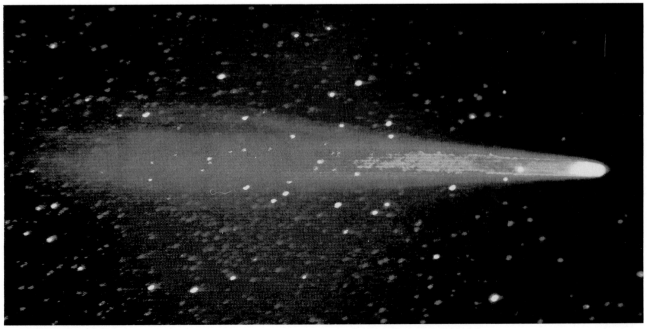

On Your Own Exercises

1. Miguel gave his little sister Jenna $3 to spend on stickers. The price per sticker varies depending on the type of sticker. Jenna wants all the stickers she buys to be of the same type.

   a. Make a table of possible values for the price per sticker p and the number of stickers n. One pair of values is shown as an example.

Price per Sticker, p	$0.25						
Number Purchased, n	12						

   b. Make a graph of the relationship between p and n, with p on the horizontal axis. Start by plotting the points in your table, adding more points if you need to, until you can clearly see the shape of the graph. Draw a smooth curve through the points.

   c. What happens to the value of n as the value of p doubles? As the value of p triples? As the value of p quadruples?

   d. What happens to the value of n as the value of p is halved? As the value of p is quartered?

   e. Write an equation for the relationship between p and n.

2. Manuel plans to serve a giant pizza at his birthday party. The amount of pizza each guest will be served depends on the number of guests.

   a. Copy and complete the table showing the fraction of the pizza each person will be served, f, for different numbers of people, n. Assume each person gets the same amount of pizza and that the entire pizza is served.

n	1	2	3	4	5	6	7	8	9	10	11	12
f												

 impactmath.com/self_check_quiz

b. Graph the relationship between n and f by plotting the points in your table and drawing a smooth curve through them.

c. Write an equation for the relationship between n and f.

3. Consider the equation $xy = 10$, where x and y can be positive or negative.

a. Solve for y in terms of x.

b. Make a table of values for this equation. Choose x values between $^-10$ and 10. Consider positive x values, negative x values, and x values between $^-1$ and 1.

c. What is the value of y when x is 0? Explain.

d. Make a graph of this relationship by first plotting the points from your table. Add points if you need to until you can see the shape, and then connect the points with a smooth curve. Think carefully about what happens to the graph near the y-axis.

In Exercises 4–9, decide whether the relationship is an inverse variation. If it isn't, tell what type of relationship it is.

4. $st = \frac{1}{4}$

5. $s = \frac{6t}{8}$

6. $s = \frac{t^2}{4}$

7. $y = 0.25x$

8. $y = \frac{0.25}{x}$

9. $x = \frac{0.25}{y}$

For Exercises 10–12, do Parts a–c.

a. Tell whether the variables could be inversely proportional. Explain how you know.

b. Plot the points on a pair of axes. Put the variable in the top row on the horizontal axis.

c. Write an equation for the relationship between the variables.

10.

m	0.5	1	2	3	4	5	10
n	60	30	15	10	7.5	6	3

11.

x	0.5	1	2	3	4	5	10
y	0.25	0.5	1	1.5	2	2.5	5

12.

t	0.25	0.5	0.75	1	1.25	1.5	2
r	0.5	0.25	0.1667	0.125	0.1	0.0833	0.0625

13. Economics Antoine wants to spend $4 on kiwi fruit at the farmer's market. The price per kiwi varies from stall to stall.

a. Is the relationship between the price per kiwi in dollars, *p*, and the number of kiwi Antoine can purchase, *n*, an inverse variation? Explain.

b. Write an equation for the relationship between *p* and *n*.

14. Mangos are very expensive this time of year at the farmer's market! They cost $4 each, but Carmen can't resist them.

a. Is the relationship between the total price Carmen pays for mangos, *p*, and the number of mangos she buys, *n*, an inverse variation? Explain.

b. Write an equation for the relationship between *p* and *n*.

15. Kiyoshi spent $8 at the farmer's market. She bought only mandarin oranges, which are $0.40 each, and nectarines, which are $0.80 each.

a. Is the relationship between the number of mandarin oranges, *m*, and the number of nectarines, *n*, an inverse variation? Explain.

b. Write an equation for the relationship between *m* and *n*.

16. Consider equations of the form $y = \frac{a}{x}$.

a. On one set of axes, make rough sketches of the graphs for the three equations below. Use *x* and *y* values from ⁻10 to 10.

  i. $y = \frac{-1}{x}$ **ii.** $y = \frac{-5}{x}$ **iii.** $y = \frac{-10}{x}$

b. Describe how the graphs of $y = \frac{a}{x}$ change as *a* decreases.

17. Consider equations of the form $y = \frac{a}{x+b}$.

a. On one set of axes, make rough sketches of the graphs for the three equations below. Use *x* and *y* values from ⁻10 to 10.

  i. $y = \frac{1}{x+1}$ **ii.** $y = \frac{1}{x+3}$ **iii.** $y = \frac{1}{x+5}$

b. Describe how the graphs of $y = \frac{a}{x+b}$ change as *b* increases.

18. Consider equations of the form $y = \frac{a}{x} + c$.

a. On one set of axes, make rough sketches of the graphs for the three equations below. Use *x* and *y* values from ⁻10 to 10.

  i. $y = \frac{1}{x} + 2$ **ii.** $y = \frac{1}{x} + 4$ **iii.** $y = \frac{1}{x} + 6$

b. Describe how the graphs change as the value of *c* increases.

Connect & Extend

19. **Physical Science** The time it takes to reach a destination depends on the speed of travel. Suppose it is 500 miles to a campsite you want to visit. Here are some ways you might get there. Some are not as practical as others, but use your imagination.

Mode of Transportation	Average Speed (mph)
Subway train	30
Car	50
Helicopter	85
Light plane	
Rocket	1,100
Horseback	
Bicycle	

a. Make estimates for the missing speeds in the table. (Don't worry about how accurate your estimates are.)

b. Copy and complete the table to show how long it would take to travel 500 miles using the estimated speed for each type of transportation.

Speed (mph)	30	50	85	1,100			
Travel Time (h)							

c. Write an equation stating the relationship between speed S and time T required to travel 500 miles.

d. Use the values in your table to sketch a smooth graph of the relationship between speed and time. Put speed on the horizontal axis.

e. What happens to the time values as the speed values increase?

f. Could the travel time for the 500-mile journey ever reach 0?

g. The speed of light is 186,000 miles per second. Determine how long the journey would take if you could ride on a light beam. Is this value shown on your graph?

h. Write an equation that gives the speed needed to travel 500 miles in T hours.

20. Compare the inverse relationship $y = \frac{3}{x}$ with the linear relationship $y = 3x$.

a. Graph and label both equations on one set of axes.

b. Consider the parts of the graphs in the first quadrant. What happens to each graph as x increases?

c. Consider the parts of the graphs in the first quadrant. As x grows closer to 0, what happens to each graph?

d. What happens when $x = 0$?

21. A large pipe organ has pipes with lengths ranging from a few centimeters to 4 meters. The table lists the pipe lengths for the E flats of all octaves, along with the frequencies of the sounds they produce. Higher sound frequencies correspond to higher pitches.

Pipe length (meters), l	4.0	2.0	1.0	0.5	0.25	0.125	0.0625
Frequency (cycles per second), f	39	78	156	312	622	1,244	2,488

a. Plot the points, using a suitable scale for the axes. Put pipe length on the horizontal axis. Draw a smooth curve through the points.

b. Use your graph to predict the pipe length that would produce an A note with a frequency of 440 cycles per second.

c. Use your graph to predict the frequency produced by a pipe 3 meters long.

d. What is the shape of the graph? What does this shape suggest about the relationship between frequency and pipe length?

e. Find an equation that fits the data fairly well. Why do you think the data may not fit exactly?

22. You found that for the equation $y = \frac{1}{x}$, there is no y value corresponding to an x value of 0. We say that the equation $y = \frac{1}{x}$ is *undefined* for $x = 0$. In Parts a–e, list any x values for which the equation is undefined.

a. $y = \frac{1}{x - 1}$

b. $y = \frac{2}{1 - 2x}$

c. $y = \frac{1}{x + 2}$

d. $y = \frac{2}{x}$

e. $y = \frac{x}{2}$

f. $y = \frac{4}{3x - 12}$

23. Consider the equation $y = \frac{1}{x^2}$.

a. Make a table of values and a graph for this equation for x values between $^-10$ and 10. Plot enough points to draw a smooth curve.

b. How is this graph similar to the graph of $y = \frac{1}{x}$?

c. How is this graph different from the graph of $y = \frac{1}{x}$?

24. A game company has spent \$200,000 to develop new game soft-ware. Although manufacturing costs vary depending on how many units of a product are produced, the company estimates that manu-facturing and shipping will be \$4 for each unit they make and sell.

a. Suppose the company produces only 20 units of the program. What is the average cost of development for each of the 20 units?

b. What is the average *total* cost (development plus manufacturing and shipping) for each of the 20 units?

c. Write an expression for the average total cost per unit for n units.

d. Complete the table for this situation.

Units	20	200	2,000	20,000	200,000	2,000,000
Average Total Cost per unit (\$)	10,004					

e. Suppose the company produces more and more units. The average cost will get closer and closer to a particular amount. What is that amount?

f. How many units must the company produce for the average total cost to be less than \$1 greater than the amount you answered in Part e? How many units are needed for the cost to be less than 1¢ greater?

**Mixed
Review**

Rearrange each linear equation into slope-intercept form, $y = mx + b$.

25. $\frac{2}{3}x - \frac{5}{8} = 2y$

26. $3y - 4x - 1 = 8x - 2y$

27. $4(x + y) - 6(5 - x) = 6$

28. $\frac{7(y + 5)}{x + 2} = 10$

Simplify each expression as much as possible.

29. $a + 4(a - 2)$

30. $2b - (8 + 2b)$

31. $90 - (5c - 1)$

32. $4d(2 + e) - 2(3 + 2d)$

33. $^-7(1 - f) + 2(7f + 2) - 9(f + 2)$

Write an equation of a line that is parallel to the given line.

34. $y = 2x - 4$ **35.** $2n = 4m$

36. $2(y - 3) = 7x + 1$ **37.** $x = ^-2$

38. Economics Taylor has $60. He is interested in buying some sports socks, which sell for $3 a pair, and some baseball caps, which are on sale for $5 each.

    a. Write an expression that shows how much s pairs of socks and c caps would cost.

    b. Taylor spent his entire $60 on socks and caps. Use your answer from Part a to express this as an equation.

    c. Graph your equation. Put number of pairs of socks on the vertical axis and number of caps on the horizontal axis.

    d. Use your graph to find all the number pairs that represent how many caps and how many pairs of socks Taylor could have bought. Be careful: he can buy only whole numbers of each item.

39. Candace's grandfather gave her four marbles: one blue, one yellow, and two green. She wants to give half of the marbles to her friend Luisa for her birthday. Candace puts all the marbles in a leather bag, and lets Luisa choose two of them without looking. What is the probability that Luisa will choose both green marbles?

40. Bryan drew this sketch of a playhouse he'd like to build. What is the area of the playhouse floor?

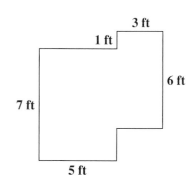

2.4 Conjectures

In the first three lessons of this chapter, you learned about quadratic and reciprocal relationships. This lesson will give you more practice with these types of relationships as you discover an important process in mathematics: making and proving *conjectures*.

VOCABULARY
conjecture

A **conjecture** is an educated guess or generalization that you haven't (yet) proved correct. Sometimes you have evidence that leads you to make a conjecture. Other times you just "get a feeling" about what will happen and make a conjecture based on very little evidence.

Remember

Consecutive integers follow one another, such as 1 and 2 or 5 and 6.

Explore

Desiree and Yutaka were looking for patterns in the products of pairs of consecutive integers.

$$5 \cdot 6 = 30$$
$$7 \cdot 8 = 56$$
$$10 \cdot 11 = 110$$

Yutaka noticed a pattern relating the products to the square of the first number.

Copy and complete the table.

First Integer	5	7	10	90	0	$^-4$	$^-1$
Second Integer	6	8	11			$^-3$	
Product	30	56	110				
First Integer Squared							

Compare the third and fourth rows of your table. Try to make a conjecture about the relationship between the product of two consecutive integers and the square of the first integer. Test your conjecture on a few more examples of your own.

Can you see a way to show that your conjecture is true?

Investigation 1 ▶ Making Conjectures

In this investigation, you will try to form conjectures about some other situations.

Problem Set A

You know that in a table for a linear relationship, as x values increase by 1, the differences in consecutive y values are constant. For example, look at this table for $y = 10 - 2x$.

x	1	2	3	4	5	6	7
y	8	6	4	2	0	$^-2$	$^-4$
Differences		$^-2$	$^-2$	$^-2$	$^-2$	$^-2$	$^-2$

Now look at the differences in y values for the quadratic relationship $y = {}^-x^2 + 2x - 1$.

x	1	2	3	4	5	6	7
y	0	$^-1$	$^-4$	$^-9$	$^-16$	$^-25$	$^-36$
Differences		$^-1$	$^-3$	$^-5$	$^-7$	$^-9$	$^-11$

The differences in the y values are not constant, but what happens if you look at *differences of the differences*?

1. Find the missing differences in this table for $y = {}^-x^2 + 2x - 1$.

x	1	2	3	4	5	6	7
y	0	$^-1$	$^-4$	$^-9$	$^-16$	$^-25$	$^-36$
Differences of the y Values		$^-1$	$^-3$	$^-5$	$^-7$	$^-9$	$^-11$
Differences of the Differences			$^-2$	?	?	?	?

To avoid confusion, the differences in the y values are called the *first differences* and the differences in the differences are called the *second differences*.

2. What do you notice about the second differences for the equation $y = {}^-x^2 + 2x - 1$?

3. Work with a partner to explore the first and second differences for two more quadratic equations. Try to make a conjecture based on your findings.

Once you have made a conjecture, the next step is to try to find a convincing argument to explain why your conjecture is true, or to find evidence that it is not true.

To show that a conjecture is true, it isn't enough to show that it *usually* works. It isn't even enough to say that the conjecture has worked in all the examples tested so far, unless you have tested *every* possible case. *You* may be convinced, but you need to be able to convince others as well.

And even if you're already convinced a conjecture is true, finding an argument to prove it is can help you understand *why* it's true.

Problem Set B

Consider the differences of squares of consecutive whole numbers.

Consecutive Numbers	Difference of Squares
1, 2	$2^2 - 1^2 = 3$
2, 3	$3^2 - 2^2 = 5$
3, 4	$4^2 - 3^2 = 7$
$\vdots$	$\vdots$
$n, n + 1$	$(n + 1)^2 - n^2 = D$

If n stands for any whole number, $n + 1$ is the next whole number. An equation for the difference of the squares D of these numbers is

$$D = (n + 1)^2 - n^2$$

1. Copy and complete the table to show the value of D for the given values of n.

n	1	2	3	4	5	6	7	8
D	3	5	7					

2. Use what you know about constant differences to determine what type of relationship $D = (n + 1)^2 - n^2$ is.

3. Use the table and your answer to Problem 2 to make a conjecture about what a simpler equation relating D to n might be.

You have made a conjecture about a simpler equation for D. Now you can try to find a convincing argument to explain why your equation must be correct. One way to do this is to use a little geometry.

You can represent the square of a whole number as a square made from tiles. The tile pattern below represents the squares of consecutive whole numbers.

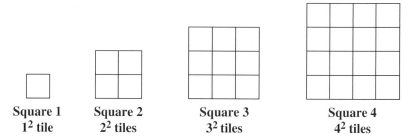

Square 1
1^2 tile

Square 2
2^2 tiles

Square 3
3^2 tiles

Square 4
4^2 tiles

The difference between the squares of two consecutive whole numbers is the number of tiles you must add to get from one square to the next. (Make sure you understand why this is true.)

4. Think about how you could add tiles to get from one square to the next in this pattern.

   a. Copy the pattern, and color the tiles that are added at each stage.

   b. How many tiles are added to go from Square 1 to Square 2? From Square 2 to Square 3? From Square 3 to Square 4?

   c. How many tiles are added to go from Square n to Square $n + 1$? Explain how you found your answer.

   d. If you did Part c correctly, your answer should prove your conjecture from Problem 3. Can you explain how?

Share & Summarize

1. Explain how finding first and second differences can help you determine whether a relationship is linear or quadratic.

2. How did you develop your conjecture for Problem 3 of Problem Set B?

3. How did you use geometry to prove your conjecture?

Investigation ▶2▶ Detective Work

Making and proving a conjecture is a bit like being a detective. You might start with a hunch, and then investigate further, looking for evidence to either support or disprove your hunch. Good detectives don't search only for evidence to support their conjectures—they try to keep an open mind and look for evidence either way, even though it might show their hunch to be wrong. Even one *counterexample*—an example for which a conjecture doesn't hold—will prove a conjecture wrong.

Problem Set C

Remember

A prime number is a number with only two whole-number factors: itself and 1.

Prime numbers are important in cryptography, the study of codes. Many mathematicians have tried to find a rule that produces prime numbers.

Consider the expression $n^2 - n + 41$ as a possibility for such a rule.

1. Evaluate $n^2 - n + 41$ for $n = 1, 2, 3,$ and 4. Is each result a prime number? (Hint: If a number is *not* prime, it must have a prime factor that is equal to or less than its square root. For example, any non-prime number less than 100 must have 2, 3, 5, or 7 as a factor.)

2. Try some other values for *n*. What do you find? Compare your results with those of other students.

3. Do you think the expression will *always* give a prime number? Explain your thinking.

4. Test the expression for $n = 41$. (Don't use your calculator.) Is the result prime?

5. Explain how you could determine the answer for Problem 4 just by looking at the expression $41^2 - 41 + 41$.

The United States Marines recruited Navajo soldiers to develop a military code that the Japanese could not break.

Here is how Dante and Kai reasoned about the sums of odd numbers.

Problem Set D

1. Convince yourself that Dante and Kai's argument is reasonable. Then try to answer Dante's question: How could you write the argument in a way that would convince someone else it must always be true?

2. Kai then wrote their argument like this:

- *If a number is even, it can be written 2k, where k is a whole number.*

- *If a number is odd, it is 1 more than an even number, so it can be written 2k + 1.*

- *So, odd + odd = (2k + 1) + (2k + 1) = 4k + 2 = 2(2k + 1), which is an even number.*

Dante disagreed. He said, "What you've shown is that 2 times a particular odd number is an even number, and that's already obvious because that's what *even* means."

a. Discuss this with a partner. Who is right? Explain.

b. If you think Dante is right, fix Kai's argument.

3. Make a conjecture about the sum of an even number and an odd number. Then write a convincing argument for why your conjecture must be true.

Problem Set E

Ben and Lucita were looking at data tables of the variables x, y, and z. Ben found that y was inversely proportional to x. Lucita found that z was also inversely proportional to x.

1. Make a conjecture about the relationship between y and z.

Now you will try to prove your conjecture.

2. Write equations for the relationships you know.

a. the relationship between x and y

b. the relationship between x and z

3. Solve each of the equations you wrote in Problem 2 for x, if necessary.

4. You now have two expressions, each equal to x. What does that mean about the relationship between the two expressions? Write an equation that shows this.

5. Now solve for y in terms of z. What type of relationship exists between y and z? Was your conjecture correct?

Share & Summarize

1. What would you say to a friend who tells you she can prove a general formula is correct and then shows you that her formula works only for a few particular examples?

2. Why is it important to discuss proofs of your conjectures with people who might not quickly agree with you about them?

On Your Own Exercises

Practice & Apply

In Exercises 1–4, make a conjecture about whether the relationship between x and y is linear, quadratic, or neither. Explain how you decided.

1.

x	1	2	3	4	5	6	7
y	$^-1$	4	15	32	55	84	119

2.

x	1	2	3	4	5	6	7
y	4	16	64	256	1,024	4,096	16,384

3.

x	1	2	3	4	5	6	7
y	6	8	10	12	14	16	18

4.

x	1	2	3	4	5	6	7
y	4	12	24	40	60	84	112

5. Write an equation for the relationship in Exercise 3.

6. In Problem Set B, you looked at $D = (n + 1)^2 - n^2$, the difference between squares of consecutive whole numbers. Now consider this equation:

$$d = (m + 2)^2 - m^2$$

In this case, d is the difference between the square of a whole number and the square of that whole number plus 2.

Numbers	Difference of Squares
1, 3	$3^2 - 1^2 = 8$
2, 4	$4^2 - 2^2 = 12$
3, 5	$5^2 - 3^2 = 16$
$\vdots$	$\vdots$
$m, m + 2$	$(m + 2)^2 - m^2 = d$

a. Copy and complete the table to show the value of d for consecutive values of m.

m	1	2	3	4	5	6
d	8	12	16			

b. Use what you know about constant differences to determine what type of relationship $d = (m + 2)^2 - m^2$ is.

impactmath.com/self_check_quiz

c. Make a conjecture about what a simpler equation for d might be. Check that your equation works for $m = 1$, $m = 2$, and $m = 3$.

d. You can use geometry to argue that your conjecture is true. Below are tile squares for 1^2 and 3^2. Think about how you add tiles to get from one square to the next. Copy the diagram, and color the tiles you would add.

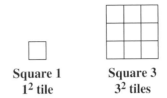

Square 1
1^2 tile

Square 3
3^2 tiles

e. Draw tile squares to represent 2^2 and 4^2, and color the tiles you would add to get from one to the other. Do the same for 3^2 and 5^2.

f. How many tiles do you add to go from the square for n^2 to the square for $(n + 2)^2$? Explain how you found your answer.

g. Does your answer from Part f prove your conjecture from Part c? Explain why or why not.

7. Héctor conjectured that when you subtract an even number from an odd number, the result is odd. He tried to prove his conjecture.

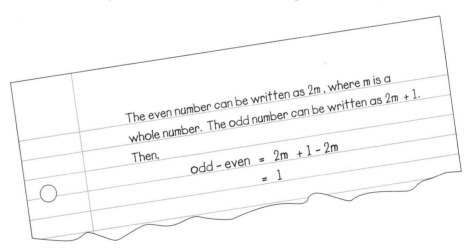

The even number can be written as $2m$, where m is a whole number. The odd number can be written as $2m + 1$.

Then,
$$\text{odd} - \text{even} = 2m + 1 - 2m$$
$$= 1$$

Héctor said, "According to my proof, an odd number minus an even number is always 1. This isn't true. What did I do wrong?"

a. What did Héctor do wrong?

b. Give a correct proof of Héctor's conjecture.

8. Make and prove a conjecture about the difference of two even numbers.

9. Make and prove a conjecture about the difference of two odd numbers.

10. You can't prove a general conjecture by checking specific examples, but just one false example—a counterexample—will disprove it. Show that the conjecture $(m + n)^2 = m^2 + n^2$ is false by giving values for m and n for which it doesn't work.

11. For each statement, find a proof or a counterexample.

a. The reciprocal of x times the reciprocal of y is the reciprocal of xy.

b. The reciprocal of x plus the reciprocal of y is the reciprocal of $x + y$.

12. Below is a table of values for the cubic equation $y = x^3$. Notice that neither the first or second differences are constant.

x	1	2	3	4	5	6	7
y	1	8	27	64	125	216	343
First Differences		7	19	37	61	91	127
Second Differences			12	18	24	30	36

a. Find the *third differences*. What do you notice?

b. Explore the first, second, and third differences for two more cubic equations. Try to make a conjecture based on your findings.

c. Make a conjecture about the differences for a *quartic equation*, which can be written in the form $y = ax^4 + bx^3 + cx^2 + dx + e$. Test your conjecture on $y = x^4$.

13. You may recall that the constant first difference for a linear equation is the value of m when the equation is written in $y = mx + b$ form. In this exercise, you will look for a relationship between a quadratic equation and the constant second difference.

a. Find the constant second difference for this table of values for $y = \frac{1}{2}x^2 + 2x$.

x	1	2	3	4	5	6	7
y	2.5	6	10.5	16	22.5	30	38.5

b. Find the constant second difference for this table of values for $y = {}^-x^2 + 2x - 1$.

x	1	2	3	4	5	6	7
y	0	$^-1$	$^-4$	$^-9$	$^-16$	$^-25$	$^-36$

In y o u r
own
words

Explain the differ-
ence between mak-
ing a conjecture
and proving some-
thing to someone.

c. Find the constant second difference for this table of values for $y = 3x^2 - 9$.

x	1	2	3	4	5	6	7
y	$^-6$	3	18	39	66	99	138

d. In Parts a–c, look for a relationship between the constant second difference and the coefficient of x^2 in the quadratic equation. Try to make a conjecture about this relationship.

e. Test your conjecture on at least two more quadratic relationships.

14. **Number Sense** *Goldbach's conjecture* states that every even integer greater than 2 can be expressed as the sum of two prime numbers. Here are some examples.

$$4 = 2 + 2 \qquad 6 = 3 + 3 \qquad 8 = 3 + 5$$

a. Test Goldbach's conjecture on 10, 12, and 100.

b. Does the conjecture seem to be true?

c. What would you need to show in order to prove the conjecture?

In Exercises 15–17, decide whether the conjecture is true or false. Try to give a convincing proof of the conjectures that are true. For false conjectures, give a counterexample.

15. The value of $n^2 - n$ is always an even number.

16. The square of every even number is a multiple of 4.

17. The difference between any two square numbers is always an even number.

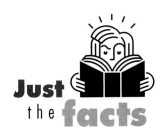
Mixed Review

Tell whether the points in each set lie on a line.

18. (0, 0); (9, 0); (0, 2)

19. (0.3, $^-2$); ($^-1$, 4.5); (10.1, $^-51$)

20. ($^-3$, $^-14.8$); (0, $^-7.3$); (1.4, $^-3.8$)

21. Gerald collected the following data in an experiment. The data are almost linear. Graph the data, and find an equation for a line of best fit.

Day	1	2	3	4	5	6	7
Height (mm)	0.38	0.92	1.33	1.82	2.35	2.88	3.36

22. Find the slope of each segment.

    a. Segment a

    b. Segment b

    c. Segment c

    d. Segment d

    e. Segment e

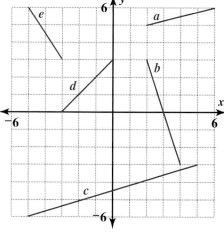

23. Economics Running Rapids charges $8.00 for the rental of a raft and $3.50 for every ride down the river in the raft.

    a. Write an expression to represent how much you would spend in a day of rafting at Running Rapids.

    b. You have a total of $30 to spend at Running Rapids. Using your expression from Part a, determine the maximum number of times you could ride down the river on the raft you have rented.

24. To simulate drawing one of three marbles from a bag, Neeraj rolled a die several times. The marbles were different colors: blue, white, and purple. Rolling 1 or 2 represented drawing the blue marble. Rolling 3 or 4 represented drawing the white marble. Rolling 5 or 6 represented drawing the purple marble.

Does this simulation represent drawing a marble and putting it back before drawing the next? Or does it represent keeping the drawn marble out of the bag before drawing the next one? How do you know?

25. The U.S. Postal Service charges extra postage for envelopes that are beyond specified size limits: $11\frac{1}{2}$ in. long and $6\frac{1}{8}$ in. high. It also charges extra if the length divided by the height is less than 1.3 or greater than 2.5. Tell whether envelopes of the given dimensions would require extra postage.

    a. 8 in. by 4 in.   **b.** 7 in. by 6 in.

    c. 11.4 in. by 5 in.   **d.** 7 in. by 7 in.

    e. 10 in. by 3 in.   **f.** 12 in. by 7 in.

Chapter Summary

Like linear relationships, *quadratic relationships* have certain characteristics. A quadratic relationship can be represented by equations of the form $y = ax^2 + bx + c$, where a, b, and c are constants, and a is not 0. The graph of a quadratic equation has a particular shape, called a *parabola*.

Inverse or *reciprocal relationships* also have a particular shape, called a *hyperbola*. You have seen inverse relationships with equations of the form $y = \frac{a}{x}$, $y = \frac{1}{x + b}$, and $y = \frac{1}{x} + c$. In addition, you saw what equations and graphs of *cubic* relationships look like.

You also made several *conjectures* in this chapter. *Proving* a conjecture not only allows you to convince others that it is true, but can also help you understand *why* it is true.

Strategies and Applications

The questions in this section will help you review and apply the important ideas and strategies developed in this chapter.

Recognizing quadratic relationships from graphs, equations, and tables

In Questions 1–9, determine whether the relationship between x and y could be quadratic, and explain how you know.

1.
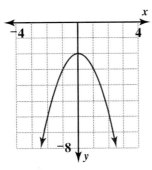

2.
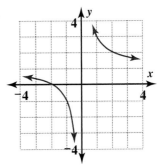

3.
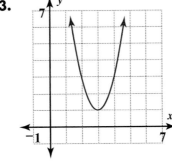

4.

x	-3	-2	-1	0	1	2	3
y	-10	-7	-4	-1	2	5	8

5.

x	-2	-1	0	1	2	3	4
y	-4	-5	-4	-1	4	11	20

6.

x	0	1	2	3	4	5	6
y	1	2	4	8	16	32	64

7. $y = 2x^3 + x^2$

8. $y = x(x + 2)$

9. $y = 2^x + 3$

Understanding the connections between quadratic equations and graphs

Match each equation with one of the graphs below.

10. $y = x^2$

11. $y = {}^-x^2$

12. $y = x^2 - 4x + 4$

13. $y = x^2 + 4$

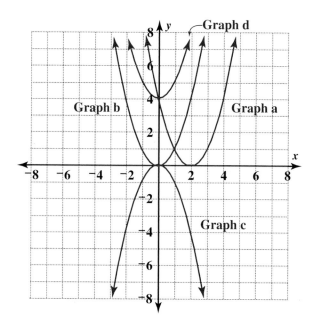

Recognizing inverse relationships from written descriptions, graphs, tables, and equations

In Questions 14–22, determine whether the relationship between x and y could be an inverse relationship, and explain how you know.

14.

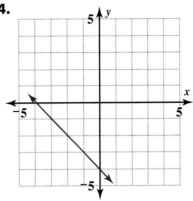

15.

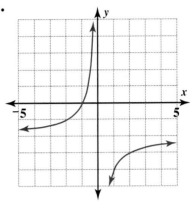

16. Carl sells hand-bound journals with blank pages. He notices that for every $5 that he increases the price x, his weekly sales y drop by 20 books.

17. Sandra has 60 square tiles. She tries to find as many rectangular arrangements of all 60 tiles as she can, recording the dimensions as length x and width y.

18.

x	1	2	4	5	6	9	10
y	60	30	15	12	10	$6\frac{2}{3}$	6

19.

x	0.5	1	1.5	2	2.5	3	3.5
y	840	420	280	210	168	140	120

20. $2xy = 7$

21. $y = \frac{x}{4}$

22. $x + 2 = \frac{1}{y}$

Solving real-world problems involving quadratic and inverse relationships

23. A radio station is broadcasting a live concert. When the audience applauds at the end of each song, the sound level rises, reaches a peak, and then falls in a way that can be approximated by a quadratic equation.

The graph shows the relationship between the noise level n (in decibels) and the time t (in seconds) after a particular song ends.

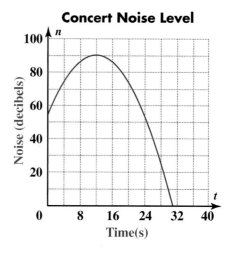

Concert Noise Level

a. At what time after the end of the song was the noise greatest? How loud did it get?

b. If a speaker tries to produce a noise that is too loud, it *distorts* the sound and produces static. The crew turns down the sound being broadcast when the noise level rises above 70 decibels and keeps it turned down until it's below 70 decibels once again. How long was the sound turned down following this song? How did you find your answer?

c. Noise levels of 50 decibels or lower are considered normal background noise. How long after the applause began did the sound subside to the level of normal background noise?

24. Jeanine was organizing an event to give her classmates a chance to read their poems before an audience. She arranged to have the school library open for 2 hours on the evening of the reading, and she wanted to determine how much time each poet would have. She realized it would depend on how many poets she invited.

a. Write an equation expressing the amount of time t in minutes available to each poet if there are n poets. Assume one poet starts immediately after another has finished.

b. If Jeanine invites 8 poets to read, how much time will each have?

c. Suppose Jeanine wants to give each poet 10 minutes. How many can she invite?

Making conjectures

25. Consider these sets of three consecutive whole numbers.

$$3, 4, 5 \qquad 5, 6, 7 \qquad 0, 1, 2 \qquad 9, 10, 11$$

Compare the products of the first and last numbers of each set to the square of the middle number.

a. Make a conjecture about what you observe.

b. Suppose the middle number is x. What are the other two numbers? Rewrite your conjecture as a mathematical statement using x.

c. Find a counterexample or a proof for your conjecture. (Hint: You might want to look at this problem geometrically.)

Demonstrating Skills

In Exercises 26–31, make a rough sketch showing the general shape and location of the graph of the equation.

26. $y = 3x^2$

27. $y = \frac{3}{x}$

28. $y = x^3$

29. $y = {}^-x^2 + 3$

30. $y = {}^-\frac{2}{x} - 1$

31. $y = (x + 1)^2$

CHAPTER 3

Exponents and Exponential Variation

Real-Life Math

It's Just a Rumor Have you ever noticed how quickly rumors can spread? For example, people who use e-mail often receive warnings about computer viruses from friends. Sometimes though, the warnings are just rumors being spread by well-meaning people. Suppose one person sent an e-mail about a virus to 10 friends. An hour later, those recipients sent the e-mail to 10 friends, and so on. In about 7 hours, 10,000,000 households with e-mail might have received the message. This is why some people say the virus warning *is* the virus!

In this chapter, you will learn that the spread of rumors and other situations can often be modeled using an exponential relationship.

Think About It How many households do you think have e-mail in the U.S.? Do you think after 7 hours most of them would have received the e-mail virus?

Family Letter

Dear Student and Family Members,

We are about to begin our study of exponents and extremely large or extremely small numbers. Exponents can be thought of as a shortcut for expressing repeated multiplication: $5 \times 5 \times 5$ is the same as 5^3. In this example, 3 is the exponent. We will study exponents and roots in greater depth than we have in the past.

We'll also explore relationships that grow (or shrink) exponentially. In this kind of relationship, the amount of change gets larger and larger (or smaller and smaller) each time. Population growth and compound interest are examples of exponential growth. If you deposit $100 in an account earning 7% interest, the balance will grow exponentially; notice that while the interest percentage is the same, the dollar amount of interest earned increases each year.

Year	Interest Earned	Account Balance
1	$7.00	$107.00
2	$7.49	$114.49
3	$8.01	$122.50
4	$8.58	$131.08
5	$9.18	$140.26
6	$9.81	$150.07

We will also learn to distinguish rational and irrational numbers and develop some strategies for handling irrationals. Rational numbers are defined as those that can be expressed as the quotient of one integer divided by another integer. Irrational numbers are those that neither repeat nor terminate when expressed as a decimal, such as $\sqrt{2}$ and π:

$$\sqrt{2} = 1.414213562\ldots \qquad\qquad \pi = 3.141592654\ldots$$

Vocabulary Along the way, we'll be learning about these new vocabulary terms:

decay factor	**real numbers**	***n*th root**
rational numbers	**irrational numbers**	**square root**
growth factor	**scientific notation**	**radical sign**

What can you do at home?

During the next few weeks, your student may show interest in exponential relationships or very large or small numbers. You might help him or her think about common occurrences of these topics, such as compound interest or the depreciation of your car as it ages. If you find examples of extremely large or small numbers in the paper—the national debt is a good example—ask your student to try to express them in both standard and exponential notation.

3.1

Exponents Revisited

In your mathematics studies in earlier years, you probably encountered exponents. In this lesson, you will take another look at exponents and how to work with expressions that involve them.

Remember

The ellipsis "..." means the list continues in the same pattern.

Explore

Consider this list of numbers.

$$2, 4, 8, 16, 32, 64, 128, \ldots$$

What could be the next two numbers in this list?

How would you describe this list?

Choose any number in the list and double it. Is the result another number in the list? Would this be true for *any* number in the list? Why or not?

Choose two numbers in the list and multiply them. Is the product also a number in the list? Would this be true for *any* two numbers in the list? Why or why not?

Investigation 1 ▶ Positive Integer Exponents

All the numbers in the list from the Explore are powers of 2. That means they can all be found by multiplying 2 by itself some number of times. Recall that you can use positive integer *exponents* to show that a number—called the *base*—is multiplied by itself.

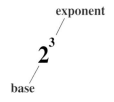

$$2^3 = 2 \cdot 2 \cdot 2 \qquad 5^4 = 5 \cdot 5 \cdot 5 \cdot 5 \qquad 120^1 = 120$$

Problem Set A

1. Look again at the list from the Explore.

$$2, 4, 8, 16, 32, 64, 128, \ldots$$

a. Use an exponent to write 128 as a power of 2.

b. Write an expression for the *n*th number in the list.

c. Is 6,002 in this list? How do you know?

d. Is 16,384 in this list? How do you know?

2. Tamika started another list of numbers using positive integer powers with 4 as the base.

a. What are the first 10 numbers in Tamika's list?

b. Is 2,048,296 in her list? How do you know?

c. The number 262,144 is in her list. Is it also a number in the list in Problem 1? How do you know?

d. The numbers 4,096 and 8,192 are in the list in Problem 1. Are they also in Tamika's list?

e. Prove It! Tamika made the conjecture that every number in her list is also in the list in Problem 1. Either explain why her conjecture is true, or give a counterexample.

Remember

A *counterexample* is an example for which a conjecture does not hold. A counterexample proves a conjecture wrong.

Problem Set B

Tell whether each statement is *sometimes true, always true,* or *never true* for positive integer values of *n*. If it is sometimes true, state for what values it is true.

1. $2^n = 2{,}048$

2. 3^n is less than 1,000,000 (that is, $3^n < 1{,}000{,}000$)

3. $4^n = 2{,}048$

4. 0.5^n is between 0 and 1 (that is, $0 < 0.5^n < 1$)

Read this poem and answer Problems 5–9, using powers to write your answers.

I pried the side from the crate;
Inside, six drawers lay in wait.
The drawers each held six yellow boxes
Whose six faces each showed six foxes.
The foxes each had six green eyes.
Alien foxes? That's a surprise!

5. How many drawers were in the crate?

6. How many boxes were in the crate?

7. What was the total number of faces on the boxes in the crate?

8. How many pictures of foxes were in the crate?

9. What was the total number of eyes on the foxes in the crate?

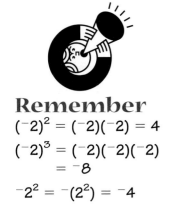

Remember

$(^-2)^2 = (^-2)(^-2) = 4$

$(^-2)^3 = (^-2)(^-2)(^-2)$
    $= ^-8$

$^-2^2 = ^-(2^2) = ^-4$

Without computing the values of the numbers in each pair, determine which number is greater. Explain your answers.

10. 3^{18} or 3^{20} **11.** $(^-2)^8$ or $(^-2)^{19}$ **12.** $^-3^{500}$ or $^-3^{800}$

13. Now consider powers of $\frac{1}{2}$.

    a. Without using your calculator, find three positive integer values of n that make $\left(\frac{1}{2}\right)^n$ less than $\left(\frac{1}{2}\right)^3$.

    b. Describe all the positive integers that would make $\left(\frac{1}{2}\right)^n$ less than $\left(\frac{1}{2}\right)^3$.

    c. How do you know your answer to Part b is correct?

14. Without computing, determine which number is greater, $\left(-\frac{1}{3}\right)^{34}$ or $\left(-\frac{1}{3}\right)^{510}$. Explain how you know your answer is correct.

V O C A B U L A R Y
scientific notation

You may recall that scientific notation makes use of powers of 10. A number is written in **scientific notation** when it is expressed as the product of a power of 10 and a number greater than or equal to 1 but less than 10. For example:

$$3{,}456 = 3.456 \times 10^3 \qquad\qquad 10{,}000{,}000 = 1 \times 10^7$$

Problem Set C

1. Copy and complete the chart without using your calculator.

Description	Number in Standard Form (approximate)	Number in Scientific Notation
Time since dinosaurs began roaming Earth (years)	225,000,000	
Projected World population in 2010		6.8×10^9
Distance from Earth to Andromeda galaxy (miles)		1.5×10^{19}
Mass of the sun (kg)	2,000,000,000,000,000,000,000,000,000,000	

For each pair of numbers, indicate which is greater.

    2. 2.34×10^5 or 1.35×10^6

    3. 3.83312×10^{31} or 8.1×10^{32}

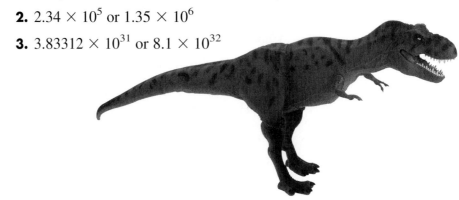

Investigation 2 Negative Integer Exponents

In Investigation 1, you worked with exponents that were positive integers. You may remember from other mathematics courses that exponents can be negative integers also.

Think & Discuss

Consider this list of numbers.

$$\ldots, \frac{1}{27}, \frac{1}{9}, \frac{1}{3}, 1, 3, 9, 27, \ldots$$

Assume the list continues in both directions.

What number could follow 27? What number could precede $\frac{1}{27}$?

Write 27, 9, and 3 using integer exponents and the same base.

Assume the pattern in the exponential forms you just wrote continues with the other numbers in the list above. Use the pattern to write the exponential forms of 1, $\frac{1}{3}$, $\frac{1}{9}$, and $\frac{1}{27}$.

The Think & Discuss may have reminded you how to use 0 and negative integers as exponents.

> For any number a not equal to 0 and any integer b:
> $$a^0 = 1 \qquad \text{and} \qquad a^{-b} = \frac{1}{a^b} = \left(\frac{1}{a}\right)^b$$

For example,

$$2^0 = 1 \qquad\qquad 8^{-3} = \left(\frac{1}{8}\right)^3 = \frac{1}{512}$$

$$0.25^0 = 1 \qquad\qquad (^-5)^{-5} = \frac{1}{(^-5)^5} = {}^-\frac{1}{3,125}$$

Problem Set D

For these problems, do not use your calculator.

1. Write each number without using an exponent.

a. 4^{-1}

b. $^-5^{-3}$

c. 1.43536326^0

2. Consider these numbers.

$$3^2 \qquad 3^{-2} \qquad \left(\tfrac{1}{3}\right)^2 \qquad \tfrac{1}{3^2} \qquad \tfrac{1}{9} \qquad 9$$

a. Sort the numbers into two groups so that all the numbers in each group are equal to one another.

b. In which group does $\left(\tfrac{1}{3}\right)^{-2}$ belong?

3. Write each number without using an exponent.

a. $\left(\tfrac{4}{5}\right)^{-2}$

b. 0.5^{-3}

4. Sort these numbers into two groups so that all the numbers in each group are equal to one another.

$$\left(\tfrac{2}{3}\right)^2 \quad \left(\tfrac{2}{3}\right)^{-2} \quad \left(\tfrac{3}{2}\right)^2 \quad \left(\tfrac{3}{2}\right)^{-2} \quad \tfrac{3^2}{2^2} \quad \tfrac{9}{4} \quad \tfrac{4}{9}$$

5. Sort these numbers into four groups so that all the numbers in each group are equal to one another.

$$10^3 \quad 10^{-3} \quad (^-10)^{-3} \quad \tfrac{1}{(^-10)^3} \quad \tfrac{1}{1,000} \quad ^-1,000 \quad \left(\tfrac{1}{10}\right)^{-3}$$

$$\left(\tfrac{1}{10}\right)^3 \quad \left(-\tfrac{1}{10}\right)^{-3} \quad \tfrac{1}{10^3} \quad 1,000 \quad \left(-\tfrac{1}{10}\right)^3 \quad \tfrac{^-1}{1,000} \quad (^-10)^3$$

6. Challenge Sort these numbers into two groups so that all the numbers in each group are equal to one another.

$$\left(\tfrac{a}{b}\right)^3 \quad \left(\tfrac{b}{a}\right)^{-3} \quad \left(\tfrac{a}{b}\right)^{-3} \quad \left(\tfrac{b}{a}\right)^3 \quad \tfrac{b^3}{a^3} \quad b^3 \div a^3 \quad a^3 \div b^3$$

7. Which of these are equivalent to a^{-n}?

$$\tfrac{1}{a^n} \qquad ^-a^n \qquad \left(\tfrac{1}{a}\right)^n \qquad 1 \div a^n$$

8. Which of these are equivalent to $\left(\tfrac{1}{a}\right)^{-n}$?

$$\tfrac{1}{a^{-n}} \qquad a^n \qquad -\left(\tfrac{1}{a}\right)^n \qquad 1 \div a^n$$

9. Which of these are equivalent to $\left(\tfrac{a}{b}\right)^{-n}$?

$$\tfrac{a}{b^n} \qquad \left(\tfrac{b}{a}\right)^n \qquad \tfrac{a^{-n}}{b^{-n}} \qquad \tfrac{b^n}{a^n}$$

Remember

Dividing by a number is equivalent to multiplying by its reciprocal.

Think & Discuss

When a fraction is raised to a negative integer power, how can you find an equivalent fraction using a positive integer power? For example, how do you find an equivalent fraction for $\left(\frac{5}{7}\right)^{-3}$?

Problem Set E

Tell whether each statement is *sometimes true, always true,* or *never true* for integer values (positive, negative, or 0) of n. If it is sometimes true, state for what values it is true.

1. $2^n = \frac{1}{2,048}$

2. 3^n is less than 1 (that is, $3^n < 1$)

3. 5^n is between 0 and 1 or is equal to 0 or 1 (that is, $0 \leq 5^n \leq 1$)

Without computing the values of the numbers in each pair, determine which number is greater. Explain how you know your answer is correct.

4. 7^{-89} or 7^{-90}

5. 3^{-15} or 6^{-15}

6. 0.4^{-5} or 0.4^{-78}

7. $^-9^{-4}$ or $^-0.5^{-4}$

8. $(^-2)^{-280}$ or $(^-2)^{-282}$

9. $(^-50)^{-45}$ or $(^-50)^{-51}$

10. 0.3^{-50} or 1.3^{-50}

As you know, you can use scientific notation with positive integer powers of 10 to express very large numbers. In the same way, you can use negative integer powers of 10 to express very small numbers. For example:

$$0.003456 = 3.456 \times 10^{-3}$$

$$0.0000001 = 1 \times 10^{-7}$$

These red blood cells have been invaded by the protozoa *Trypanosoma,* the cause of West African sleeping sickness. Red blood cells measure about 5×10^{-3} cm. Can you estimate the length of *Trypanosoma?*

Problem Set F

1. Complete the chart without using your calculator.

Description	Number in Standard Form (approximate)	Number in Scientific Notation
Average mass of a hydrogen atom (grams)	0.0000000000000000000000016735	
Diameter of the body of a Purkinje cell (meters)		8×10^{-5}
Diameter of some fats in the body (meters)		5×10^{-10}
Average mass of an oxygen atom (grams)	0.000000000000000000000026566	

For each pair of numbers, indicate which is greater.

2. 2.34×10^{-5} or 1.35×10^{-6}

3. 3.83312×10^{-31} or 8.1×10^{-32}

Just the facts

A Purkinje cell is a type of brain cell called a *neuron*. The human brain contains about 30 million of them.

Share & Summarize

1. Explain what a^{-b} means, assuming b is a positive integer.

2. How would you decide whether a^{-5} is greater than or less than a^{-7}? Assume a is positive and not equal to 1.

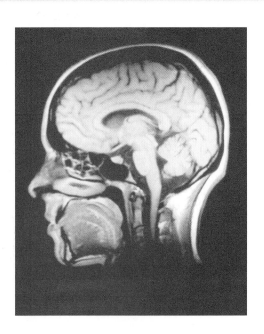

You may recall from earlier mathematics classes the following *exponent laws*, which can make calculations with exponents much simpler. In these laws, the bases *a* and *b* cannot be 0 if they are in a denominator or if they are raised to a negative exponent or to 0.

Product Laws

$$a^b \cdot a^c = a^{b+c}$$

$$a^c \cdot b^c = (ab)^c$$

Quotient Laws

$$\frac{a^b}{a^c} = a^{b-c}$$

$$\frac{a^c}{b^c} = \left(\frac{a}{b}\right)^c$$

Power of a Power Law

$$(a^b)^c = a^{bc}$$

EXAMPLE

Ben explains how he remembers the first product law.

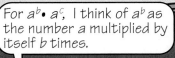

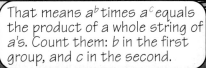

In the next problem set, you will think about ways to explain why some of the other exponent laws are true. Although these laws are true for all integer values of the exponents, you will focus on either positive or negative integers to make your job easier. Use Ben's line of reasoning as a guide.

Problem Set G

1. **Prove It!** In this problem, you will show that the first quotient law, $\frac{a^b}{a^c} = a^{b-c}$, works for positive integer exponents. Assume $a \neq 0$.

   a. First show that the law works when $b > c$.

   b. Now show that the law works when $b < c$.

2. **Prove It!** Show that the second product law, $a^c \cdot b^c = (ab)^c$, works for negative integer exponents. Hint: Let $c = {}^-x$ for a positive integer x.

Problem Set H

1. Copy this multiplication chart. Without using your calculator, find the missing expressions. Write all entries as powers.

×	3^{-2}	3^x	3^4
3^4			
3^a			
${}^-3^2$			

2. Copy this division chart. Without using your calculator, find the missing expressions by dividing the first expression in that row by the first expression in that column. For example, the first unshaded cell represents the quotient $2a^4 \div a^5$, or $\frac{2a^4}{a^5}$, which is equivalent to $2a^{-1}$. Write all entries as powers or products of powers.

÷	a^5	a^{-2}	$(a^3)^2$
$2a^4$			
a^{-3}			
$(2a)^5$			

3. Copy this chart. Without using your calculator, fill in the missing expressions. Write all entries as powers or products of powers.

×		$2a$		
b^{-4}	b^4			
a^8			$a^{10}b^{-4}$	a^8b^{-8}
			b	
	$(2ab)^4$			

Problem Set ▌

For these problems, do not use your calculator.

1. Rewrite each expression using a single base.

  a. $(a^{-m})^0$ (Assume $a \neq 0$.)

  b. $[(^-d)^3]^4$

  c. Challenge $(^-10^{-4})^{-5}$

2. Rewrite each expression using a single base.

  a. $(2^3 \cdot 2)^2$

  b. $(a^m)^n \div (a^{-m})^n$

  c. $4^3 \cdot n^3 \div (^-16n)^3$

3. Find at least two ways to write each expression as a product of two expressions.

  a. $32n^{10}$ **b.** m^7b^{-7}

Share & Summarize

Without using a calculator, rewrite this expression as simply as you can. Show each step of your work, and record which exponent law, if any, you used for each step.

$$\frac{2^6 n^3}{(16n^2)^3}$$

Investigation 4 ▶ Exponent Laws and Scientific Notation

Lucita and Tala were discussing how to multiply 4.1×10^4 by 3×10^6.

We can start by rearranging things a little.

4.1 times 3 is 12.3, and we can use a product rule for the powers of 10.

$(4.1 \times 10^4)(3 \times 10^6) = (4.1 \times 3)(10^4 \times 10^6)$
$= 12.3 \times 10^{4+6}$
$= 12.3 \times 10^{10}$

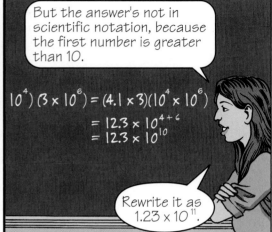

But the answer's not in scientific notation, because the first number is greater than 10.

$10^4)(3 \times 10^6) = (4.1 \times 3)(10^4 \times 10^6)$
$= 12.3 \times 10^{4+6}$
$= 12.3 \times 10^{10}$

Rewrite it as 1.23×10^{11}.

Think & Discuss

How would you divide two numbers written in scientific notation? For example, what is $(2.12 \times 10^{14}) \div (5.3 \times 10^6)$? Write your answer in scientific notation.

Problem Set J

For these problems, don't use your calculator unless otherwise indicated.

1. There are about 4×10^{11} stars in our galaxy and about 10^{11} galaxies in the observable universe.

a. If every galaxy has as many stars as ours, how many stars are there in the observable universe? Show how you found your answer.

b. If only 1 in every 1,000 stars in the observable universe has a planetary system, how many planetary systems are there? Show how you found your answer.

Remember

1 billion is
1,000,000,000.

c. If 1 in every 1,000 of those planetary systems has at least one planet with conditions suitable for life as we know it, how many such systems are there? Show how you found your answer.

d. At the end of the 20th century, the world population was estimated at about 6 billion people. Compare this number to your answer in Part c. What does your answer mean in terms of the situation?

2. *Escherichia coli* is a type of bacterium that is sometimes found in swimming pools. Each *E. coli* bacterium has a mass of 2×10^{-12} gram. The number of bacteria increase so that, after 30 hours, one bacterium has been replaced by a population of 4.8×10^{8} bacteria.

a. Suppose a pool begins with a population of only 1 bacterium. What would be the mass of the population after 30 hours?

b. A small paper clip has a mass of about 1 gram. The paper clip has how many times the mass of the 4.8×10^{8} *E. coli* bacteria? Show how you found your answer.

3. The speed of light is about 2×10^{5} miles per second.

a. On average, it takes light about 500 seconds to travel from the sun to Earth. What is the average distance from Earth to the sun? Write your answer in scientific notation.

b. The star Alpha Centauri is approximately 2.5×10^{13} miles from Earth. How many seconds does it take light to travel between Alpha Centauri and Earth?

c. Use your answer to Part b to estimate how many years it takes for light to travel between Alpha Centauri and Earth. You may use your calculator.

4. These data show current estimates of the energy released by the three largest earthquakes recorded on Earth. The *joule* is a unit for measuring energy, named after British physicist James Prescott Joule.

There are 88 recognized constellations, or groupings of stars that form easily identified patterns. This is the constellation Orion, "the Hunter."

Largest Recorded Earthquakes

Location	Date	Energy Released (joules)
Tambora, Indonesia	April 1815	8×10^{19}
Santorini, Greece	about 1470 B.C.	3×10^{19}
Krakatoa, Indonesia	August 1883	6×10^{18}

a. The Santorini earthquake was how many times as powerful as the Krakatoa earthquake?

b. The Tambora earthquake was how many times as powerful as the Krakatoa earthquake?

5. A scientist is growing a culture of cells. The culture currently contains 2×10^{12} cells.

a. The number of cells doubles every day. If the scientist does not use any of the cells for an experiment today, how many cells will she have tomorrow?

b. If she uses 2×10^9 of the 2×10^{12} cells for an experiment, how many will she have left? Show how you found your answer. Be careful—this problem is different from the others you've done!

Problem Set K

Copy these multiplication and division charts. Without using your calculator, find the missing expressions. Write all entries using scientific notation. For the division chart, divide the row label by the column label.

$\times$	4×10^{28}	
$^{-}2 \times 10^{12}$		4×10^5
6×10^{-20}		
8×10^a		

$\div$	2×10^6	
$^{-}4 \times 10^{12}$		2×10^5
8×10^{-10}		
8×10^a		

Share & Summarize

Jordan has written his calculation for three problems involving scientific notation. Check his work on each problem. If his work is correct, write "correct." If it is incorrect, write a note explaining his mistake and how to solve the problem.

1. $(2 \times 10^5) \cdot 2 = 2 \cdot 2 \times 10^5 = 4 \times 10^5$

2. $(6 \times 10^{-5}) \cdot (2 \times 10^{-7}) = 6 \cdot 2 \times 10^{-5} \cdot 10^{-7} = 1.2 \times 10^{-12}$

3. $(3 \times 10^{12}) - (3 \times 10^{10}) = 3 \times 10^{12-10} = 3 \times 10^2$

Investigation ▶ Modeling Our Solar System

MATERIALS
- masking tape
- ruler

Just the facts

Every 248 years, the orbits of Pluto and Neptune cross and Pluto becomes the eighth planet instead of the ninth.

In this investigation, you will examine the relative distances between objects in our solar system.

Make a Prediction

The average distances of planets from the sun are often written in scientific notation.

Planet	Average Distance from Sun (miles)
Mercury	3.6×10^7
Venus	6.7×10^7
Earth	9.3×10^7
Mars	1.4×10^8
Jupiter	4.8×10^8
Saturn	8.9×10^8
Uranus	1.8×10^9
Neptune	2.8×10^9
Pluto	3.7×10^9

Imagine lining up the planets with the sun on one end and Pluto on the other so that each planet is at its average distance from the sun. How would the planets be spaced? Here is one student's prediction:

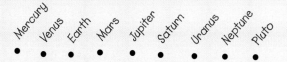

1. Now make your own prediction. Without using your calculator, sketch a scale version of the planets lined up in a straight line from the sun. Don't worry about the sizes of the planets.

Create a Model

To check your prediction, some of the members of your class will represent parts of the solar system in a large-scale model. The scale model will allow you to compare the planets' average distances from the sun, although it will not model the relative sizes of the planets.

In a large space, use masking tape to mark a line along which you will make your model. The sun will be at one end of the line; Pluto will be at the other.

As a class, answer the next three questions to determine the locations of the planets in your model.

2. Measure the length of the line. This will be the distance between the sun and Pluto in your model.

3. What is the actual distance between the sun and Pluto in miles?

4. What number would you multiply distances (in feet) in your model by to find the actual distance (in miles)?

5. How can you calculate the distances of the planets from the sun in your model?

6. Use your answer to Question 5 to estimate the distance between each planet and the sun in your scale model. Copy the table, fill in the proper unit in the last column, and then record the scaled distances.

Planet	Average Distance from Sun (miles)	Average Distance from Sun in Scale Model (_____)
Mercury	3.6×10^7	
Venus	6.7×10^7	
Earth	9.3×10^7	
Mars	1.4×10^8	
Jupiter	4.8×10^8	
Saturn	8.9×10^8	
Uranus	1.8×10^9	
Neptune	2.8×10^9	
Pluto	3.7×10^9	

Your teacher will assign your group to a planet or the sun. Your group should create a sign stating the name of your planet (or the sun) and its average distance from the sun, in miles, in scientific notation.

Decide as a class at which end of the line the sun will be, and determine where your celestial body belongs in the scale model. Choose one member of your group to represent your celestial body by standing on the line with the sign.

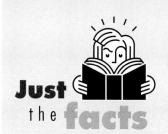

Just the facts

The Nine Planets is one of many Web sites that contains information about our solar system. This site uses a common convention for scientific notation in which the symbols "$\times 10$" are replaced by the letter "e"; the exponent follows as a full-size numeral. Using this convention, 3.6×10^7 is written 3.6 e7.

Check Your Prediction

Answer these questions as a class.

7. Which two planets are closest together? What is the actual distance between them?

8. Which two adjacent planets are farthest apart? (*Adjacent* means next to each other.) What is the actual distance between them?

9. Compare the two distances in Questions 7 and 8. How many times farther apart are the planets in Question 8 than the planets in Question 7?

10. Did your sketch give a reasonably accurate picture of the distances?

11. Was there anything about the spacing of the planets that surprised you? If so, what?

12. Was there anything about the spacing of the planets that did *not* surprise you? If so, what?

13. The moon is an average of 2.4×10^5 miles from Earth.

a. Without computing, estimate where the moon would be on your model, and ask your teacher to stand there.

b. Now compute the exact scaled distance of the moon from Earth in your model.

c. Was your estimation about the location of the moon correct? If not, what was different?

What Did You Learn?

14. Draw a scale version of the planets that shows the relative distances between the planets. Don't worry about representing the sizes of the planets, but do your best to get the distances between planets right.

15. Below are three number lines marked with numbers in scientific notation. Which number line has numbers in the correct places?

a.

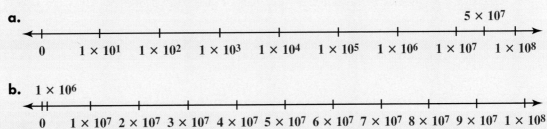

b.

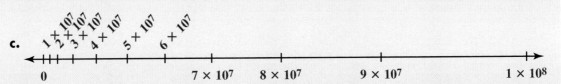

c.

On Your Own Exercises

Practice & Apply

1. **Social Studies** One of these numbers is in standard notation, and one is in scientific notation. One is the world population in 1750; the other is the world population in 1950.

$$2.56 \times 10^9 \qquad\qquad 725,000,000$$

Which number do you think is the world population in 1750? In 1950? Explain your reasoning.

2. For what values of n, if any, will n^2 be equal to or less than 0?

3. For what values of n, if any, will n^3 be equal to or less than 0?

Given that n represents a positive integer, decide whether each statement is *sometimes true, always true,* or *never true.* If it is sometimes true, state for what values it is true.

4. $4^n = 65,536$

5. 4^n is less than 1,000,000 (that is, $4^n < 1,000,000$)

6. n^2 is negative (that is, $n^2 < 0$)

7. 0.9^n is greater than or equal to 0, and at the same time 0.9^n is less than or equal to 1 (that is, $0 \le 0.9^n \le 1$)

8. For what positive values of x will x^{20} be greater than x^{18}?

9. For what positive values of x will x^{18} be greater than x^{20}?

10. For what positive values of x will x^{18} be equal to x^{20}?

11. For what negative values of x will x^{20} be greater than x^{18}?

12. For what negative values of x will x^{18} be greater than x^{20}?

13. For what negative values of x will x^{18} be equal to x^{20}?

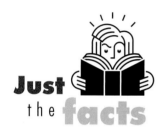

Just the facts

If each of the estimated 9×10^6 residents of New York City produces 4 pounds of trash per day, that's about 1.3×10^{10} pounds of garbage every year!

14. Challenge In Investigation 1, you explored positive integer powers of 2 and of 4.

n	1	2	3	4	5	6	7	8	9
2^n	2	4	8	16	32	64	128	256	512
4^n	4	16	64	256	1,024	4,096	16,384	65,536	262,144

Now think about positive integer powers of 8.

a. List the first five positive integer powers of 8.

b. Name three numbers that are on all three lists—that is, three numbers that are powers of 2, 4, and 8.

c. List three numbers greater than 16 that are powers of 2 but are not powers of 8.

d. List three numbers greater than 16 that are powers of 4 but are not powers of 8.

e. Describe the powers of 2 that are also powers of 8.

f. Describe the powers of 4 that are also powers of 8.

15. For what positive values of x will x^{-20} be greater than x^{-18}?

16. For what positive values of x will x^{-18} be greater than x^{-20}?

17. For what values (positive or negative) of x will x^{-18} be equal to x^{-20}?

18. The sixth power of 2 is 64; that is, $2^6 = 64$.

a. Write at least five other expressions, using a single base and a single exponent, that are equivalent to 64.

b. Write the number 64 using scientific notation.

Sort each set of expressions into groups so that the expressions in each group are equal to one another. Do not use your calculator.

19. m^2 $\left(\frac{1}{m}\right)^2$ m^{-2} $\left(\frac{1}{m}\right)^{-2}$ $\frac{1}{m^2}$ $1 \div m^2$

20. 3^x $\left(\frac{1}{3}\right)^x$ $\left(\frac{1}{3}\right)^{-x}$ $\frac{1}{3^x}$ 3^{-x} $1 \div 3^x$

21. Prove It! Prove that the second quotient law, $\frac{a^c}{b^c} = \left(\frac{a}{b}\right)^c$, works for positive integer exponents c. Assume b is not equal to 0.

22. Challenge Prove that the power of a power law, $(a^b)^c = a^{bc}$, works for positive integer exponents b and c.

Rewrite each expression using a single base and a single exponent.

23. $2^7 \cdot 2^{-4} \cdot 2^x$

24. $(-4^m)^6$

25. $m^7 \cdot 28^7$

26. $(-3)^{81} \cdot (-3)^{141}$

27. $\frac{55^{-8}}{9^{-8}}$

28. $\left(\frac{m^{84}}{m^{12}}\right)^x$

29. $3^{-5} \cdot 8^5$

30. $n^a \div n^{\frac{a}{3}}$

31. $(22^2 \cdot 22^5)^0$

Rewrite each expression as simply as you can.

32. $4a^4 \cdot 3a^3$

33. $m^{-3} \cdot m^4 \cdot b^7$

34. $\frac{10n^{-15}}{5n^5}$

35. $(4x^{-2})^6$

36. $(-m^2 n^3)^4$

37. $(a^m)^n \cdot (b^3)^2$

38. $(x^{-2})^3 \cdot x^5$

39. $\frac{12b^5}{4b^{-2}}$

40. $\frac{(x^4 y^{-5})^{-3}}{(xy)^2}$

Copy each chart. Without using your calculator, find the missing expressions. Write all entries as powers or products of powers. For the division chart, divide the row label by the column label.

41.

×	2^{10}	2^{-x}	-2^x	
-2^{-3}				
		2^{a-x}	$(2n)^a$	
2^{2a}				

42.

÷	4^{-2}	4^x	-4^x	n^7
-4^7				
4^a				
4^7				

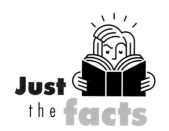

Just the facts

Sirius, also called the Dog Star, is a double star orbited by a smaller star called Sirius B, or the Pup.

43. Physical Science The speed of light is about 2×10^5 miles per second. At approximately 5×10^{13} miles from Earth, Sirius appears to be the brightest of the stars. How many seconds does it take light to travel between Sirius and Earth? How many years does it take?

44. Social Studies The population of the world in the year 1 A.D. has been estimated at about 200,000,000. By 1850, this number had grown to about 1 billion, and by 2000, the population was close to 6×10^9.

   a. The 1850 population was how many times the population in 1 A.D.?

   b. The 2000 population was how many times the 1850 population?

   c. Over which time period did the world population grow more: the 1,850 years from 1 A.D. to 1850, or the 150 years from 1850 to 2000?

45. Copy this division chart. Without using your calculator, find the missing expressions by dividing the row label by the column label. Express all entries in scientific notation.

÷			3×10^{x}
3×10^{-20}	6×10^{-29}		
6×10^{14}		3×10^{134}	
			$5 \times 10^{a-x-1}$

Connect  **Extend**

46. Social Studies According to the 1790 census, the population of the United States in 1790 was 3,929,214. You can approximate this value with powers of various numbers; for example, 2^{21} is 2,097,152 and 2^{22} is 4,194,304. Using powers of 2, the number 2^{22} is the closest possible approximation to 3,929,214.

What is the closest possible approximation using powers of 3? Powers of 4? Powers of 5?

47. Which of these sets of numbers share numbers with the powers of 2? Explain how you know.

a. positive integer powers of 6

b. positive integer powers of 7

c. positive integer powers of 16

48. Fine Arts A piano has eight C keys. The *frequency* of a note determines how high or low it sounds. Moving from the left of the keyboard to the right, each C note has twice the frequency of the one before. For example, "middle C" has a frequency of about 261.63 vibrations per second. The next higher C has a frequency of about 523.25 vibrations per second.

If the first C key has a frequency of x, what is the frequency of the last C key?

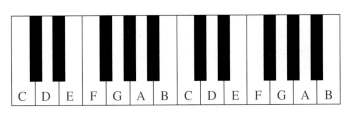

49. Economics Julián's mother offered him $50 a month in allowance. Julián said he would rather have his mother pay him 1 penny the first day of the month, 2 pennies the second day, 4 the third day, 8 the fourth day, and so on. His mother would simply double the number of pennies she gave him each day until the end of the month. His mother said that sounded fine with her.

a. Would Julián receive more money with an allowance of $50 a month or using his plan? Explain why.

b. If Julián's plan produces more money, on what day would he receive more than $50 a month?

c. With his plan, how much money would Julián receive the last day of June, which has 30 days?

d. Challenge With his plan, how much would Julián receive in all for the month of June? Filling in a table like this one might help you answer this question.

Day	Amount Received Each Day	Total Amount
1	$0.01	$0.01
2	$0.02	$0.01 + 0.02 = $0.03
3	$0.04	$0.03 + 0.04 = $0.07
4		

50. Sports A particular tennis tournament begins with 64 players. If a player loses a single match, he or she is knocked out of the tournament. After one round, only 32 players remain; after two rounds, only 16 remain; and so on.

Six students have conjectured a formula to describe the number of players remaining, p, after r rounds. Which rule or rules are correct? For each rule you think is correct, show how you know.

- Terrill: $p = \frac{64}{2^r}$
- Mi-Yung: $p = 64 \cdot 2^{-r}$
- Antonia: $p = 64 \cdot \frac{1}{2^r}$
- Peter: $p = 64 \cdot \left(\frac{1}{2}\right)^r$
- Damon: $p = 64 \cdot 0.5^r$
- Tamika: $p = 64 \cdot (-2)^r$

51. This list of numbers continues in the same pattern in both directions.

$$\ldots, \tfrac{1}{5}, 1, 5, 25, 125, 625, \ldots$$

Héctor wanted to write an expression for this list using n as a variable. To do that, though, he had to choose a number on the list to be his "starting" point. He decided that when $n = 1$, the number on the list is 5. When $n = 2$, the number is 25.

a. Using Héctor's plan, write an expression that will give any number on the list.

b. What value for n gives you 625? 1? $\tfrac{1}{5}$?

Without computing the value of each pair of numbers, determine which number is greater. For each problem, explain why.

52. 2^{80} or 4^{42} **53.** $3^{-1,600}$ or 27^{-500} **54.** 12^{20} or 4^{45}

55. A pastry shop sells a square cake that is 45 cm wide and 10 cm thick. A competitor offers a square cake of the same thickness that is 2 cm wider. The first baker argues that the area of the top of the rival cake is $(45 + 2)^2$ cm^2 and is therefore only 4 cm^2 larger than the one he sells.

How do you think the first baker misused one of the rules for calculating with exponents? What is the actual difference in areas?

56. Astronomy Earth travels around the sun approximately 6×10^8 miles each year. At approximately what speed must Earth travel in miles per second? Give your answer in scientific notation.

57. Life Science The diameter of the body of a Purkinje cell is 8×10^{-5} m.

a. If a microscope magnifies 1,000 times, what will be the scaled diameter, in meters, as viewed in the microscope?

b. What is the scaled diameter, in centimeters, as viewed in the microscope?

Mixed Review

State whether the data in each table could be linear, and tell how you know.

58.

a	$^-4$	$^-3$	$^-2$	$^-1$	0	1
b	$^-32$	$^-13.5$	$^-4$	$^-0.5$	0	0.5

59.

c	$^-4$	$^-3$	$^-2$	$^-1$	0	1
d	$^-12.1$	$^-9.6$	$^-7.1$	$^-4.6$	$^-2.1$	0.4

Each table represents a linear relationship. Write an equation to represent each relationship.

60.

a	-4	-3	-2	-1	0	1
b	-8.8	-6.6	-4.4	-2.2	0	2.2

61.

e	-4	-3	-2	-1	0	1
f	-15	-13.75	-12.5	-11.25	-10	-8.75

Rewrite each equation in $y = mx + b$ form, and tell whether the relationship represented by the equation is increasing or decreasing.

62. $4(x - y) = {}^-3$

63. $\dfrac{4 - 3x}{2y} = 1$

64. $-\dfrac{1}{3}x = {}^-4 - \dfrac{2}{3}y$

65. $\dfrac{2}{y} = \dfrac{1}{3 + 2x}$

66. Economics Julia found three Web sites that sell 4-inch-square stickers of her favorite band's logo. The three sites sell the stickers for different prices, and charge different amounts for shipping.

Site 1: Stickers are 75¢ each; shipping is $4 for any size order.

Site 2: Stickers are 60¢ each; shipping is $5.50 for any size order.

Site 3: Stickers are $1.25; shipping is included.

a. For each site, write an equation to represent the charge C for ordering any number of stickers s.

b. Graph your three equations on axes like these. Label each graph with its site number.

c. Use your graph to answer this question: If Julia wants to order 16 stickers, which site will charge her the least?

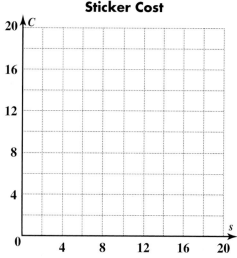

Sticker Cost

d. Use your graph to answer this question: If Julia wants to order 10 stickers, which site will charge her the least?

67. Consider the parabola at left.

a. What is its vertex?

b. What is its line of symmetry?

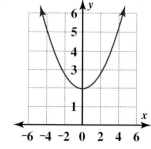

Exponential Relationships

You have probably heard people comment that something changing quickly is "growing exponentially." In common speech, *exponential change* is often used to mean *rapid change.* In mathematics, it has a more precise meaning. In this lesson, you will look at several kinds of exponential relationships and review the precise meaning for the term *exponential change.*

Explore

Lewis and Clark Middle School is holding its annual canned-food drive, sponsored by the mathematics teachers. The teachers set up a point system to reward students who bring cans of food. Each student can choose from two plans for receiving points.

Canned Food Drive

EARN valuable points based on the number of cans donated!

Total Cans	Total Points Plan A	Plan B
1	1,000	2
2	2,000	4
3	3,000	8
4	4,000	16

Describe the way the points in each plan change as the number of cans collected rises.

Which plan would you choose? Is that always the best plan, no matter how many cans you bring?

Plan A represents a kind of growth with which you are very familiar: linear growth. With each additional can, 1,000 points are added to the total. The rate of growth is constant.

For Plan B, the number of points is multiplied by 2 for each additional can. Quantities that are repeatedly multiplied by a number greater than 1 are said to *grow exponentially.* You can also say they show *exponential growth.* The number by which they are multiplied is called the **growth factor.** The growth factor for Plan B is 2.

VOCABULARY
growth factor

Investigation Exponential Growth

The kingdom of Tonga is a group of more than 150 islands in the Pacific Ocean. In 1994, the population of Tonga was estimated to be 100,000 people. Suppose the population has been increasing by about 2% each year since 1994.

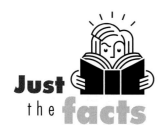

Think & Discuss

Make a table that shows the estimated population of Tonga for the 5 years after 1994.

Assume the population continues to grow by 2% per year. How can you predict the population 25 years after 1994?

Years after 1994	Estimated Population
0	100,000
1	
2	
3	
4	
5	

Here's how Kai and Lydia thought about how to find the population of Tonga after 25 years.

Problem Set A

1. What number would you multiply the 1995 population estimate of Tonga by to get the 1996 population estimate?

2. Estimate the 1996 population. Compare your answer to the 1996 population value you found in the Think & Discuss (2 years after 1994).

3. Make a table like the one below. In the second column, enter the computations you would do if you used Kai and Lydia's calculation method. For the third column, rewrite the computations in the second column using exponential notation.

Years after 1994	Estimated Population of Tonga	
	Written as a Product	Using Exponential Notation
0	100,000	$100{,}000 \cdot 1.02^0$
1	$100{,}000 \cdot 1.02$	$100{,}000 \cdot 1.02^1$
2	$100{,}000 \cdot 1.02 \cdot 1.02$	$100{,}000 \cdot 1.02^2$
3		
4		
5		

4. If the population of Tonga continues to increase by 2% each year, what will the population be in the year 2019, or 25 years after 1994? Show how you found your answer.

5. What is the growth factor in this situation?

6. Write an equation to show what the population p would be n years after 1994.

7. Suppose the population of Tonga increased by 5% each year, instead of 2%. What would the growth factor be? Explain how you know.

8. If the population of Tonga increased by 20% each year, what would the growth factor be? Write an equation to show what the population p would be n years after 1994.

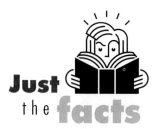

The endangered
orangutan has an
average life span of
35 years. An adult
female gives birth to a
single baby, on average,
every 3 to 8 years.

Problem Set B

Many living things grow exponentially during the early part of their lives. Later their growth slows and eventually stops. The actual growth rate for a particular organism depends on many things. For example, most bacteria grow more quickly in a warm environment than a cold one.

These estimates describe the early growth for four living organisms.

Organism	Growth in Weight
Orangutan	20% per month
Wheat plant	5% per day
Kitten	10% per week
Duckweed	50% per week

Below are four equations for estimating the weight of each organism at various times. K represents the starting weight of the organism, and W represents its weight at time t.

For each problem, do Parts a and b.

a. Match the equation to the organism it represents.

b. Tell what value of t would give the weight of the organism after 1 month. (Assume there are 4 weeks or 30 days in 1 month.)

1. $W = K \cdot 1.1^t$ **2.** $W = K \cdot 1.2^t$

3. $W = K \cdot 1.5^t$ **4.** $W = K \cdot 1.05^t$

Share & Summarize

1. Which of these tables could represent exponential growth? Explain how you know.

a.

x	1	2	3	4
y	3	12	48	192

b.

x	1	2	3	4
y	4	8	16	28

c.

x	1	2	3	4
y	2	4	6	8

d.

x	1	2	3	4
y	$^-60$	$^-30$	$^-20$	$^-15$

2. Suppose a population of c people grows by $R\%$ each year. Explain how you can estimate the size of the population after t years.

Investigation ▶ 2 ▶ Growing Even More

In Lesson 3.1, you solved a problem about *Escherichia coli* bacteria. Now you will examine how a population of this bacteria grows over time.

MATERIALS

graphing calculator

Problem Set C

These data show the population over time of the *E. coli* bacteria in a sample of water from a particular swimming pool.

Hour	Number of Bacteria
0 (start)	50
1	100
2	200
3	400
4	800

1. Assume the pattern in the table continues. What would be the population in Hour 5?

2. Write an equation giving the population p in the nth hour.

3. What is the growth factor in this situation? How does this appear in the equation you wrote?

4. Three students tried to describe this relationship in words.

   • Evan: "The population begins at 50 and increases by 50 each hour."

   • Tamika: "The population doubles every hour."

   • Jesse: "The population begins at 50 and doubles every hour."

   a. One description doesn't fit the data. Whose is it, and what's wrong?

   b. Look at the two descriptions that *do* fit the data. For each one, could you produce another series of bacteria populations that matches the description but is different from the data given in the table? Explain.

5. How does the equation you wrote in Problem 2 use the starting population?

6. Graph your equation from Problem 2 on your calculator. Use the table to help choose reasonable window settings.

   a. Make a sketch of the graph. Label the minimum and maximum values on each axis.

   b. Your graph should show points for non-integer values of n. Thinking about the context of this situation, does it make sense for n to have non-integer values? Explain.

7. About when does the population of bacteria in this pool exceed 100,000? Explain how you found your answer.

Problem Set D

The table gives data on the bacteria population in a sample taken from another pool. The sample was placed under a heat lamp and left undisturbed for 5 hours before the size of the population was first measured.

Hour	Bacteria
0 (start)	—
1	—
2	—
3	—
4	—
5	25,000
6	125,000
7	625,000

Assume the bacteria population is growing exponentially.

1. What is the growth factor?

2. Assuming the growth pattern didn't change, what was the approximate population of the bacteria when the sample was taken from the pool (at Hour 0)? Explain how you found your answer.

3. Write an equation showing the population p of the bacteria in this sample in the nth hour. How do you use the starting population and the growth factor in your equation?

4. Look back at the graph you made for Problem 6 of Problem Set C.

    a. Using the same window settings, graph both that equation and the new one you wrote for Problem 3 above. Make a sketch of the graphs, being sure to label the axes. Label the graphs with their equations.

    b. Compare the two graphs.

Share & Summarize

1. How can you write an equation for an exponential relationship by looking at a table of data? Assume the table's input values are consecutive integers.

2. A scientist has c cells of a certain type of bacteria. Each cell grows and splits so that every cell has been replaced by x cells 1 hour later. Write an expression giving the number of cells after t hours.

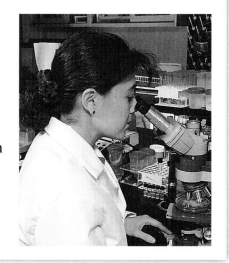

Investigation 3 ▶ Exponential Decay

In Investigations 1 and 2, you examined exponential growth relationships. In each situation, quantities were repeatedly multiplied by a number *greater than 1*. In this investigation, you will see what happens when quantities are repeatedly multiplied by a positive number *less than 1*.

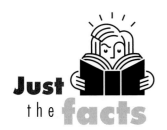
Problem Set E

The brightness of light can be described with a unit called a *lumen*. A certain type of mirror reflects $\frac{3}{5}$ of the light that hits it. Suppose a light of 2,000 lumens is shined on a series of several mirrors.

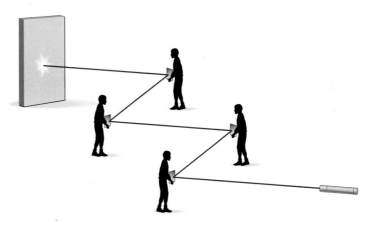

1. Copy and complete the table to indicate how much light would be reflected by the *n*th mirror for values of *n* ranging from 0 to 3. (Mirror 0 represents the original light.)

Mirror Number	Reflected Light (lumens)
0	2,000
1	
2	
3	

2. How much light is reflected by the *n*th mirror in the series?

3. Graph this relationship on your calculator. Put the mirror number on the horizontal axis and the amount of reflected light (in lumens) on the vertical axis. Sketch your graph, and label the minimum and maximum values on each axis.

4. The final mirror in a particular series reflects about 12 lumens of light. How many mirrors are in this series?

5. The final mirror in another series reflects only about 0.12 lumen of light. How many mirrors are in this series?

VOCABULARY
decay factor

6. The graph shows the amount of light reflected by another series of mirrors. The amount of light reflected by this type of mirror is different from the amount reflected by the other mirrors you have investigated. The intensity of the light being shone on the first mirror is also different.

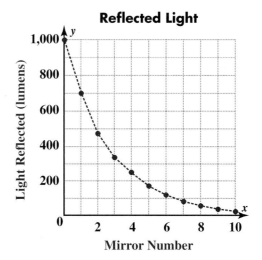

Reflected Light

a. What is the intensity of the light being shone on the first mirror?

b. How much light was reflected by the first mirror?

c. What fraction of light does this mirror reflect?

When quantities are repeatedly multiplied by a positive number less than 1—as was the case in Problem Set E—the quantities *decay exponentially.* Another way to say this is that they *decrease exponentially.* The number the quantities are repeatedly multiplied by is called the **decay factor.** The decay factor for the first mirror problem in Problem Set E is $\frac{3}{5}$, or 0.6.

After you purchase an item, its value may decrease year by year as it becomes worn out or out of fashion. This is called *depreciation.*

Think & Discuss

Suppose your school purchases a computer for $1,500, and the computer depreciates by 20% each year. That means that at the end of each year the computer is worth 20% less than its value at the beginning of the year.

What will be the value of the computer at the end of year 1?

Here's how Marcus and Ben thought about depreciation.

Marcus recalled the method demonstrated in Investigation 1. For exponential growth, you add the change—which is a percentage increase—to the original value. With depreciation, though, the change is a percentage decrease, so you have to subtract:

$$\text{Value at end of year} = \$1{,}500 - 20\% \text{ of } \$1{,}500$$
$$= \$1{,}500 - 0.2(\$1{,}500)$$

Marcus realized he could have just multiplied by a single number:

$$\$1{,}500 - 0.2(\$1{,}500) = \$1{,}500(1 - 0.2)$$
$$= \$1{,}500(0.8)$$

Ben reasoned that, at the end of the year, the computer's value will be 80% of its current value, since $100\% - 20\% = 80\%$. To compute the value of the computer at the end of the year, simply take 80% of $1,500:

$$80\% \text{ of } \$1500 = \$1{,}500(0.8)$$
$$= \$1{,}200$$

Both boys found a calculation—multiplying by 0.8—that lets them compute the value at the end of the year using the value at the beginning of the year.

MATERIALS

graphing calculator

Problem Set F

1. Find the value of the computer at the end of the second year and at the end of the third year.

2. Assume the computer continues to depreciate by 20% a year. What would its value be at the end of the 25th year? Explain how you found your answer.

3. Write an equation giving the computer's value V at the end of year n.

4. Does the computer's value decrease exponentially? If so, what is the decay factor?

5. Graph the equation from Problem 3 on your calculator.

   a. Sketch the graph, remembering to label the axes.

   b. When will the computer be worth half its present value, that is, $750?

   c. When will the computer be worth one-fourth its present value, that is, $375?

6. Suppose the computer's value depreciates by 17% instead of 20%. What would you multiply $1,500 by to get the value at the end of the first year? Explain how you know.

7. Suppose the computer's value depreciates by 40%. What would the decay factor be?

8. This equation describes the depreciation of a designer telephone, with v representing the value in dollars of the telephone n years from now.

$$v = 219(0.75)^n$$

a. What is the value of the telephone now?

b. What will the value of the telephone be in one year?

c. What is the decay factor for this relationship?

d. Describe this exponential relationship in words. Include the initial value of the telephone, the decay factor, and the time frame in your description.

Share & Summarize

Suppose $y = ab^x$ describes the exponential decay of y as x increases.

1. What does a represent?

2. What does b represent? How do you know?

Investigation  Identifying Exponential Growth and Decay

You have seen several exponential relationships and examined how to represent them with words, tables, equations, and graphs. Now you will see representations of different relationships and identify which are exponential.

Problem Set G

1. For each equation, state whether it describes an exponential relationship. If it does, tell whether that relationship involves growth or decay.

a. $y = 100 \cdot 3^x$
b. $y = 100 \cdot 0.3^x$
c. $y = 32 \cdot x$
d. $y = 20 \cdot \left(\frac{1}{4}\right)^x$
e. $y = 0.3(1.5^x)$
f. $y = 3x^2$

2. Consider these graphs.

i.
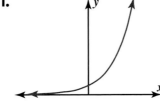

ii.
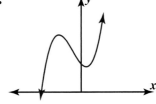

iii.
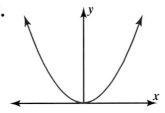

iv.
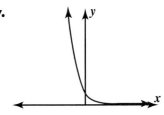

a. Which of these graphs could *not* describe an exponential relationship? Explain.

b. Assume that the graphs that *might* describe an exponential relationship actually *do*. For each, would the relationship involve growth or decay?

3. One of these is a graph of $y = 0.5^x$. The other is of $y = \frac{1}{x+1}$, which describes an inverse relationship. The graphs have different scales.

i.
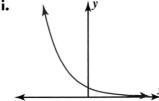

ii.
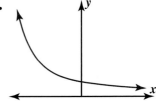

a. Just by looking at the graphs, can you tell which describes an exponential relationship?

b. One of the important characteristics of an inverse relationship is that its graph will have a vertical line that the graph cannot cross. For $y = \frac{1}{x+1}$, that vertical line is $x = {}^-1$. There is no y value when x is $^-1$ because you can't divide a number by 0.

Is there any x value for which the exponential relationship $y = 0.5^x$ has no y value?

4. For each table, state whether the relationship described could be exponential. If it could be, tell whether it would involve growth or decay. Explain your reasoning.

a.

x	y
1	6
2	9
3	12
4	15

b.

x	y
1	6
2	9
3	13.5
4	20.25

c.

x	y
111	56
112	28
113	14
114	7

5. A scientist is growing a culture of amoebas. Amoebas reproduce by splitting in half, so that one amoeba becomes two. Suppose that each amoeba in this culture splits twice in a day, so that after one day a single amoeba has become four amoebas. Assume that none of the amoebas die.

Is the relationship between the number of days since the scientist began growing the culture and the number of amoebas exponential? If so, what is the growth factor? If not, explain how you know.

6. Sydney is hiking on the Appalachian Trail. She hikes 15 miles every day. Is the relationship between the number of days since Sydney began hiking and the number of miles she has hiked exponential? If so, what is the growth factor? If not, explain how you know.

In the next problem set, you will compare exponential relationships with different growth and decay factors.

Problem Set **H**

Imagine that you have just discovered a table of data from a scientist who is experimenting with four cultures of bacteria. Each culture contains cells that were treated differently. Next to the table, the scientist noted that all of the populations changed exponentially, but with different growth or decay factors. Unfortunately, the scientist spilled coffee on the page, so many of the entries aren't legible.

1. Copy the scientist's table, and fill in the missing pieces of data. Remember that each culture changed exponentially.

Bacteria Count

Days	Culture 1	Culture 2	Culture 3	Culture 4
0	100	100	100	100
1	300		70	
2		25		
3				12,500
4				

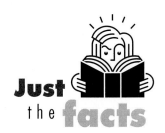

Just the facts

The Appalachian Trail, the longest marked hiking trail in the world, runs more than 2,000 miles from Maine to Georgia.

MATERIALS

graphing calculator

2. Which cultures *grew* exponentially? What is the growth factor?

3. Which cultures *decayed* exponentially? What is the decay factor?

4. Now you will compare equations for the four cultures.

   a. For each culture, write an equation giving the population p of the culture d days after the experiment began.

   b. How are your equations similar?

   c. How are your equations different?

5. Now you will compare graphs for the four cultures.

   a. Start with the cultures that *grew*. Graph the equations for those cultures in the same window of your calculator, with the horizontal axis showing Days 0 to 3. Sketch the graphs, and label each with the appropriate equation. Be sure to label the axes as well.

   b. Now consider the cultures that *decayed*. Graph them in the same window, with the horizontal axis showing Days 0 to 3. Sketch the graphs.

   c. How are the graphs for the cultures that grew different from those for the cultures that decayed?

   d. How are the graphs for the cultures that grew similar to those for the cultures that decayed?

   e. How can you tell which of two growth factors is greater just by looking at the graphs of the equations? How can you tell from the graphs which of two decay factors is greater?

Share & Summarize

1. How can you tell from a table when the relationship between two quantities might be exponential and not some other type of relationship? Assume the values for one quantity are consecutive integers.

2. How can you tell from a graph when the relationship between two quantities might be exponential?

3. How are graphs and equations representing exponential decay different from those representing exponential growth?

On Your Own Exercises

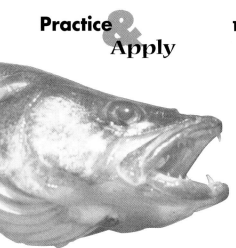

Practice & Apply

1. **Life Science** Kyle has estimated that the population of fish in the lake on his family's farm is 1,000 this year.

   a. If the population increases by 24% per year, what number can Kyle multiply this year's population by to estimate the fish population for next year?

   b. Estimate the number of fish in the lake next year.

   c. If the population increases by 6.9% per year, what number can Kyle multiply this year's population by to estimate the fish population for next year?

   d. Estimate the number of fish in the lake next year, given the information in Part c.

2. **Social Studies** The 1995 population of Delhi, a city in India, was estimated to be 10,000,000. From 1990 to 1995, the population increased by about 4% per year. Assume that the population continues to grow by 4% per year after 1995.

   a. Copy and complete the table.

Population of Delhi

Years after 1995	Estimated Population in Standard Notation	Estimated Population in Exponential Notation
0	10,000,000	$10,000,000 \cdot 1.04^0$
1	10,400,000	$10,000,000 \cdot 1.04^1$
2		
3		
4		
5		

   b. If the population of Delhi continues to increase by 4% per year, what will it be in the year 2020, which is 25 years after 1995? Write your answer in standard notation.

   c. Write an equation to represent the population p at n years after 1995.

   d. What is the growth factor in this situation?

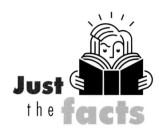

Life Science The data in each table represent how a certain population of bacteria grows over time. Identify the growth factor for each table, assuming the growth is exponential.

3.

Hours from Start	Population
0	2
1	10
2	50

4.

Hours from Start	Population
14	15,000,000
15	60,000,000
16	240,000,000

5.

Hours from Start	Population
25	3×10^8
26	6×10^8
27	1.2×10^9

6. Look at the table in Exercise 3. What will be the population of the bacteria in hour n?

7. The population of frogs in Kyle's family's lake is 1,000 this year.

   a. If the population decreases by 24% a year, what number can Kyle multiply this year's population by to estimate the frog population in the lake next year?

   b. If the population decreases by 6.9% a year, what number can Kyle multiply this year's population by to estimate the frog population in the lake next year?

8. This equation describes how the value of Geoffrey's new car, v, changes over time measured in years, t.

$$v = 18,000(0.6)^t$$

Tala thinks the car depreciates by 60% per year. Tamika thinks it depreciates by 40% per year. Who is right? Explain.

For each equation, state whether the relationship between x and y is exponential. If it is, tell whether growth or decay is involved and name the growth or decay factor.

9. $y = 4x$　　　　**10.** $y = 4^x$　　　　**11.** $y = 5(4^x)$

12. $y = 5x^4$　　　　**13.** $y = 5 \cdot 0.25^x$　　　　**14.** $y = 5$

15. Life Science When you take medicine to combat a headache or to lower a fever, the medicine enters your bloodstream. Your body eliminates the medicine in such a way that each hour a particular fraction is removed.

Suppose you have 200 mg of medicine in your bloodstream. Every hour, $\frac{1}{3}$ of the drug still in your bloodstream is eliminated.

a. How many milligrams will remain in your bloodstream at the end of the first hour?

b. How many milligrams will remain at the end of n hours?

c. How much of the medicine will have been eliminated at the end of n hours?

16. Below are five graphs of exponential relationships and five equations that represent them. Match each equation to a graph.

a. $y = 2(1.1^x)$

b. $y = 2(0.2^x)$

c. $y = 0.1(1.3^x)$

d. $y = 0.1(1.5^x)$

e. $y = 2(0.9^x)$

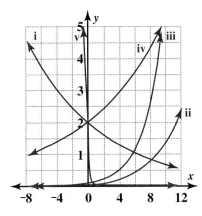

Connect & Extend

17. Economics When you invest your money with *compound interest,* a percentage of the money you invested is added to your investment after a given amount of time. The percentage added is the interest.

Suppose you invest $200 at 5% interest per year.

a. How much interest will you receive at the end of the first year?

b. How much money will you have altogether at the end of the first year?

c. How did you find your answer to Part b?

When interest is compounded yearly, the interest added is counted as part of your investment the next time the percentage is calculated.

d. The interest you earn the second year is 5% of your answer to Part b. How much interest will you receive the second year?

e. Complete the table.

Years	Account Value
0	$200
1	
2	
3	
4	
5	

f. How much money would you have in the account in 50 years if you invested $200 at an interest rate of 5%, compounded yearly? How much money would you have in *n* years?

g. When she was 20, Ella invested $600 at an interest rate of 5%, compounded yearly. When Jan was 45, he invested $2,000, also at a rate of 5%, compounded yearly.

How much will each person have in the bank at age 60?

18. Social Studies When an illness such as the flu starts to spread, the number of sick people can often be modeled using an exponential relationship.

Below are tables created for three communities showing how the number of people who had a particular illness changed over time. All three tables represent exponential growth and have the same growth factor.

Table A

Days Since Illness Was Identified	Number of Sick People
0	128
1	192
2	288

Table B

Days Since Illness Was Identified	Number of Sick People
5	200
6	300
7	450

Table C

Days Since Illness Was Identified	Number of Sick People
4	648
5	972
6	1,458

a. What is the growth factor in the three tables?

b. Could Tables A and B be describing data from the same community? Why or why not?

c. Could Tables A and C be describing data from the same community? Why or why not?

19. A scientist recorded data on the growth of a population of bacteria, but later splattered orange juice on her notebook.

Hour	Population
0	
1	600
2	
3	21,600
4	

a. Assume the population grows exponentially, and complete the scientist's table.

b. What is the growth factor for the bacteria population?

c. How many bacteria will there be in hour n?

20. Life Science The population of birds at a wildlife reserve is projected to decrease over time according to this equation, where x is the time in years from now.

$$y = 1,000(0.7)^x$$

a. What is the current population of birds?

b. If the bird population continues to decrease according to this equation, will the birds ever disappear from the reserve? Explain.

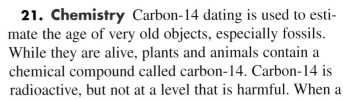

21. Chemistry Carbon-14 dating is used to estimate the age of very old objects, especially fossils. While they are alive, plants and animals contain a chemical compound called carbon-14. Carbon-14 is radioactive, but not at a level that is harmful. When a plant or an animal dies, the amount of carbon-14 in its remains decreases exponentially.

Each type of radioactive material has a *decay rate*. This rate determines how long it will take for the amount of radioactive material in something to decrease to half. This period of time is the *half-life*.

Imagine that a scientist is working with four samples of radioactive compounds. These equations describe different rates of decay for the four compounds, where a is the amount and t is time in centuries:

- Sample 1: $a = 100 \cdot 0.8^t$
- Sample 2: $a = 100 \cdot 0.7^t$
- Sample 3: $a = 100 \cdot 0.6^t$
- Sample 4: $a = 100 \cdot 0.5^t$

a. Match each sample to one of the tables below.

i.

Time (centuries)	Amount of Compound (grams)
0	100.0
1	50.0
2	25.0
3	12.5

ii.

Time (centuries)	Amount of Compound (grams)
3	51.2
4	41.0
5	32.8
6	26.2

iii.

Time (centuries)	Amount of Compound (grams)
2	36.0
3	21.6
4	13.0
5	7.8

iv.

Time (centuries)	Amount of Compound (grams)
7	8.2
8	5.8
9	4.0
10	2.8

b. What is the starting value for each sample?

c. What is the decay factor for each sample?

d. Match each sample to one of the graphs below.

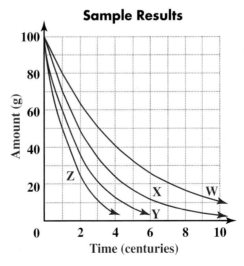

Sample Results

e. For each sample, approximate how long it takes for the amount of the compound to fall to half of its original value. This is the half-life of the compound.

22. Dion and Lucita are looking at this table of data, which represents the exponential growth of bacteria in a certain swimming pool.

Hour	Population
0	10
1	20
2	40
4	80
5	160
$\vdots$	$\vdots$
24	167,772,160
25	335,544,320

Dion says the equation $p = 10 \cdot 2^h$ describes the table, where p is the population and h is the number of hours.

a. Is Dion's equation correct?

Lucita says that the equation $p = 10 \cdot 16{,}777{,}216^d$ describes the table, where p is the population and d is the number of days.

b. Express 1 hour ($h = 1$) as a fraction of a day. Do the same for 2 hours, 3 hours, and 24 hours.

c. Using a calculator, test Lucita's equation. Find the population for the number of days corresponding to 1 hour, 2 hours, 3 hours, and 24 hours. Does her equation seem to be correct?

d. Resolve Dion and Lucita's dilemma. If one or both of the equations is incorrect, explain why. If both of the equations are correct, explain how this can be.

23. Challenge Examine this table.

x	y
0.1	$^-1$
1	0
10	1
100	2
1,000	3
10,000	4

a. Plot the points in the table. Draw a dashed curve through them.

b. In all of the exponential relationships you've seen, y grew or decayed exponentially as x increased. In this relationship, x is growing exponentially as y increases. What is the growth factor?

Mixed Review

Geometry State the area of the square with the given side length.

24. side length 3 m

25. side length 22 cm

26. Find an equation of the line that passes through the points $(^-2, 0)$ and $(^-6, ^-6)$.

27. Find an equation of the line with slope $^-2.5$ that passes through the point $(2, ^-4)$.

28. Match each equation to a graph.

a. $y = -x^2 + 2$

b. $y = (x - 2)^2$

c. $y = -x^2$

d. $y = -(x + 2)^2$

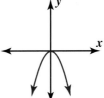

Graph F

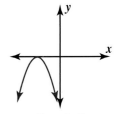

Graph G

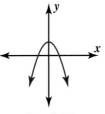

Graph H

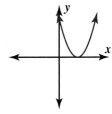

Graph J

Remember

Lines always have a constant difference in their y values. That is, when the x values change by a certain amount, such as 1 or 2, the y values also change by a certain amount.

For the linear equations in Exercises 29–31, answer Parts a and b.

a. What is the constant difference between the y values as the x values increase by 1?

b. What is the constant difference between the y values as the x values decrease by 3?

29. $y = \frac{x}{2}$

30. $2x - 2 = y$

31. $\frac{y}{3x} = 3$

32. Sports A popular cliff for divers has a height of 18 meters from the water's surface. Once a diver has left the cliff, the height in meters of the diver, h, after t seconds can be approximated with this equation:

$$h = 18 - 4.9t^2$$

a. Create a table of heights for several values of t. When you choose values for your table, think about what values would make sense in this situation.

b. Use your table to sketch a graph of the relationship between height and time.

c. Use your graph to predict how long it will take a diver to reach the halfway point of the dive.

d. Use your graph to predict how long it will take a diver to hit the surface of the water.

3.3 Radicals

You have worked quite a bit with numbers raised to positive and negative integer exponents. You know how to evaluate 2^3, 4^{-3} and even $(^-100)^{100}$.

But what if you were asked, "What number squared is 441?" This is the same as knowing the exponent (in this case, 2, because you're squaring), and the result (in this case, 441) but not knowing the base. In this lesson, you will examine problems like this.

Explore

Why do you think these numbers are called "square numbers" or "perfect squares"? What do they have to do with a square?

16   64   100

Without using the $\sqrt{}$ key on your calculator, try to find the side length of a square whose area is 12.5316.

You can take a number that is not an integer, like 5.5, and make it the side length of a square. A square with a side length of 5.5 has an area of 30.25. Even though 30.25 is not a perfect square, it's still equal to a number multiplied by itself.

Investigation Square Roots

To solve an equation like $2x = 6$, you can think about "undoing" the multiplication by 2 that gives the product 6. To undo multiplication, you divide: $6 \div 2 = 3$, so x must be 3.

Think again about the question, "What number squared is 441?" This can be written as the equation $x^2 = 441$. To find the answer, you can think about "undoing" the process of squaring a number. That is, you have to find the **square roots** of 441.

VOCABULARY
square root

Thinking about solutions to an equation gives you a new way to look at square roots, as the students in this class discovered.

Both 5 and $^-5$ are square roots of 25. This is important to remember when you want to solve $x^2 = 25$, because there are two answers: $x = 5$ and $x = ^-5$.

V O C A B U L A R Y
radical sign

The symbol $\sqrt{}$ is called a **radical sign.** It's used to mean the *positive* square root of a number, so $\sqrt{25} = 5$. To indicate the *negative* square root of a number, write a negative sign in front: $^-\sqrt{25} = ^-5$.

Problem Set A

Solve these problems without using your calculator.

1. $\sqrt{49}$ **2.** What are the square roots of 100?

3. $^-\sqrt{225}$ **4.** What are the square roots of 2.25?

5. $\sqrt{0}$ **6.** What is the negative square root of 0.01?

7. How many square roots does a positive number have?

8. How many square roots does 0 have?

9. How many square roots does a negative number have?

10. Decide whether the statement below is true or false. Explain your answer.

$$^-4 = \sqrt{16}$$

Problem Set B

Solve these problems without using a calculator.

1. $(\sqrt{49})^2$

2. $(\sqrt{2.25})^2$

3. $(\sqrt{81})^2$

4. $(^-\sqrt{100})^2$

5. $(^-\sqrt{4})^2$

6. $^-(\sqrt{0.04})^2$

7. What happens when you square the square root of a number n? Why?

Use a calculator to check your answers to these problems.

8. $\sqrt{100^2}$

9. $\sqrt{(^-4)^2}$

10. $\sqrt{0.04^2}$

11. $\sqrt{(^-0.04)^2}$

12. Consider the value of $\sqrt{n^2}$.

   a. Assume n is positive, and write an expression equivalent to $\sqrt{n^2}$ that has no radical signs. Hint: Consider your answer to Problem 8.

   b. Assume n is negative, and write an expression equivalent to $\sqrt{n^2}$ that has no radical signs. Hint: Consider your answers to Problems 9 and 11.

   c. If possible, write a single expression equivalent to $\sqrt{n^2}$ that has no radical signs, for any value of n (positive, negative, or 0). Hint: Consider your answers to Problems 10 and 11.

Problem Set C

Solve these equations, if possible, using whatever method you prefer. If an equation has no solution, write "no solution" and explain why.

1. $\sqrt{x} = 7$

2. $\sqrt{x} + 8 = ^-6$

3. $0 = 2.3 + \sqrt{x}$

4. $\sqrt{x + 2} = 5$

5. $x = \sqrt{^-16}$

6. $x^2 = 64$

Share & Summarize

1. Juana, a sixth grader, said, "My older sister loves to tell me about her math class. When she learned about square roots, she showed me some examples, like $\sqrt{4} = 2$. At first, I thought finding a square root meant the same thing as finding half of something."

   Write a sentence or two explaining the difference between finding half of something and finding the square root of something.

2. Are there any numbers for which taking half gives the same result as taking the square root? If so, what are they?

Investigation 2 ▶ Simplifying Radical Expressions

What happens when you try to add, subtract, multiply, or divide numbers that involve radical signs? The next problem set will help you think about how and when you can combine terms with radical signs.

Problem Set D

1. Does the square root of a sum equal the sum of the square roots? That is, does $\sqrt{x + y} = \sqrt{x} + \sqrt{y}$?

   Complete the following table, testing several examples. For the last column, choose your own positive x and y values to test. Then make a conjecture about whether the two expressions are equivalent.

(x, y)	$(0, 0)$	$(4, 4)$	$(36, 16)$	$(25, \frac{1}{4})$	$(\_\_, \_\_)$
$\sqrt{x} + \sqrt{y}$					
$\sqrt{x + y}$					

2. Now do the same for multiplication: Is $= \sqrt{x \cdot y} = \sqrt{x} \cdot \sqrt{y}$? Complete the table, and then make a conjecture.

(x, y)	$(0, 0)$	$(5, 5)$	$(9, 25)$	$(0.64, 100)$	$(\_\_, \_\_)$
$\sqrt{x} \cdot \sqrt{y}$					
$\sqrt{x \cdot y}$					

3. Complete the table, and make a conjecture about the differences of square roots.

(x, y)	$(0, 0)$	$(4, 4)$	$(81, 49)$	$\left(\frac{25}{9}, \frac{16}{9}\right)$	$(\_\_, \_\_)$
$\sqrt{x} - \sqrt{y}$					
$\sqrt{x - y}$					

4. Complete the table, and make a conjecture about dividing square roots.

(x, y)	$(0, 2)$	$(3, 3)$	$(4, 16)$	$\left(\frac{4}{9}, 2.25\right)$	$(\_\_, \_\_)$
$\sqrt{x} \div \sqrt{y}$					
$\sqrt{x \div y}$					

Expressions that contain one or more numbers or variables under a radical sign are called *radical expressions*. These are all radical expressions.

$$3\sqrt{x} \qquad 2.3m\sqrt{3m} \qquad x + 3\sqrt{x}$$

Computing with radical expressions can get difficult when the expressions are complicated. Sometimes it helps to *simplify* radical expressions to make them easier to work with. A radical expression is simplified when

- numbers under radical signs have no square factors
- the number of radical signs in the expression is as small as possible

In Problem Set D, you probably conjectured that

$$\sqrt{x} \cdot \sqrt{y} = \sqrt{x \cdot y}$$

As long as both x and y are nonnegative—that is, positive or 0—this conjecture is true.

You might also have conjectured that

$$\sqrt{x} \div \sqrt{y} = \sqrt{x \div y}$$

This rule is true for nonnegative x and positive y. These rules will help you learn how to simplify radical expressions.

To simplify a number under a square root sign, look for factors of the number that are perfect squares. Rewrite the number as a product in which at least one factor is a perfect square, and take the square roots of the perfect squares.

To simplify $\sqrt{24}$, rewrite it as $\sqrt{4 \cdot 6}$ or $\sqrt{4} \cdot \sqrt{6}$. Since $4 = 2 \cdot 2$, $\sqrt{4} \cdot \sqrt{6}$ is equivalent to $2 \cdot \sqrt{6}$ or $2\sqrt{6}$.

To simplify $\sqrt{18x^4}$, rewrite it as $\sqrt{9x^4 \cdot 2}$ or $\sqrt{9x^4} \cdot \sqrt{2}$. Since $9x^4 = 3x^2 \cdot 3x^2$, $\sqrt{9x^4} \cdot \sqrt{2}$ is equivalent to $3x^2\sqrt{2}$.

To simplify $\sqrt{30}$, you could rewrite it as $\sqrt{15 \cdot 2}$, $\sqrt{10 \cdot 3}$, or $\sqrt{5 \cdot 6}$. However, since none of the factors of 30 are perfect squares, $\sqrt{30}$ is already simplified.

Perfect squares can be hard to find. For example, to simplify $\sqrt{48}$, you could rewrite it as $\sqrt{8} \cdot \sqrt{6}$. But since neither 8 or 6 is a perfect square, you might think $\sqrt{48}$ is already simplified.

However, if you rewrite $\sqrt{48}$ as $\sqrt{4} \cdot \sqrt{12}$, you *can* simplify:

$$\sqrt{48} = \sqrt{4} \cdot \sqrt{12}$$
$$= 2 \cdot \sqrt{12}$$

Then simplify $\sqrt{12}$ further:
$$= 2 \cdot \sqrt{4 \cdot 3}$$
$$= 2 \cdot \sqrt{4} \cdot \sqrt{3}$$
$$= 2 \cdot 2 \cdot \sqrt{3}$$
$$= 4 \cdot \sqrt{3} \text{ or } 4\sqrt{3}$$

Problem Set E

Simplify if possible. Assume all variables are positive.

1. $\sqrt{75}$ **2.** $\sqrt{60}$

3. $\sqrt{42}$ **4.** $\sqrt{\frac{1}{8}}$

5. $\sqrt{50x^3}$ **6.** $\sqrt{72a^4b^5}$

Now see whether you can "unsimplify" expressions. Fill each blank with an expression different from the given one. Assume all variables are nonnegative.

7. $2\sqrt{3}$ is the simplified form of _____.

8. $6\sqrt{2}$ is the simplified form of _____.

9. $5y\sqrt{3}$ is the simplified form of _____.

10. $\frac{1}{4}x^2\sqrt{3x}$ is the simplified form of _____.

When adding or subtracting terms with radical signs, the terms with radical signs behave in ways similar to expressions with variables.

EXAMPLE

To simplify expressions with variables, combine *like terms*.

$$3x + 4y - 2x + 5y = 3x - 2x + 4y + 5y$$
$$= (3 - 2)x + (4 + 5)y$$
$$= 1x + 9y$$
$$= x + 9y$$

To simplify expressions with radicals, combine *like radical terms*.

$$3\sqrt{2} + 4\sqrt{5} - 2\sqrt{2} + 5\sqrt{5} = 3\sqrt{2} - 2\sqrt{2} + 4\sqrt{5} + 5\sqrt{5}$$
$$= (3 - 2)\sqrt{2} + (4 + 5)\sqrt{5}$$
$$= 1\sqrt{2} + 9\sqrt{5}$$
$$= \sqrt{2} + 9\sqrt{5}$$

Sometimes you must simplify the terms before you can combine them:

$$\sqrt{20} + \sqrt{45} = \sqrt{4 \cdot 5} + \sqrt{9 \cdot 5} = 2\sqrt{5} + 3\sqrt{5}$$
$$= (2 + 3)\sqrt{5}$$
$$= 5\sqrt{5}$$

Problem Set F

For Problems 1–4, decide whether the two expressions are equivalent. Explain your reasoning or show your work.

1. $\sqrt{2} + \sqrt{2} + \sqrt{2}$ and $\sqrt{6}$

2. $3\sqrt{2} + \sqrt{3}$ and $3\sqrt{5}$

3. $\sqrt{50} + \sqrt{98}$ and $12\sqrt{2}$

4. $-\frac{1}{2}\sqrt{80}$ and $\sqrt{45} + \sqrt{20} - 7\sqrt{5}$

5. Write an expression equivalent to $2\sqrt{3}$ that includes addition or subtraction.

6. Below are four radical expressions.

   i. $^{-}3\sqrt{3} + \sqrt{48}$ **ii.** $2\sqrt{12} - \sqrt{27}$

   iii. $12\sqrt{2} - 7\sqrt{2}$ **iv.** $3\sqrt{32} - \sqrt{98}$

   a. Simplify each expression.

   b. Are any of the expressions equivalent to each other? If so, which?

Share & Summarize

When she is trying to simplify radical terms, Susan gets stuck if she can't immediately find a factor that is a perfect square. For example, she said, "When I try to simplify $\sqrt{60}$, I come up with $\sqrt{6} \cdot \sqrt{10}$. Neither of those are perfect squares, so I can't figure out how to simplify it."

1. Simplify $\sqrt{60}$.

2. Describe a method Susan could use to find the right factors. Your method should work for any problem Susan tries.

Investigation 3 ▶ nth Roots

In Investigation 1, you "undid" the process of squaring numbers. Now you will undo the process of taking numbers to higher powers.

Think & Discuss

Think about why "cubic numbers" or "perfect cubes" have those names. Give an example of a cubic number, and show how it is related to a cube.

Some number cubed is equal to 8. What is that number? How many answers can you find?

Some number cubed is equal to ⁻64. What is that number? How many answers can you find?

Is there a number squared that equals ⁻64? If so, what is it?

Some number cubed is equal to ⁻0.027. What is that number?

When you answer questions like "What does N equal, if $N^3 = {}^-64$?" you are undoing the process of cubing a number. To answer this question, suppose someone has cubed a number and told you the result, and you want to find the original number.

The *cube root* of 64 is 4, because 4 *cubed* is 64. That is, $4^3 = 4 \cdot 4 \cdot 4 = 64$. The radical sign is used to indicate cube roots by adding a "3":

$$\sqrt[3]{64} = 4$$

Every number has one cube root. For square roots, the radical sign by itself always represents the positive square root. For cube roots, however, the cube root of a positive number is always positive, and the cube root of a negative number is always negative. So, the symbol $\sqrt[3]{}$ represents the one cube root a number has.

$$\sqrt[3]{8} = 2 \qquad \sqrt[3]{-8} = -2$$

Problem Set G

Evaluate without using a calculator.

1. $\sqrt[3]{8}$

2. $\sqrt[3]{-125}$

3. $\sqrt[3]{0.000001}$

4. $(\sqrt[3]{8})^3$

5. $(\sqrt[3]{-125})^3$

6. $(\sqrt[3]{37})^3$

7. What happens when you cube the cube root of any number? Why?

Solve each equation using whatever method you choose.

8. $\sqrt[3]{x} = -\frac{1}{8}$

9. $6 = \sqrt[3]{2n}$

10. $\sqrt[3]{z + 5} = -2$

Just the facts

Using the set of complex numbers, every number except 0 has three cube roots.

You can also take higher roots. The *fourth root* of 625 is 5, because 5 to the fourth power is 625. That is, $5^4 = 5 \cdot 5 \cdot 5 \cdot 5 = 625$. Another fourth root of 625 is $^-5$, because $^-5 \cdot {}^-5 \cdot {}^-5 \cdot {}^-5 = 625$.

The radical sign for the *positive* fourth root is $\sqrt[4]{}$, so $\sqrt[4]{625} = 5$. However, if you want the fourth roots of 625, you need to consider both $^-5$ and 5.

Think & Discuss

What do you think a *fifth root* is? Give an example.

What do you think a *sixth root* is? Give an example.

What's another name for a *second root*?

What's another name for a *third root*?

Fifth roots, sixth roots, and so on are similar to square roots and cube roots.

In general, $\sqrt[n]{}$ denotes the **nth root** of x.

- When n is even, $\sqrt[n]{x}$ denotes the positive nth root. The negative nth root is $-\sqrt[n]{x}$. When n is even, x must be nonnegative.

- When n is odd, $\sqrt[n]{x}$ is positive if x is positive and negative if x is negative.

The nth root of 0 is always 0, no matter what n is.

Notice that the square root of x is written $\sqrt{x}$, not $\sqrt[2]{x}$. If there is no number, the radical is assumed to represent a square root.

VOCABULARY
nth root

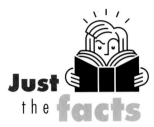

Just the facts

The nth root of x can also be expressed like this:

$$x^{\frac{1}{n}}$$

Using complex numbers, every number except 0 has n nth roots.

Problem Set H

Evaluate without using a calculator.

1. $\sqrt[5]{32}$

2. Find the fourth roots of 256.

3. Find the seventh root of $^-128$.

4. $\sqrt[6]{729}$

5. Are there any numbers for which the second, third, fourth, fifth, and sixth roots are the same number? If so, list them.

6. Are there numbers with both a positive and a negative fourth root? If so, give an example. If not, explain why not.

7. Is it possible to find a number with only a positive fourth root? If so, give an example. If not, explain why not.

8. Is it possible to find a number with both a positive and a negative fifth root? If so, give an example. If not, explain why not.

Order each set of numbers from least to greatest.

9. $\sqrt[5]{7}$ $\sqrt[3]{7}$ $\sqrt[4]{7}$ $\sqrt{7}$ $\sqrt[6]{7}$

10. $\sqrt[4]{\frac{1}{4}}$ $\sqrt[81]{\frac{1}{4}}$ $\sqrt{\frac{1}{4}}$ $\sqrt[80]{\frac{1}{4}}$

Share & Summarize

1. Describe the general relationship between finding an nth root of some number and raising some number to the nth power.

2. Is it possible to find a number with both a positive and a negative ninth root? If so, give an example. If not, explain why not.

Investigation ▶4▶ Irrational Numbers

V O C A B U L A R Y
rational numbers

Integers and fractions such as $\frac{1}{2}$ and $\frac{7}{10}$ are *rational numbers*. The word *rational* comes from the word *ratio,* which should help you remember the definition of a rational number. A **rational number** is a number that can be written as the ratio, or quotient, of two integers.

EXAMPLE

The number 0.5 is a rational number because it can be written as a ratio of two integers, such as $\frac{1}{2}$ and $\frac{7}{14}$.

The number ⁻1.8 is a rational number because it can be written as a ratio of two integers, such as $-\frac{9}{5}$ and $-\frac{27}{15}$.

The number 3 is also a rational number. It can be written as $\frac{3}{1}$ or $\frac{9}{3}$, for example.

V O C A B U L A R Y
irrational numbers
real numbers

Numbers that *cannot* be written as the ratio of two integers are called **irrational numbers.** The rational numbers and irrational numbers together form the set of **real numbers.** All real numbers can be located on the number line.

Think & Discuss

Below are seven numbers. Try to decide whether each number is rational by looking for a ratio of integers that is equal to the number.

$$3 \qquad 0.7 \qquad {}^{-}2.55 \qquad \sqrt{20} \qquad 4.5678 \qquad \sqrt{25} \qquad 7.\overline{4}$$

Think about numbers with one nonzero digit to the right of the decimal point, such as 0.7. Are all such numbers rational?

Think about numbers that have two nonzero digits to the right of the decimal point, such as 2.55. Are all such numbers rational?

Are numbers with 3 nonzero digits to the right of the decimal point rational? Are numbers with 10 nonzero digits to the right of the decimal point rational? Are numbers with n nonzero digits to the right of the decimal point rational?

Can numbers whose nonzero digits to the right of the decimal point go on forever be rational? Explain.

Remember
A bar written over one or more numerals means those numerals repeat. For example,
$$2.1\overline{5} = 2.1555\ldots$$
$$9.\overline{32} = 9.323232\ldots$$

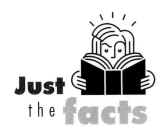

Just the facts

A group called the Pythagoreans in ancient Greece had developed a way of life based on the idea that rational numbers are the essence of all things. When the existence of irrational numbers was proved, they took extreme measures to cover it up!

The Parthenon, in Athens, Greece, was built during the fifth century B.C.

In the Think & Discuss, you may not have been sure whether $\sqrt{20}$ is rational or irrational. You will now explore this number.

Problem Set I

1. Using your calculator, evaluate $\sqrt{20}$. What does the calculator display for the decimal form of $\sqrt{20}$?

2. Multiply your answer to Problem 1 by itself, using your calculator. What do you get?

3. Now imagine multiplying your answer for Problem 1 by itself *without* using a calculator. What would be the rightmost digit of the product? (For example, the rightmost digit of 4.36 is 6, and the rightmost digit of 121.798 is 8.)

4. Could the number your calculator displayed in Problem 1 really be equal to $\sqrt{20}$? How might your findings be explained?

Calculators can be deceptive because they have only enough screen space to show a certain number of digits. If a number has more digits than a calculator can handle, some of the digits at the end simply don't appear on the screen.

So, what about $\sqrt{20}$? It has more digits than a calculator will display. That means it could be one of three types of numbers:

- It could have a set number of nonzero digits to the right of the decimal point.

  Such numbers are called *terminating decimals* because the digits eventually stop, or terminate. Since you can find a ratio of integers equal to a terminating decimal, these numbers are rational.

- It could have nonzero digits to the right of the decimal point that continue forever in a repeating pattern; $0.3\overline{24}$ is an example of such a number.

  Such numbers are called *nonterminating, repeating decimals.* They are also rational, although this fact will not be shown here. For example, $0.\overline{3}$ is equal to $\frac{1}{3}$.

- It could have nonzero digits to the right of the decimal point that continue forever without ever repeating.

  Such numbers are called *nonterminating, nonrepeating decimals.* As it turns out, *any* such number is irrational.

The number $\sqrt{20}$ is a nonterminating, nonrepeating decimal, so it is irrational.

You have already seen many irrational numbers. In fact, any number that has a radical sign in its simplified form is irrational.

You have worked with this kind of irrational number when representing distances and lengths. For example, the length of the hypotenuse of this triangle is irrational.

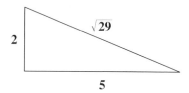

Not all irrational numbers are roots, though. One famous irrational number is π.

Problem Set J

Work with a partner to decide whether each number below is rational or irrational. If a number is rational, write it as a ratio of two integers. If it is irrational, explain how you know.

1. $3\frac{1}{2}$

2. $\sqrt{3} \cdot \sqrt{45}$

3. 4.627803

4. $\sqrt[3]{32} \div \sqrt[3]{4}$

5. $\sqrt{40} - \sqrt{10}$

6. $\sqrt{22}$

7. $\sqrt{8} \cdot \sqrt{2}$

8. 0

9. $1.\overline{6}$

10. $\frac{\sqrt{2}}{2}$

Share & Summarize

Tell whether the members of each group are *always rational, sometimes rational,* or *never rational.* Explain your reasoning or give examples.

1. integers

2. cube roots

3. radical expressions

4. products of two irrational numbers

On Your Own Exercises

Practice  **Apply**

In Exercises 1–10, find the indicated roots without using a calculator.

1. the square roots of 1.21

2. $\sqrt{1.21}$

3. the square roots of $\frac{4}{49}$

4. $\sqrt{\frac{4}{49}}$

5. the square roots of 0.0064

6. $\sqrt{0.0064}$

7. $(\sqrt{26})^2$

8. $(\sqrt{0.09})^2$

9. $\sqrt{(^-3)^2}$

10. $\sqrt{x^2}$

Solve each equation. If it has no solution, write "no solution."

11. $\sqrt{x-3} = 9$

12. $5\sqrt{x} = 25$

13. $\sqrt{\frac{x}{7}} = 3$

14. $\sqrt{x} = 36$

15. $\sqrt{x+2} + 8 = 1$

16. $\sqrt{x-20} = ^-18$

Tell whether each computation is correct or incorrect.

17. $\sqrt{2} \cdot \sqrt{3} = \sqrt{6}$

18. $\sqrt{4} + \sqrt{15} = \sqrt{19}$

19. $\sqrt{3} - \sqrt{0} = \sqrt{3}$

20. $\sqrt{20} \div \sqrt{45} = \frac{2}{3}$

Simplify each radical expression. If it is already simplified, say so.

21. $\dfrac{\sqrt{20} + \sqrt{80}}{\sqrt{20}}$

22. $\sqrt{17} - \sqrt{30}$

23. $\sqrt{8} \cdot \sqrt{12}$

24. Challenge $\sqrt{x+2} + \sqrt{4x+8}$

25. Below are four radical expressions. Assume x is positive.

    i. $\sqrt{800x^3}$

    ii. $4x\sqrt{50x}$

    iii. $3\sqrt{32x^3}$

    iv. $5\sqrt{8x^3} + 2x\sqrt{2x} + 2\sqrt{32x^3}$

  a. Simplify each expression.

  b. Which expressions are equivalent to each other?

Decide whether the expressions in each pair are equivalent. Explain.

26. $3\sqrt{5}$ and $\sqrt{5} + \sqrt{5} + \sqrt{5}$

27. $\sqrt{32} - \sqrt{18}$ and $\sqrt{14}$

In Exercises 28–35, find the indicated roots without using a calculator.

28. the cube root of $^-216$

29. $\sqrt[3]{^-216}$

30. the sixth roots of 64

31. $\sqrt[6]{64}$

32. the fifth root of $^-243$

33. $\sqrt[5]{^-243}$

34. Challenge the eighth roots of x^{16}

35. Challenge $\sqrt[8]{x^{16}}$

Order each set of numbers from least to greatest.

36. $\sqrt[9]{^-41}$ $\quad$ $\sqrt[3]{^-41}$ $\quad$ $\sqrt[11]{^-41}$ $\quad$ $\sqrt[5]{^-41}$ $\quad$ $\sqrt[7]{^-41}$

37. $\sqrt[3]{-\frac{1}{3}}$ $\quad$ $\sqrt[311]{-\frac{1}{3}}$ $\quad$ $\sqrt[5]{-\frac{1}{3}}$ $\quad$ $\sqrt[105]{-\frac{1}{3}}$

Prove that each number is rational by finding a pair of integers whose ratio, or quotient, is equal to the number.

38. 3.56

39. $^-0.000230$

40. $^-1.\overline{6}$

41. $5\frac{3}{8}$

Tell whether each number is rational or irrational. If it is rational, find two integers whose ratio is equal to it. If it is irrational, explain how you know.

42. $^-3.\overline{3}$

43. $4\frac{1}{3}$

44. $^-\sqrt{28}$

45. $^-5.8237$

46. $\sqrt[4]{32}$

47. $\sqrt{45} - \sqrt{10} \cdot \sqrt{2}$

48. For this problem, assume x is positive.

a. Name three values of x for which $\sqrt{x}$ is less than x.

b. Name three values of x for which $\sqrt{x}$ is greater than x.

c. In general, how can you tell whether $\sqrt{x}$ is greater than x?

49. Prove It! Evan said, "If $x^2 = y^2$, then $x = y$."

a. Try several values for x and y to investigate Evan's conjecture.

b. Is Evan's conjecture true? If it is, explain why. If not, give a counterexample.

50. For what values of x and y is $\sqrt{x} + \sqrt{y} = \sqrt{x+y}$ true?

51. For what values of x and y is $\sqrt{x} - \sqrt{y} = \sqrt{x-y}$ true?

52. Prove It! In Investigation 2, you saw that for nonnegative values of x and y, $\sqrt{x} \cdot \sqrt{y} = \sqrt{x \cdot y}$. In Parts a–d, you will prove this conjecture.

To make things easier, define two new variables: $u = \sqrt{x}$ and $v = \sqrt{y}$.

a. What is the product uv in terms of x and y?

b. In terms of x and y, what are u^2 and v^2 equivalent to?

c. Fill in the blank, using x and y:

$$(uv)^2 = u^2v^2 = \underline{\hspace{2cm}}$$

d. Use your answer to Part c to write an expression in terms of x and y that is equivalent to uv.

In y o u r
own
words

Write a letter to a younger student explaining the difference between rational and irrational numbers. Include the definitions of both kinds of numbers and several examples of each.

e. Parts a and d both asked you to write the product uv in terms of x and y, so your answers to those parts are equivalent. Have you shown that $\sqrt{x} \cdot \sqrt{y} = \sqrt{(x \cdot y)}$?

f. Challenge Use Parts a–d as a guide to prove that

$$\sqrt{x} \div \sqrt{y} = \sqrt{x \div y}$$

if x is nonnegative and y is positive. If it helps, rewrite both sides as fractions: $\frac{\sqrt{x}}{\sqrt{y}} = \sqrt{\frac{x}{y}}$.

53. Geometry Isosceles right triangles have two legs that are the same length. If the length of one leg of an isosceles right triangle is a, what is the length of the hypotenuse? Simplify your answer.

54. List these numbers from least to greatest.

$$0 \quad 1 \quad {}^{-}1 \quad \sqrt[51]{2} \quad \sqrt[51]{-2} \quad \sqrt[51]{0.2} \quad \sqrt[51]{-0.2}$$

55. Consider $\sqrt[8]{n}$.

a. If n is greater than 1, is $\sqrt[8]{n}$ greater than or less than n?

b. If n is greater than 0 but less than 1, is $\sqrt[8]{n}$ greater than or less than n?

56. Challenge Consider $\sqrt[n]{x}$.

a. If $\sqrt[n]{x}$ is positive, what can you say for sure about x? Explain.

b. If $\sqrt[n]{x}$ is negative, what can you say for sure about n and x? Explain.

57. You have worked quite a bit with integer exponents. Exponents can also be fractions. When $\frac{1}{n}$ is used as an exponent, it means "take the nth root." So, for example,

$$81^{\frac{1}{2}} = \sqrt{81} = 9 \qquad (^{-}27)^{\frac{1}{3}} = \sqrt[3]{-27} = {}^{-}3 \qquad 64^{\frac{1}{4}} = \sqrt[4]{64} = 4$$

The laws of exponents apply to fractional exponents just as they do to integer exponents.

Evaluate each expression without using a calculator. In Parts e–h, use the laws of exponents to help you.

a. $1.44^{\frac{1}{2}}$ **b.** $125^{\frac{1}{3}}$ **c.** $(^{-}32)^{\frac{1}{5}}$ **d.** $^{-}32^{\frac{1}{5}}$

e. $(3^{\frac{1}{3}})^3$ **f.** $\left(\frac{9}{25}\right)^{\frac{1}{2}}$ **g.** $(64)^{-\frac{1}{3}}$ **h.** $16^{\frac{3}{4}}$

Remember

Two figures are similar if they are the same shape.

58. Geometry If two triangles are similar, you can multiply the side lengths of one triangle by some number to get the side lengths of the other triangle. For example, in similar Triangles A and B below, you can get the side lengths of Triangle B by multiplying the side lengths of Triangle A by 2. This number is called the *scale factor.*

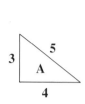

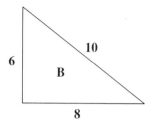

a. Imagine all the triangles that are similar to Triangle C.

If you scale Triangle C by a factor of n, what will be the side lengths of the new triangle?

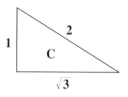

b. Is it possible to find a triangle similar to Triangle C whose sides are all irrational numbers? If so, give an example by stating the scale factor and the new side lengths. If not, explain why not.

c. Is the area of the Triangle C rational or irrational? Explain.

d. **Challenge** Is it possible to find a triangle similar to Triangle C whose area is rational? If so, give an example by stating the scale factor and the new side lengths. If not, explain why not.

59. Challenge In Parts a and b, tell whether the number is rational or irrational. If it is rational, find two integers whose ratio is equal to it. If it is irrational, explain how you know.

a. $6\sqrt{3}$

b. 2π (Hint: If it is rational, it is equal to $\frac{a}{b}$ for some integers a and b. What might π be equal to in terms of a and b?)

c. In general, if you multiply a nonzero rational number n by an irrational number m, will your result be rational or irrational? Explain how you know.

60. Identify all the pairs of equivalent fractions in this list.

$$\frac{1}{3} \quad \frac{7}{12} \quad \frac{2}{9} \quad \frac{8}{10} \quad \frac{28}{48} \quad \frac{20}{25} \quad \frac{6}{27} \quad \frac{12}{36}$$

Write each fraction in lowest terms.

61. $\dfrac{48}{80}$ **62.** $\dfrac{140}{196}$ **63.** $\dfrac{198}{231}$ **64.** $\dfrac{140}{315}$

65. Without graphing, decide which of these equations represent parallel lines. (Assume that q is on the horizontal axis.) Explain.

a. $2p = 3q + 5$ **b.** $p = 3q^2 + 5$

c. $p = 1.5q - 7.1$ **d.** $p = 3q + 3$

Determine whether the three points in each set are collinear.

66. $(1, {}^-1.6), (5, 0), (6, 0.4)$ **67.** $(7, {}^-4), (3, {}^-1), ({}^-2, 2.75)$

Write a quadratic equation for each table.

68.

x	y
$^-3$	$^-9$
$^-2$	$^-4$
$^-1$	$^-1$
0	0
1	$^-1$
2	$^-4$
3	$^-9$

69.

x	y
$^-3$	59
$^-2$	54
$^-1$	51
0	50
1	51
2	54
3	59

70.

x	y
$^-3$	18
$^-2$	8
$^-1$	2
0	0
1	2
2	8
3	18

71. Jonas and Julia conducted an experiment in which they dropped a rubber ball from various heights and measured how high the ball bounced. Here are the data they collected.

Drop Height (in.)	6	7	10	14	16	18	20	24	26	30
Bounce Height (in.)	2	2.5	4	5	6	6.5	7	8.5	10	12

a. Make a graph of the data, with drop height on the horizontal axis.

b. Draw a line that goes through most of the data points, and use your graph to write an equation of the line.

c. Now find the mean of the drop heights and the mean of the bounce heights from the data in the table.

d. Plot the point that has these two means as its coordinates, and adjust the line you drew in Part d to go through this point. Write an equation for your new line.

Chapter Summary

This chapter began with a review of integer exponents and the laws of exponents. You also studied exponential relationships, both growth and decay.

You then worked with square roots of numbers. You generalized the concept of square roots for cube roots, fourth roots, and even higher roots. You also learned the difference between rational and irrational numbers.

Strategies and Applications

The questions in this section will help you review and apply the important ideas and strategies developed in this chapter.

Understanding integer exponents

1. Suppose y is a positive integer.

  a. Explain what x^y means.

  b. Explain how x^y and x^{-y} are related.

2. Suppose r is a number not equal to 0 or 1. Which is greater, r^{11} or r^{21}, in each of the following cases? Explain your answers.

  a. r is greater than 1.

  b. r is between 0 and 1.

  c. r is between $^-1$ and 0.

  d. r is less than $^-1$.

Understanding scientific notation

3. Which is greater, 2.3×10^{32} or 3.2×10^{23}? Explain.

4. An atom of an element is composed of protons, neutrons, and electrons. The resting mass of a proton is about 1.7×10^{-24} gram. The resting mass of an electron is about 9.1×10^{-28} gram. Answer these questions without using your calculator.

  a. A proton has how many more grams of mass than an electron?

  b. A proton has how many times the mass of an electron?

Recognizing and describing exponential relationships

5. Determine which of these equations describe an exponential relationship. For each exponential relationship, indicate whether it is growth or decay.

$$y = 3(x^4) \qquad y = 3(4^x) \qquad y = 3x + 4$$

$$y = \frac{1}{4x} \qquad y = \left(\frac{1}{4}\right)^x \qquad y = x^4$$

For Questions 6–8, do Parts a–c.

a. Identify whether the relationship described is exponential.

b. Explain your reasoning.

c. If the relationship is exponential, write an equation describing it.

6. A checkers tournament with 2,048 contestants has been set up in several rounds. At each round, every contestant plays an opponent. Only the winners continue to the next round. Consider the relationship between the round number (Round 1, Round 2, and so on) and the number of contestants playing in that round.

7.

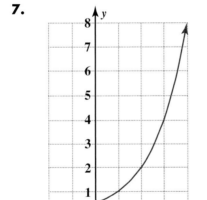

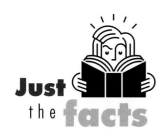
8. *Modems* are devices that allow computers to "talk" to other computers over a phone line. Over the years, technical improvements have allowed modems to communicate at faster and faster rates.

Each time modem speeds increased, Elisa bought a new one. Her first modem had a speed of only 1,200 kilobytes per second (kps). The table shows the speed of each of her modems. Consider the relationship between the modem number and the speed of the modem.

Modem Number	Speed (kps)
1	1,200
2	2,400
3	4,800
4	9,600
5	14,400
6	28,800
7	33,600
8	57,600

Understanding the laws of exponents

9. Prove It! By completing Parts a and b, prove that the power to a power law works when one exponent is a positive integer and the other is a negative integer. Assume a is a positive number and b and c are positive integers.

a. Show that $(a^b)^{-c} = a^{-bc}$.

b. Now show that $(a^{-b})^c = a^{-bc}$.

Understanding roots

10. Explain why there are two values of x for which $x^2 = 16$, but only one value of x for which $x = \sqrt{16}$. What are the values of x in each case?

11. Why is $^-3$ the fifth root of $^-243$?

Identifying rational and irrational numbers

12. Write each number as a ratio of integers, if possible, and identify the number as rational or irrational.

a. $0.\overline{4}$

b. 0.4

c. $\sqrt{0.4}$

d. $\sqrt{0.4} \cdot \sqrt{10}$

Demonstrating Skills

Evaluate or simplify each number without using your calculator.

13. 0.4^3

14. $\left(\frac{2}{3}\right)^4$

15. 8^{-3}

16. $\left(\frac{2}{3}\right)^{-4}$

17. $\sqrt{121}$

18. $\sqrt{32}$

19. $\sqrt[3]{-1{,}000}$

20. $\sqrt[3]{0.027}$

21. $\sqrt[5]{32}$

22. $2\sqrt[7]{(-3)^{14}}$

23. $\frac{1}{3}\sqrt{27}$

24. $\sqrt[x]{13^x}$

Write each expression using a single exponent and a single base. Assume x is not zero.

25. $a^3 \cdot b^3$

26. $(2x)^4 \cdot (2x)^{-7}$

Simplify each expression as much as possible. When necessary, assume the variable is not negative.

27. $\sqrt{52}$

28. $^{-}\sqrt{70}$

29. $\sqrt{18x^3}$

30. $3m\sqrt{4m^6}$

CHAPTER 4

Solving Equations

Real-Life Math

Get with the Program Mathematical programming is a technique used in a wide variety of fields. While much of the mathematics you have studied has been around for centuries, mathematical programming didn't exist until the late 1940's, just after World War II.

Mathematical-programming problems are also called optimization problems because they involve looking for an optimal solution—one that gives a maximum or minimum value for a variable such as profit, cost or time. These problems usually involve several equations and inequalities, with many variables. Petroleum companies use mathematical programming to find the best ways to blend gasoline. This technique is also used to determine the minimum amount of food needed to ensure that astronauts get the proper nutrition on space shuttle missions, because of limited space.

Think About It List some of the variables that a manufacturing company might need to consider in order to maximize its profits.

Family Letter

Dear Student and Family Members,

Our next chapter in mathematics is about solving equations. This is one of the most important and most frequently used mathematical skills, one with many applications in the sciences, social studies, and everyday life. Solving equations can help us figure out how many gallons of gas you will need to drive 500 miles. Scientists launching a satellite, economists making predictions, and citizens working out income taxes all have to solve complex equations.

We have learned several methods for solving equations in earlier grades. In this chapter, we will review and extend these methods to solve inequalities and systems of two equations with two variables. We will also learn how to use graphing calculators to find approximate solutions to equations and pairs of equations.

Situations involving inequalities are common in daily life. For example, if you go to the store with \$5.00 and want to buy markers at \$1.95 each and ink pads at \$0.59 each, the inequality

$$1.95m + 0.59p \leq 5.00$$

tells which combinations of markers and pads you will be able to buy. In this example, 1 marker and 5 pads, or 2 markers and 2 pads, will satisfy the inequality.

We will also learn to solve systems of equations with two variables. In such problems, we have to find a pair of values that satisfies *both* equations. For example:

> In a game of basketball, Corrine scored 10 times for a total of 23 points. Some of the shots were 2-point shots, and some were 3-point shots. How many of each type did Corinne score?

To answer this question, we set up the system of equations below, where x represents the number of 2-point baskets and y stands for the number of 3-pointers:

$$x + y = 10 \quad \text{and} \quad 2x + 3y = 23$$

Vocabulary Along the way, we'll be learning about these new vocabulary terms:

elimination	**substitution**
inequality	**system of equations**

What can you do at home?

Ask your student to show you some of the equations he or she is solving, and methods for solving them. Encourage your student to think about ways these skills can be used outside of school.

Revisiting Equations

Solving mathematical problems often involves writing and solving equations. You have probably already encountered several strategies for solving equations.

Think & Discuss

Describe some strategies you might use to solve this equation.

$$6\left(\frac{2(2x-6)}{4}\right) = 12$$

Here are three equation-solving methods you have probably used.

- *Guess-check-and-improve:* Guess the solution, check your guess by substituting it into the equation, and use the result to improve your guess if you need to.

- *Backtracking:* Start with the output value and work backward to find the input value.

- *Doing the same thing to both sides:* Apply the same mathematical operation to both sides of the equation until the solution is easy to see.

In this lesson, you will review two of these methods: backtracking and doing the same thing to both sides.

Investigation 1 Reviewing Equation-Solving Methods

Suppose you want to solve an equation that consists of an algebraic expression on one side and a number on the other, such as this equation:

$$2\left(\frac{2n}{6} - 1\right) = 16$$

Backtracking may be a good solution method for this type of equation.

Solve $2\left(\frac{2n}{6} - 1\right) = 16$ by backtracking.

Think of *n* as the input and 16 as the output. Make a flowchart to show the operations needed to get from the input to the output.

Input   **Output**

This flowchart shows that you multiply the input by 2, divide the result by 6, subtract 1, and then multiply by 2. The output value is 16. To backtrack, start from the output and work backward, *undoing* each operation, until you find the input.

The value in the fourth oval is multiplied by 2 to get 16, so this value must be 8.

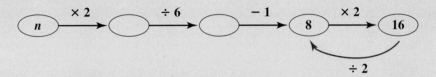

The number 1 is subtracted from the value in the third oval to get 8, so this value must be 9.

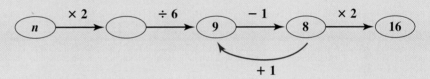

The value in the second oval is divided by 6 to get 9, so this value must be 54.

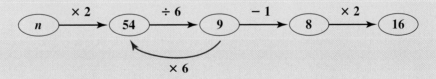

The input is multiplied by 2 to get 54, so the input must be 27.
So, the solution of $2\left(\frac{2n}{6} - 1\right) = 16$ is 27.

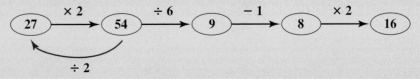

Problem Set **A**

Solve each equation by backtracking. Check each solution by substituting it into the original equation.

1. $3\left(\frac{n}{2} - 1\right) = 15$

2. $\frac{2(n + 1) - 3}{4} = 5$

3. $2\left(\frac{n}{4} - 3\right) + 1 = 5$

For fairly simple equations, you may be able to backtrack in your head. Here's how you might think about solving $\frac{n}{3} + 1 = 4$:

This equation means "divide by 3 and then add 1," which gives 4. To backtrack, "subtract 1" from 4 to get 3, and then "multiply by 3" to get 9.

Solve each equation mentally if you can. If you have trouble, use pencil and paper. Check each solution by substituting it into the equation.

4. $6(3m - 4) = 12$

5. $\frac{3m}{4} + 4 = 12$

6. $\frac{3(n + 4)}{6} = 12$

Some equations cannot be solved directly by backtracking.

Think & Discuss

Héctor tried to solve $3a + 4 = 2a + 7$ by backtracking, but he couldn't figure out how to make a flowchart. Why do you think he had trouble?

How could you solve $3a + 4 = 2a + 7$?

Equations that cannot easily be solved by backtracking can often be solved by doing the same thing to both sides.

EXAMPLE

Solve $3a + 4 = 2a + 7$ by doing the same thing to both sides.

$$3a + 4 = 2a + 7$$
$$a + 4 = 7 \qquad \text{after subtracting } 2a \text{ from both sides}$$
$$a = 3 \qquad \text{after subtracting 4 from both sides}$$

Problem Set **B**

Solve each equation by doing the same thing to both sides. Check your solutions.

1. $3a - 4 = 2a + 3$

2. $11b - 6 = 9 + 6b$

3. $7 - 5x = 12 - 3x$

4. $3y + 7 = 7 - 2y$

Sometimes you need to simplify an equation before you can solve it.

Problem **C**

1. Since one side of this equation is a number, Tala thought she could solve it by backtracking.

$$4(3x + 2 - 2x + 2) = 20$$

a. Tala drew a flowchart for this equation but couldn't figure out how to backtrack to solve it. Why do you think she had trouble?

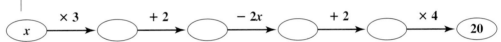

b. Tala thought the equation might be easier to solve if she simplified it so x appeared only once. Simplify the equation, and then solve it by backtracking.

Simplify each equation and then solve it using any method you choose. Check your solutions.

2. $5(2a - 3) + 2a + 3 = 0$

3. $^-6(8 - 3n) + 2n = ^-8$

4. $2k + 2\left(\frac{k}{4} - 3\right) - k + 1 = 25$

Just t h e **facts**

Flowcharts are often used by builders to organize the many steps involved in a construction project.

Problem Set D

Now try applying equation-solving techniques to solve word problems.

1. Mikayla said, "I'm thinking of a number. When I subtract 3 from my number, multiply the result by 8, and then divide this result by 3, I get 32. What is my number?"

2. Neva and Owen collect action figures. Owen has five times as many action figures as Neva. Owen also has 60 more than Neva. Write and solve an equation to find the number of action figures Neva and Owen each have.

3. The sum of four consecutive whole numbers is 150.

    a. Let x represent the first number. Write expressions to represent the other three numbers.

    b. Write and solve an equation to find the four numbers.

4. The sum of three consecutive *even* numbers is 78.

    a. Let n represent the first even number. Write expressions for the next two even numbers.

    b. Write and solve an equation to find the three numbers.

5. A video distributor has 800 copies of a new video to distribute among four stores. She plans to give the first two stores the same number of copies. The third store will receive three times as many videos as either of the first two. The fourth store will receive 20 more than either of the first two.

    a. How many copies will each store receive?

    b. The distributor has 700 copies of another video that she wants to distribute using the same rules. How many copies do you think she should give to each store? Explain your answer.

Share & Summarize

1. Could you solve all the backtracking problems in Problem Set A by doing the same thing to both sides?

2. Write an equation, similar to those in Problem Set A, that can be solved by backtracking. Solve your equation first by backtracking and then by doing the same thing to both sides, recording each step in your solutions. What connection do you notice between the two methods?

Lab Investigation ▶ Clock Systems

The arithmetic and algebra you are familiar with involve the *real number system*. This system involves all the numbers on the number line and mathematical operations such as addition, subtraction, multiplication, and division.

In this investigation, you will explore number systems called *clock systems*. Although clock systems also involve subtraction, multiplication, and division, you will work only with addition.

Telling Time

Our method of telling time uses hours from 1 through 12. A number system based on the way we tell time can be called the *clock-12 system*.

Suppose it is 4 o'clock now.

1. Consider times in the next 8 hours.

   a. What time will it be in 2 hours? In 4 hours? In 7 hours?

   b. How did you find your answers to Part a?

2. Now consider times *beyond* 8 hours from now.

   a. What time will it be in 10 hours? In 12 hours? In 20 hours?

   b. How did you find your answers to Part a?

   c. What time will it be in 21 hours? In 35 hours?

3. Lydia said she could find the time h hours after 4 o'clock by computing $4 + h$, and then subtracting 12 until the answer is between 1 and 12.

   a. Explain why her method works.

   b. Can you find a quicker way to find the time h hours after 4 o'clock? If so, describe it. Then illustrate your method by using it to find the time 50 hours after 4 o'clock.

4. Your answers to Part a of Questions 1 and 2 should have been in the range from 1 to 12.

   a. Explain why.

   b. The clock-12 system involves only integers. How many numbers does this system have?

Just the facts

The first clocks, used in ancient Egypt, were shadow clocks—an early version of the sundial.

Adding in the Clock-12 System

You can write addition equations in the clock-12 system, just as you can in the real number system. For example, in Part a of Question 2, you found that, in the clock-12 system, $4 + 10 = 2$.

5. Copy and complete the table to show *all* the possible sums in the clock-12 system.

+	12	1	2	3	4	5	6	7	8	9	10	11
12												
1												
2												
3												
4												
5												
6												
7												
8												
9												
10												
11												

In the clock-12 system, 12 behaves like 0 in the real number system: when you add 12 to a number, the result is equal to the original number. Because of this property of 12, it is convenient to rename it as 0. Cross out the 12s in your table and replace them with 0s.

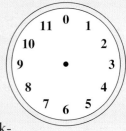

6. Find each sum in the clock-12 system. (The clock-12 system now consists of whole numbers from 0 to 11.)

   a. $4 + 8$ **b.** $0 + 5$ **c.** $6 + 11$

7. Find the diagonal of 0s in your table. Which pairs of numbers add to 0?

8. Because they add to 0, the pairs of numbers you identified in Question 7 are *additive inverses*.

   a. What is the additive inverse of 5 in the clock-12 system?

   b. What is the additive inverse of 6 in the clock-12 system?

Solve each equation in the clock-12 system.

   9. $x + 4 = 3$ **10.** $y + 5 = 5$

   11. $1 + x = 0$ **12.** $4 + p = p$

Adding in the Clock-6 System

This clock face is for the clock-6 system.

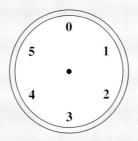

The clock-6 system works in a similar way to the clock-12 system. For example:

$$1 + 3 = 4 \qquad 2 + 4 = 0 \qquad 3 + 5 = 2$$

13. Create an addition table for the clock-6 system.

+	0	1	2	3	4	5
0						
1						
2						
3						
4						
5						

14. Which pairs of numbers add to 0 in the clock-6 system?

Find each sum in the clock-6 system.

15. $4 + 5$

16. $2 + 3$

17. $5 + 1$

18. $(3 + 4) + (0 + 1)$

19. What is the additive inverse of 4 in the clock-6 system?

Solve each equation in the clock-6 system. Remember, you can use only the numbers from 0 to 5 for your solutions.

20. $x + 5 = 0$

21. $3 + y = 3$

Using the idea of an additive inverse, Marcus thought of a way to solve equations in the clock-6 system.

Use Marcus's method to solve each equation. Show your solutions.

22. $1 + j = 0$ **23.** $k + 2 = 1$ **24.** $m + 3 = 1$

What Have You Learned?

25. Addition in another clock system would work in a manner similar to the clock-12 and clock-6 systems.

    a. Choose another clock system to investigate. Draw the clock face for your system.

    b. Make a table showing all the possible sums for your system.

    c. What is the additive inverse of 2 in your clock system?

26. Create and solve three addition equations (for example, $x + 3 = 1$) for your clock system.

27. Write an explanation to another student of how to solve addition equations in your system.

28. Explain how addition in a clock system is different from addition in the real number system.

On Your Own Exercises

Practice & Apply

Solve each equation by backtracking. (Backtrack mentally if you can.) Check your solutions.

1. $5(n - 2) = 45$ **2.** $2(n - 5) = 7$ **3.** $\frac{n}{2} + 5 = 7$

4. $3(4m - 6) = 12$ **5.** $\frac{3m - 3}{6} = 12$ **6.** $3\left(\frac{m}{6} - 4\right) = 12$

Solve each equation by doing the same thing to both sides.

7. $2x + 3 = x + 5$ **8.** $7y - 4 = 4y - 13$

Simplify and solve each equation.

9. $4 - 2(^-5a - 10) = 30$ **10.** $\frac{b - 2}{4} = \frac{6}{5}$

In Exercises 11–14, write and solve an equation to find the number of coins each friend has.

11. Ken has three more coins than twice the number Javier has. Khalid has five fewer coins than Javier. They have 50 coins altogether.

12. Da-Chun has five times as many coins as Austin. Da-Chun also has 16 more than Austin.

13. Kai has two less than double the number of coins Ty has. Kai also has 23 more than Ty.

14. Emilio has three fewer than double the number Jacob has. Latisha has 20 more than three times the number Jacob has. They have 65 altogether.

15. Lindsey said, "I'm thinking of a number. If I multiply it by 5, subtract 4, and then multiply the result by 2, I get 62. What is my number?"

16. Hannah delivers boxes of pizza dough to five pizzerias every Thursday. She normally delivers the same number of boxes to each store. The weekend before the big game, she receives calls from the five restaurants requesting various quantities of dough.

- Sam's Pizzeria wants twice the usual number of boxes.

- Pizza House wants three extra boxes.

- Pizza Pit wants six extra boxes.

- Paul's Pizza Parlor will be closing for the game and wants to cut its regular order in half.

- Pizza Heaven wants the usual amount.

Hannah totals the orders and says, "I don't know if I have the 64 boxes of dough I need." How many boxes of dough does each store usually receive on Thursday?

17. Write two equations that can't be solved directly by backtracking. Explain why backtracking won't work.

18. Evan made this toothpick pattern. He described the pattern with the equation $t = 5n - 3$, where t is the number of toothpicks in Stage n.

Stage 1   Stage 2   Stage 3

a. Explain how each part of the equation is related to the toothpick pattern.

b. How many toothpicks would Evan need for Stage 10? For Stage 100?

c. Evan used 122 toothpicks to make one stage of his pattern. Write and solve an equation to find the stage number.

d. Is any stage of the pattern composed of 137 toothpicks? Why or why not?

e. Is any stage of the pattern composed of 163 toothpicks? Why or why not?

f. Evan has 250 toothpicks and wants to make the largest stage of the pattern he can. What is the largest stage he can make? Explain your answer.

19. Recall that the absolute value of a number is its distance from 0 on the number line. You can solve equations involving absolute values. For example, the solutions of the equation $|x| = 8$ are the two numbers that are a distance of 8 from 0 on the number line, 8 and $^-8$.

Solve each equation.

a. $|a| = 2.5$

b. $|2b + 3| = 8$

c. $|9 - 3c| = 6$

d. $\frac{|5d|}{25} = 1$

e. $|^-3e| = 15$

f. $20 + |2.5f| = 80$

20. Tamika and Lydia were making hair ribbons to sell at a crafts fair. Tamika cut seven segments from one length of ribbon and had 2 feet left over. Lydia said, "I'm cutting segments twice as long as yours. If your length of ribbon had been just 1 foot longer, I could have cut four segments from it."

From their conversation, determine how long Tamika's and Lydia's segments were.

21. Nine square tiles are used to cover a floor that covers an area of 36 square feet.

    a. Write and solve an equation to find the dimensions of the tiles.

    b. Draw and label a picture to show how these nine tiles could be used to cover a floor that measures 3 feet by 12 feet. Assume that you are able to cut tiles in half if you need to.

Mixed Review

22. Identify the three pairs of equivalent equations.

    a. $p = 2q - 4$ **b.** $p - 2q = 4$

    c. $p - 2q - 4 = 0$ **d.** $^-2p = 8 + 4q$

    e. $^-p - 4 = 2q$ **f.** $0.5p = q - 2$

23. List four numbers that are greater than $^-2$ and less than 1.

Each table represents a linear relationship. Tell whether each relationship is increasing or decreasing.

24.

x	y
1	2
$^-2$	3.5
2	1.5
$^-3$	4
0	2.5
$^-1$	3

25.

x	y
1	3.2
$^-4$	2.2
0	3
$^-1$	2.8
$^-3$	2.4
$^-2$	2.6

26.

x	y
$^-4$	0.5
0	$^-0.5$
$^-1$	$^-0.25$
$^-3$	0.25
$^-2$	0
1	$^-0.75$

27. Write 64, 256, and 1,024 using integer exponents and the same base.

28. Sort these expressions into two groups so that the expressions in each group are equal to one another.

$$m^3 \qquad \left(\tfrac{1}{m}\right)^3 \qquad m^{-3} \qquad \left(\tfrac{1}{m}\right)^{-3} \qquad \frac{1}{m^3} \qquad m \div m^4$$

4.2 Inequalities

The equals sign, =, expresses a comparison between two quantities: it states that two quantities are *equal*. The table lists other mathematical symbols you have used to express comparisons.

Symbol	What It Means	Examples
$<$	is less than	$5 < 7 \qquad 4 + 8 < 15 \qquad {}^-17 < {}^-4 \cdot 3$
$>$	is greater than	$7 > 5 \qquad 80 > 20 + 9 \qquad 3^4 > 2^4$
$\leq$	is less than or equal to	$5 \leq 7 \qquad 6 \cdot {}^-5 \leq {}^-30 \qquad \frac{1}{3} \leq \frac{1}{2}$
$\geq$	is greater than or equal to	$7 \geq 5 \qquad 10^8 \geq 10^7 \qquad \frac{1}{3} \geq \frac{3}{9}$

VOCABULARY
inequality

Mathematical statements that use these symbols to compare quantities are called **inequalities.** The inequalities above compare only numbers. You have also used inequalities to compare expressions involving variables.

- The statement $x > 7$ means that the value of the variable x is greater than 7. Some possible values for x are 7.5, 12, 47, and 1,000.

- The statement $n - 3 \leq 12$ means that the value of "n minus 3" is less than or equal to 12. Possible values for n are 15, 1.2, 0, and $^-$5.

Think & Discuss

How is the meaning of $7 < x$ similar to or different from that of $x > 7$?

It is sometimes convenient to combine two inequalities like this:

$$7 < x < 10$$

What do you think $7 < x < 10$ means?

How could you write $7 < x < 10$ as two inequalities joined by the word *and*?

List three values of x that satisfy the inequality $7 < x < 10$.

Thermostats are used to control the temperature in a room or a building. How do you think a thermostat might use inequalities to do this?

Investigation 1 ▶ Describing Inequalities

The problems in this investigation will give you more practice working with inequalities.

Problem Set A

1. List six whole numbers that satisfy the inequality $n + 1 < 11$.

2. Consider this statement: k is less than or equal to 14 and greater than 9.

   a. Write this statement in symbols.

   b. List all the whole numbers for which the inequality in Part a is true.

3. Consider the inequality $15 \leq m \leq 18$.

   a. Write the inequality in words.

   b. List all the whole numbers for which the inequality is true.

   c. Are there other numbers—not necessarily whole numbers—for which the inequality is true? If so, give some examples. How many are there?

4. If you consider only whole-number values for n, four out of the five statements below represent the same values. Which inequality below is *not* equivalent to the others?

$$10 < n < 20 \qquad 11 \leq n \leq 19 \qquad 11 \leq n < 20$$

$$11 < n < 19 \qquad 10 < n \leq 19$$

List six values—not necessarily whole numbers—that satisfy each inequality or pair of inequalities.

5. $10.2 < p < 14.7$

6. $q \geq 12$ and $q > 15$
   (The values of q must make *both* inequalities true.)

7. $20 \geq |r| \geq 17.75$

8. $3s > 12$

9. $t > 0$ and $t^2 \leq 16$

Remember

The absolute value of a number is its distance from 0 on the number line:

$|3| = 3 \qquad |{-}3| = 3$

Now you will write inequalities to represent situations that are described in words.

Problem Set B

Write an inequality or a pair of inequalities to describe each situation. Make your answers as specific and complete as possible.

1. A box of matches contains at least 48 but fewer than 55 matches. Write an inequality for the number of matches n in the box.

2. Tamika expects from 100 to 120 people to buy tickets to the talent show. Each ticket costs $5. Write an inequality to represent the total amount of money m she expects from ticket sales.

3. Four friends—Sandy, Mateo, Felisa, and Destiny—are comparing their heights. Sandy, who is 155 cm tall, is the shortest. Mateo is 165 cm tall. Felisa is not as tall as Mateo. Destiny is at least as tall as Mateo. Write inequalities to represent Felisa's and Destiny's heights, F and D.

4. The Completely Floored store sells square tiles in a variety of sizes, with side lengths ranging from 2 cm to 20 cm. Write an inequality to represent the range of possible areas a, in square centimeters, for the tiles.

5. Rondell's walk to school takes 15 minutes, give or take 2 minutes. Write an inequality for the time t it takes Rondell to walk to school.

6. The bed of José's trailer is 120 cm above the ground. Each layer of cartons he loads onto the trailer adds an additional 40 cm to the height. His loaded trailer must be able to pass under a footbridge whose underside is 4 meters high.

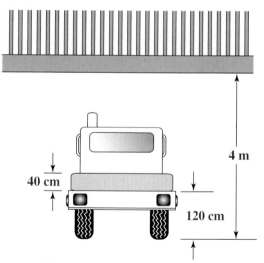

  a. Write an expression for the height of the trailer when it holds n layers of cartons.

  b. Use your expression to write an inequality relating the height of the loaded trailer to the height of the footbridge.

7. The families of Chelsea and her four friends are throwing a party for the girls' sixteenth birthdays. They estimate they will spend $7 per person for food and $3 per person for beverages, plus $200 to hire a disk jockey. The total amount they have budgeted for the party is $500.

  a. Let p represent the number of people they can invite. Write an expression for the costs of the party for p people.

  b. Use your expression to write an inequality relating the costs, the number of people, and the families' budget.

8. An architect estimated that the maximum floor area for a square elevator in a particular building is 4 m². Building regulations require that the minimum area be 2.25 m². Write an inequality for the possible side lengths s for the elevator's floor.

Share & Summarize

Make up a situation that can be represented by an inequality. Express your inequality in symbols.

Investigation 2 Solving Inequalities

Most of the equations you have solved have had one or two solutions. Inequalities, however, can have many solutions. In fact, they often have an infinite number!

In Problem 8 of Problem Set A, you found some of the solutions of the inequality $3s > 12$. For example, the values 4.5, 7, and 10 all satisfy $3s > 12$. In fact, any value greater than 4 will satisfy this inequality, but you certainly can't list them all.

For this reason, the solutions of an inequality are usually given as another inequality. For example, you can express all the solutions of $3s > 12$ by writing $s > 4$.

Problem Set C

1. Solve each equation or inequality. That is, find the value or values of *a* that make the equation or inequality true.

   a. $3a - 10 = 35$ **b.** $3a - 10 > 35$ **c.** $3a - 10 < 35$

2. This is how Tamika thought about Problem 1.

a. Test at least four other values for *a*. Does Tamika's conclusion seem correct?

b. Use this graph of $y = 3a - 10$ to explain why Tamika's conclusion is correct.

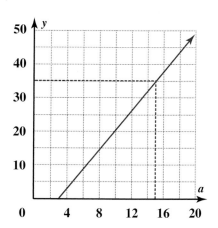

Use Tamika's method to solve each inequality.

3. $5a + 7 < 42$

4. $\frac{b}{7} + 1 \geq 6$

5. $4c - 3 > 93$

6. $^-6(d + 1) \geq 24$

Think & Discuss

To solve a linear inequality of the form $mx + b < c$ or $mx + b > c$ for *x*, where *a*, *b*, and *c* are constants, you can solve the related equation $mx + b = c$ and then test just *one* value greater than or less than the solution. Explain why this method works.

Ben solved the inequality $3a - 10 < 35$ by doing the same thing to both sides.

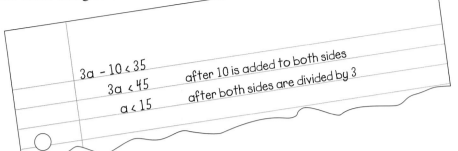

$$3a - 10 < 35$$
$$3a < 45 \quad \text{after 10 is added to both sides}$$
$$a < 15 \quad \text{after both sides are divided by 3}$$

Problem Set D

Use Ben's method (doing the same thing to both sides) to solve each inequality. Then solve the inequality by using Tamika's method (solving the related equation and testing one value greater than or less than the solution). If you get different solutions, tell which solution is correct.

1. $7x + 2 \geq 100$ **2.** $^-7x + 2 \leq 100$

3. Consider these inequalities.

$$5 < 6 \qquad ^-4 > ^-10 \qquad 16 \geq 2 \qquad ^-7 < 4$$

a. Multiply both sides of each inequality by a negative number. Is the resulting inequality true? If not, how can you change the inequality symbol to make it true?

b. Now *divide* both sides of each inequality above by a negative number. Is the resulting inequality true? If not, how can you change the inequality symbol to make it true?

c. Use what you discovered in Parts a and b to explain why Ben's and Tamika's methods sometimes give different results.

d. How can you alter Ben's method to always give the right solution?

Solve each inequality.

4. $2b > 15$ **5.** $^-2c > 15$ **6.** $12 + 6d \leq 54$

7. $12 - 6e \leq 54$ **8.** $^-3(f - 12) < ^-93$ **9.** $\frac{5g}{^-7} + 1 \geq ^-\frac{23}{7}$

10. Solve the inequality $0 < 1 - u < 1$. Explain your solution method.

Share & Summarize

Suppose one of your classmates has been absent during this investigation. Explain to him or her, in writing, how to solve a linear inequality, and describe some common mistakes to avoid.

Investigation 3 ▶ Graphing Inequalities

You can use number-line graphs to show values that satisfy inequalities.

> ### EXAMPLE
>
> This number line shows the solutions of the inequality $^-3 \leq n < 1$. The filled-in circle indicates that $^-3$ is included in the solution. The open circle indicates that 1 is *not* included. The heavy line shows that all numbers between $^-3$ and 1 are included.
>
>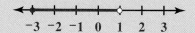
>
> Below is the graph of $x \geq 1$. The arrow indicates that the solution extends to the right, beyond the part of the number line shown.
>
>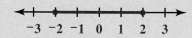
>
> This is a graph of $|z| \leq 2$. It shows that values of z less than or equal to 2 and greater than or equal to $^-2$ are solutions of the inequality.
>
>

Problem Set E

1. Consider the inequality $0 < b < 4.7$.

    a. List all the *integer* values that satisfy the inequality.

    b. Use your answer to Part a to help you graph *all* the values that satisfy $0 < b < 4.7$.

2. In Parts a–d, you will consider the inequalities $x < 10$ and $x > 5$. Draw a separate number-line graph for each part.

    a. Graph all x values for which $x < 10$.

    b. Graph all x values for which $x > 5$.

    c. Graph all x values for which $x < 10$ *and* $x > 5$.

    d. Graph all x values for which $x < 10$ *or* $x > 5$.

    e. Explain how the words *and* and *or* affect the graphs in Parts c and d.

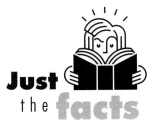

Just the facts

Size regulations for taking certain fish species from public waters can often be stated as an inequality. In Florida in 1999, you could keep a black drum fish you had caught only if it was not shorter than 14 in. and not longer than 24 in.

MATERIALS

graph paper

3. In Parts a–c, you will consider the inequalities $c \geq 10$ and $c \leq 5$. Draw a separate graph for each part.

    a. Graph all c values for which $c \geq 10$ and $c \leq 5$.

    b. Graph all c values for which $c \geq 10$ or $c \leq 5$.

    c. Explain how the words *and* and *or* affect the graphs in Parts a and b.

4. Consider inequalities that involve absolute values.

    a. Graph all m values that satisfy the inequality $|m| \geq 2.5$.

    b. Graph all t values that satisfy the inequality $|t| < 3$.

5. Now consider the inequality $x^2 < 16$.

    a. List the *integer* values that satisfy the inequality.

    b. Use your answer to Part a to help you graph *all* x values that satisfy the inequality $x^2 \leq 16$.

    c. Express the solution of $x^2 \leq 16$ as an inequality.

Solve each inequality, and graph the solution on a number line.

6. $\frac{2p}{5} < 10$
                      7. $^-2(k - 5) \leq 10$

You can show solutions of a two-variable equation such as $y = 2x + 3$ by making a graph of the equation. Any point (x, y) on the graph satisfies the equation. You will now learn how to graph inequalities with two variables.

Problem Set F

1. Graph the equation $y = x$.

2. List the coordinates of three points that are above the line $y = x$ and three points that are below the line.

3. Which inequality, $y > x$ or $y < x$, describes the coordinates of points that are above the line $y = x$?

4. Which inequality, $y > x$ or $y < x$, describes the coordinates of points that are below the line $y = x$?

5. Based on your answers to Problems 3 and 4, predict whether each given point will be above, below, or on the line $y = x$. Test each prediction by plotting the point.

    a. $(5, 11)$
    b. $(7, 3)$
    c. $(0, ^-6)$
    d. $(1, 1)$

6. List three (x, y) pairs that make the inequality $y > 3x$ true.

7. Graph the equation $y = 3x$. Then plot the points you listed in Problem 6. Where do they appear on your graph?

You can use what you learned in Problem Set F to graph inequalities.

EXAMPLE

Graph the inequality $y > 3x$.

First graph the *equation* $y = 3x$. Use a dashed line because the points on the line do not make the inequality true. Then shade the area containing the points that make the *inequality* $y > 3x$ true. You saw in Problem Set F that these are the points above the line.

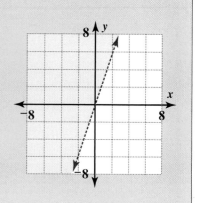

MATERIALS

graph paper

Problem Set G

Graph each inequality.

1. $y < 2x + 3$ **2.** $y > x + 2$ **3.** $y \le {}^-4x$

Challenge Graph each of these nonlinear inequalities.

4. $y \ge x^2$ **5.** $y < x^2 - 1$

Share & Summarize

1. Consider these statements.

$$x \ge a \text{ and } x \le b \qquad x \ge a \text{ or } x \le b$$

a. Describe how the number-line graph of "$x \ge a$ and $x \le b$" is different from the number-line graph of "$x \ge a$ or $x \le b$."

b. Give values of a and b for which the number-line graph of "$x \ge a$ or $x \le b$" includes *all* numbers.

c. Give values of a and b for which the number-line graph of "$x \ge a$ and $x \le b$" includes no numbers.

2. Explain the steps involved in graphing an inequality with two variables. Illustrate with an example different from those given in the text.

Practice & Apply

1. List six whole numbers that satisfy the inequality $n - 2 > 5$.

2. Consider the inequality $^-4 < x \leq 0$.

   a. List all the integers that satisfy the inequality.

   b. List three non-integers that satisfy the inequality.

List five values that satisfy each inequality. Include negative and positive values, if possible.

3. $^-2n \geq 6$ **4.** $p^2 < 4$ **5.** $6 < x < 7$

6. $y > 5$ and $y > 12$ **7.** $m - 3 > 9$ **8.** $1 \leq ^-x \leq 5$

9. $|q| < 5$ **10.** $0 \leq |b| \leq 6$ **11.** $|s| - 5 \geq 6$

12. Isabel said it would take her at least an hour and a half, but no more than 2 hours, to finish her homework. Write an inequality to express the number of hours, h, Isabel thinks it will take to do her homework.

13. Sareeta earns $0.12 for each pamphlet she delivers. She thinks she can deliver from 500 to 750 pamphlets on Sunday. Write an inequality for the number of dollars, d, Sareeta expects to earn on Sunday.

14. Dan's Delivery Service charges $9 to ship any package weighing 5 pounds or less. Mr. Valenza wants to send a box containing tins of cookies to his daughter in college. Each tin weighs 0.75 pound, and the packing materials weigh about 1 pound.

   a. Write an expression, using t to represent the number of tins in the box, to represent the weight of the box.

   b. Use your expression to write an inequality representing the number of tins Mr. Valenza can send for $9.

15. Wednesday nights are special at the video arcade: customers pay $3.50 to enter the arcade and then only $0.25 to play each game. Roberto brought $7.50 to the arcade and still had some money when he left. Write an inequality for this situation, using n to represent the number of games Roberto played.

16. Solve each equation or inequality.

   a. $1 - x = 0$

   b. $1 - x < 0$

   c. $1 - x > 0$

17. Solve each equation or inequality.

    a. $1 - d = 1$ **b.** $1 - d < 1$ **c.** $1 - d > 1$

Solve each inequality.

18. $5(e - 2) > 10$

19. $\frac{f}{2} + 5 > 10$

20. $\frac{g + 2}{5} > 10$

21. $^-3 \leq h - 2 \leq 1$

22. Consider the inequality $^-7 \leq x < ^-1.3$.

    a. List all the integer values that satisfy the inequality.

    b. Graph *all* the values that satisfy the inequality.

23. Consider the inequalities $m < ^-3$ and $m \geq 0$.

    a. Graph all m values for which $m < ^-3$ and $m \geq 0$.

    b. Graph all m values for which $m < ^-3$ or $m \geq 0$.

24. Consider the inequalities $y \leq 3$ and $y \geq ^-4$.

    a. Graph all y values for which $y \leq 3$ and $y \geq ^-4$.

    b. Graph all y values for which $y \leq 3$ or $y \geq ^-4$.

25. Graph all r values for which $|r - 3| \geq 3$.

26. Consider the inequality $x^3 \leq 27$.

    a. Express the solution of $x^3 \leq 27$ as an inequality.

    b. Graph the solution on a number line.

Graph the solution of each inequality on a number line.

27. $^-\frac{3p}{4} < 6$

28. $12 - 5q \geq 32$

Graph each inequality.

29. $y < 3x + 7$

30. $y \leq ^-3x - 7$

31. $y > ^-2x + 4$

32. How would you change this graph of $y = x^2 + 3$ to represent $y > x^2 + 3$?

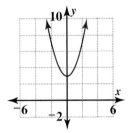

33. Graph the inequality $y \geq {}^-x^2$.

34. Sports In U.S. amateur boxing, fighters compete in classes based on their weight. Five of the 12 official weight classes are listed below. Copy and complete the table to express the weight range for each class as an inequality.

Weight Class	Weight Range (pounds)	Inequality
Super heavyweight	over 201	
Heavyweight	179–201	
Welterweight	140–147	$140 \leq w \leq 147$
Featherweight	120–125	
Light flyweight	under 107	

35. Megan is writing a computer game in which a player stands on the balcony of a haunted house and drops water balloons on ghosts below. The player chooses where the balloon will land and then launches it.

Since water splatters, Megan's game gives the player points if a ghost is anywhere within a square centered where the balloon lands. The square extends 15 units beyond the center in all four directions. That is, if both of the ghost's coordinates are 15 units or less from the center, the player has scored a hit.

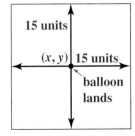

Suppose the balloon lands at (372, 425). The nearest ghost has the coordinates (x, y), and it counts as a hit by the game. Use inequalities to describe the possible values for x and y. (Hint: You will need two inequalities, one for x and one for y. Should you say "and" or "or" between them?)

When you draw a graph, you have to decide the range of values to show on each axis. Each exercise below gives an equation and a range of values for the x-axis. Use an inequality to describe the range of values you would show on the y-axis, and explain how you decided. (It may help to try drawing the graphs.)

36. $y = 2x + 7$ when $0 \leq x \leq 10$

37. $y = 2x - 10$ when $^{-}5 \leq x \leq 5$

38. $y = x^2 + 1$ when $^{-}5 \leq x \leq 5$

39. $y = ^{-}2x$ when $^{-}5 < x < 0$

40. Physical Science The pitch of a sound depends on the frequency of the sound waves. High-pitched sounds have higher frequencies than low-pitched sounds. Most animals can hear a much greater range of frequencies than they can produce. The table shows the range of frequencies various animals can produce and hear.

Animal	Frequencies of Sounds Produced (hertz)	Frequencies of Sounds Heard (hertz)
Human being	$85 \leq f \leq 1,100$	$20 \leq f \leq 20,000$
Bat	$10,000 \leq f \leq 120,000$	$1,000 \leq f \leq 120,000$
Dog	$452 \leq f \leq 1,080$	$15 \leq f \leq 50,000$
Grasshopper	$7,000 \leq f \leq 100,000$	$100 \leq f \leq 15,000$

Source: *World Book Encyclopedia*, Vol. 18. Chicago: World Book Inc., 1997.

a. Make a number-line graph showing the range of frequencies a grasshopper can produce but cannot hear.

b. Make a number-line graph showing the range of frequencies a dog can hear that a human being cannot produce.

c. Make a number-line graph showing the range of sounds a bat can make that a dog can hear.

d. Make a number-line graph showing the range of frequencies both a dog and a grasshopper can produce.

41. Challenge Use a graph to solve the inequality $x^2 < x$. Explain how you found your answer.

42. Make a graph showing the values for which $y < 3x + 2$ and $y > x + 5$.

Write each equation in the form $y = mx + b$.

43. $-2y = 14x + \frac{1}{2}(6x + 12)$

44. $\frac{1}{5}(10x + 5) - 2 + 9x - 3y = y$

45. $\frac{6(x - 7)}{2(3 - y)} = 0.4$

Rewrite each expression as simply as you can.

46. $0.5a^3 \cdot 3a^3$ **47.** $m^7 \cdot m^{-5} \cdot b^5$ **48.** $(2x^{-2})^2$

49. $(-m^2n)^4$ **50.** $(a^m)^n \cdot (b^3)^0$ **51.** $\frac{(x^2y^{-3})^{-2}}{(xy)^4}$

52. Geoff raised $65 for cancer research for every mile he walked in a recent fund-raiser, in addition to the $100 he was donating himself. Write an equation for the relationship between the number of miles Geoff walked, m, and total amount he raised, r.

53. Match each graph to an equation.

    a. $y = x^2 - 3$

    b. $y = x^2 + 3$

    c. $y = x^2$

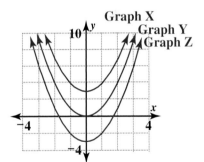

54. Examine how this pattern grows from one stage to the next.

Stage 0   Stage 1   Stage 2

    a. Copy and complete the table.

Stage, n	0	1	2	3
Line Segments, s	2	8		

    b. What kind of relationship is this?

    c. Write an equation to describe the relationship between the number of line segments in a stage, s, and the stage number, n.

Using Graphs and Tables to Solve Equations

Some equations can be solved using several methods, so you can choose the easiest or quickest method, or the one you like the most. You have already solved equations using backtracking, guess-check-and-improve, and doing the same thing to both sides.

In this lesson, you will learn to find approximate solutions of equations by using a graphing calculator. You will also learn how to use graphs to determine how many solutions an equation has.

▶ **MATERIALS**

graphing calculator

Explore

Graph $y = 3x + 3$ on your calculator. Use the standard viewing window ($^-10 \le x \le 10$ and $^-10 \le y \le 10$).

Use the graph to estimate the solution of each of these equations. Explain what you did to find your answers.

$$3x + 3 = 0 \qquad 3x + 3 = ^-6$$

Graph $y = 3x^2 - 3$ in the same viewing window as $y = 3x + 3$.

Use your graph to estimate the solution of each of these equations. Explain how you found your answers.

$$3x^2 - 3 = 0 \qquad 3x^2 - 3 = ^-6$$

Use the two graphs to solve the equation $3x + 3 = 3x^2 - 3$. Explain what you did to find your answer.

Investigation ▶ 1 ▶ Finding Values from a Graph

If objects such as tennis balls are tossed or projected straight upward, those with faster starting speeds will go higher. The pull of gravity will slow each object until it momentarily reaches a maximum height, and then the object will begin to fall. The formula

$$h = vt - 16t^2$$

approximates an object's height above its starting position at a chosen time. The variable h represents the height in feet, t represents the time in seconds, and v represents the starting speed, or initial *velocity,* in feet per second. This formula assumes the starting height is at ground level, and it ignores such complications as air resistance.

Just the **facts**

Galileo Galilei (1564–1642) first uncovered the mathematical laws governing falling bodies and the height of thrown objects.

Suppose someone throws a tennis ball toward the ground. The ball bounces straight upward, with a speed of 30 feet per second (about 20.5 miles per hour) as it leaves the ground. In this case, $v = 30$, so the height as the ball bounces is given by

$$h = 30t - 16t^2$$

where t is the time in seconds after the ball leaves the ground.

You can use the formula to find the height h of the ball for various times t. For example, after one-half second, or at time $t = 0.5$, the formula gives

$$h = 30(0.5) - 16(0.5)^2$$
$$h = 15 - 4$$
$$h = 11$$

So, the ball is 11 feet above the ground 0.5 second after it leaves the ground.

The graph shows the relationship between h and t for the equation $h = 30t - 16t^2$. You can see from the graph that the ball is at a height of 11 meters when the time is 0.5 second.

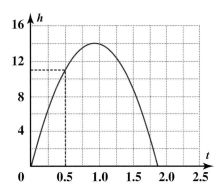

Think & Discuss

After looking at the graph of $h = 30t - 16t^2$, Evan said the ball did not bounce straight upward. Why do you think he said this?

Is he correct? Explain.

MATERIALS

graphing calculator

Problem Set A

1. You can display the graph of $h = 30t - 16t^2$ on your calculator. When you enter the equation, use x for t and y for h.

a. Graph the equation $y = 30x - 16x^2$ in the standard viewing window ($^-10 \leq x \leq 10$ and $^-10 \leq y \leq 10$). Describe the graph.

b. Find a new viewing window that allows you to identify points along the graph of $y = 30x - 16x^2$. Make a sketch of the graph. Label the minimum and maximum values on each axis.

c. Estimate the coordinates of the highest point on the graph.

2. You can use the Trace feature on your calculator to estimate when the ball hits the ground.

a. What is the value of h when the ball is on the ground?

b. Estimate the values of t that give this value of h.

c. Write the equation you would need to solve to find the value of t when the ball is on the ground.

d. Check your estimates from Part b by substituting them into your equation. Are your estimates exact solutions?

3. What does the fact that the equation has two solutions tell you about the flight of the ball?

4. The Zoom feature on your calculator allows you to look more closely at any part of a graph. To see how this works, start with your graph from Problem 1.

You already used the Trace feature to estimate when the ball hits the ground the second time. Now you can use the Zoom feature to get a better estimate. Select Zoom In from the Zoom menu, and focus on where the curve crosses the *x*-axis.

Once again, use the Trace feature on this new graph to try to get a more accurate estimate of a solution of $30t - 16t^2 = 0$. Check your estimate by substituting it for *t* in the equation $30t - 16t^2 = 0$. If your estimate is not an exact solution, zoom in once more and refine your estimate again.

5. For the formula $h = 30t - 16t^2$, write an inequality that shows the values *t* can have. Explain your inequality.

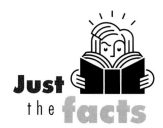

Just the facts

Because they must be extremely accurate, equations for calculating the height and speed of a rocket account for such factors as air resistance and the decreasing mass of the rocket as it burns fuel.

Problem Set B

1. How could you use this graph of $h = 30t - 16t^2$ to estimate the solution of $30t - 16t^2 = 3$?

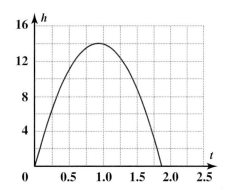

2. Evan said he could tell immediately from the graph that there are exactly two solutions of the equation $30t - 16t^2 = 3$. How did he know?

3. Use your calculator's Zoom and Trace features to find approximate solutions of $30t - 16t^2 = 3$ to the nearest hundredth.

4. What do the solutions of $30t - 16t^2 = 3$ tell you about the ball's flight?

5. Bharati said the equation $30t - 16t^2 = 20$ has no solutions. How could she tell this from the graph?

6. Write another equation of the form $30t - 16t^2 = ?$ that has no solutions.

7. How many solutions does $30t - 16t^2 = 10$ have?

8. Is there a value of c that would make the equation $30t - 16t^2 = c$ have exactly one solution? If so, explain how to find it using the graph, and give the value.

9. For the formula $h = 30t - 16t^2$, write an inequality that shows the values h can have. Explain your inequality.

Share & Summarize

1. How can you use a graphing calculator to help estimate solutions of equations? Illustrate your answer by writing an equation and finding solutions to the nearest hundredth.

2. How many solutions are possible for an equation of the form $30x - 16x^2 = c$, where c is a constant?

Investigation 2 Using Tables to Solve Equations

In Investigation 1, you estimated solutions of equations by making graphs on a calculator. Now you will use your calculator's Table feature to make the guess-check-and-improve process more systematic.

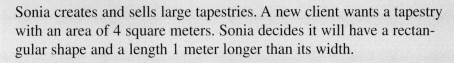

Think & Discuss

Sonia creates and sells large tapestries. A new client wants a tapestry with an area of 4 square meters. Sonia decides it will have a rectangular shape and a length 1 meter longer than its width.

• If x represents the width of the tapestry in meters, what equation can Sonia solve to find the length and width of the tapestry?

• Sonia used guess-check-and-improve to try to find the value of x. She first tried a width, x, of 1 meter and found the area would be 2 m². Then she found that for a width of 2 m, the area would be 6 m². What might Sonia choose as her third guess for x?

• Use guess-check-and-improve to find the value of x to the nearest tenth of a meter. Keep track of your guesses and the resulting areas.

• Now find the x value to the nearest hundredth of a meter.

• How do you know the value you found for the width gives an area closer to 4 m² than any other value to the nearest hundredth?

• How could you find a value of x that gives an even closer approximation to an area of 4?

Just the facts

An art form that has been around for centuries, tapestries are colorful, richly woven cloths. Today they are often used to decorate the walls of homes and offices.

In the Think & Discuss on page 245, you chose a first guess for x and then improved your guess one step at a time. Graphing calculators have a Table feature that allows you to examine many guesses at a glance.

MATERIALS

graphing calculator

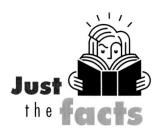

Just the facts

The Greek letter delta, Δ, is used to represent the increment between consecutive values in a sequence. On many calculators, Δ is used to indicate the amount of change in the variable x from one row of a table to the next.

EXAMPLE

Try Sonia's problem using the Table feature. (Follow along with this Example using your calculator.)

- On the Table Setup screen, set the starting x value to 0 and the increment to 1.

- Enter Sonia's equation in the form $y = x(x + 1)$ or $y = x^2 + x$.

- Look at the Table screen to see which value or values of x give values of y closest to 4.

The table at right shows part of what may appear on your screen.

X	Y$_1$
0	0
1	2
2	6
3	12
4	20
5	30
6	42

From the table, you can see that x must be between 1 and 2 since the corresponding y values go from 2 to 6.

To search for values between 1 and 2, use the Table Setup feature to show x values in smaller increments.

- On the Table Setup screen, set the starting x value to 1 and the increment to 0.1.

When you return to the Table screen, you should see something like the table at right.

You can now see that the solution has an x value between 1.5 and 1.6.

X	Y$_1$
1	2
1.1	2.31
1.2	2.64
1.3	2.99
1.4	3.36
1.5	3.75
1.6	4.16

In Problem Set C, you will use the method shown in the Example to find a more accurate estimate for x.

Problem Set C

x	y
1.5	3.75
1.51	
1.52	
1.53	
1.54	
1.55	
1.56	
1.57	
1.58	
1.59	
1.6	

1. In the Example, the table shows that the width of Sonia's tapestry must be between 1.5 and 1.6 meters. You can use Table Setup again to improve the accuracy of x to the hundredths place.

   a. What values of x and what increment should you use to get a more exact estimate for x?

   b. Adjust the values of x and the increment in Table Setup and then view the table once more. You will have to scroll down to see all the values you need. Copy and complete the table at left.

   c. Which two values of x is the solution between? Which of these gives an area closer to 4? Explain.

   d. What are the width and length of the tapestry to the nearest centimeter?

2. You can use Table Setup again to improve the accuracy of your estimate for x to the thousandths place (the nearest millimeter).

   a. What should you enter as the starting x value and the increment in Table Setup?

   b. Between which two values of x does the exact answer lie?

   c. Which of the two values gives a better approximation? How do you know?

3. Another client wants a tapestry with an area of 6 square meters and a length 2 meters greater than its width.

   a. Write an equation representing the area of the tapestry, using x to represent the width.

   b. Use the Table and Table Setup features to find the width to the nearest centimeter. For each new table you examine, record the two values of x that give areas closest to 6.

Woven of wool and silk in the late 15th century, this is one of a series of six tapestries entitled *The Lady and the Unicorn*.

You will now use the Table feature to approximate solutions of quadratic equations. You will learn how to find exact solutions of quadratic equations in Chapter 7.

MATERIALS

graphing calculator

Problem Set D

Each equation below has two solutions. Use your calculator's Table feature to approximate the solutions to the nearest hundredth (you may have to scroll through the table). Check each answer by substituting it into the equation.

1. $2m(m - 5) = 48$

2. $3t(t - 5) = 48$

3. Since some quadratic equations have two solutions, there may be a second solution of each equation you have explored for the area of Sonia's tapestries.

    a. How could you use a graphing calculator table to determine whether there is another solution of $x(x + 1) = 4$?

    b. Determine whether there is a second solution of $x(x + 1) = 4$. If so, estimate it to the nearest hundredth. Check the solution by substitution.

    c. Determine whether there is a second solution of $x(x + 2) = 6$. If so, estimate it to the nearest hundredth. Check the solution by substitution.

    d. Could Sonia use the solutions you found in Parts b and c to design a tapestry? Why or why not?

Share & Summarize

Think about how you might solve these equations.

    i. $2w - 3 = 3w - 7$ **ii.** $x^3 - 2x^2 = 5$

    iii. $3y(2y + 3) - 15 = 0$ **iv.** $5 = 3z + 1$

1. Which equations would you solve by doing the same thing to both sides or backtracking? Explain.

2. Which equations would you solve using your calculator's Table feature? Explain.

On Your Own Exercises

Practice & Apply

1. **Physical Science** A ball is launched straight upward from ground level with an initial velocity of 50 feet per second. Its height h in feet above the ground t seconds after it is thrown is given by the formula $h = 50t - 16t^2$.

   a. Draw a graph of this formula with time on the horizontal axis and height on the vertical axis. Show $0 \leq t \leq 4$ and $0 \leq h \leq 40$.

   b. What is the approximate value of t when the ball hits the ground?

   c. About how high does the ball go before it starts falling?

   d. After approximately how many seconds does the ball reach its maximum height?

2. Suppose the height in feet of a ball bouncing vertically into the air is given by the formula $h = 30t - 16t^2$, where t is time in seconds since the ball left the ground. A graph of this equation is shown below.

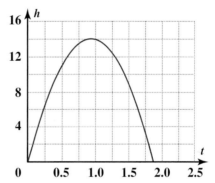

   a. On its way up, the ball reached a particular height at 0.5 second. It reached this height again on its way down. How could you determine at what time the ball reached this height on the way down?

   b. What is the value of this time to the nearest tenth?

3. In Investigation 1, you found that the maximum value of $30t - 16t^2$ is approximately 14. The exact value is 14.0625. Write an equation or an inequality showing the values of h for which $h = 30t - 16t^2$ has each given number of solutions.

   a. two   **b.** none   **c.** one

4. Draw a graph of $y = x^2 - 4x + 2$. Use your graph to find a value for d so that the equation $x^2 - 4x + 2 = d$ has each given number of solutions.

   a. two   **b.** one   **c.** none

5. Solve the equation $(x + 2)(x - 3) = 14$ by constructing a table of values. Use integer values of x between $^-6$ and 6.

6. Solve the equation $k^2 - k + 3 = 45$ by constructing a table of values. Use integer values of x between $^-9$ and 9.

7. Examine this table.

t	$t(t - 3)$
$^-2$	10
$^-1$	4
0	0
1	$^-2$
2	$^-2$
3	0
4	4
5	10
6	18
7	28

   a. Use the table to estimate the solutions of $t(t - 3) = 5$ to the nearest integer.

   b. If you were searching for solutions by making a table with a calculator, what would you have to do to find solutions to the nearest tenth?

   c. Find two solutions of $t(t - 3) = 5$ to the nearest tenth.

8. The equation $x^3 + 5x^2 + 4 = 5$ has a solution between $x = 0$ and $x = 1$. Find this solution to the nearest tenth.

9. This is a graph of $y = 0.1x^2 + 0.2x + 1$.

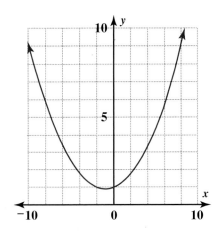

   a. Between which two pairs of integer values for x do the solutions of $0.1x^2 + 0.2x + 1 = 4$ lie?

   b. Find both solutions of $0.1x^2 + 0.2x + 1 = 4$ to the nearest hundredth.

10. Physical Science The formula $h = vt - 16t^2$ approximates a projected object's height above its starting position t seconds after it begins moving straight upward with an initial velocity of v ft/s.

For a thrown ball, the starting position will usually be some distance above the ground, such as 5 feet. So the height of the ball above the ground will be given by this formula, where s is the initial height (at $t = 0$) and v is the initial velocity:

$$h = vt - 16t^2 + s$$

a. Write an equation describing the height h of the ball after t seconds if it is thrown upward with a starting velocity of 30 feet per second from 5 feet above the ground.

b. What equation would you solve to find how long it takes the ball to reach the ground?

c. What equation would you solve to find how long it takes the ball to return to its starting height?

11. An object that is dropped, like one thrown upward, will be pulled downward by the force of gravity. However, its initial velocity will be 0. If air resistance is ignored, you can estimate the object's height h at time t with the formula $h = s - 16t^2$, where s is the starting height in feet.

a. If a baseball is dropped from a height of 100 ft, what equation would you solve to determine the number of seconds that would pass before the baseball hits the ground?

b. Solve your equation.

Just the facts

To accurately determine the height of a falling skydiver, air resistance—which depends on the diver's speed and cross-sectional area—must be calculated.

12. A rock climber launched a hook from the base of a cliff. The edge of the cliff is 100 feet above the climber. Use the formula for the height of a thrown object, $h = vt - 16t^2$, to answer these questions. (Assume that the hook is thrown straight up and, if it travels high enough, catches the edge of the cliff during its descent.)

a. Can the hook reach the top of the cliff if its initial velocity is 70 feet per second, or 70 fps? Explain.

b. Find an initial velocity that would be sufficient to allow the hook to reach the cliff's edge.

c. At an initial velocity of 100 fps, how far above the edge of the cliff will the hook rise?

d. At an initial velocity of 100 fps, how long will it take until the hook catches the top of the cliff on its way back down?

13. A cartoon cat pushes a piano out of a fourth-story window that is approximately 50 feet above the sidewalk

a. An object falls a distance of $s = 16t^2$ feet in t seconds. For how long will the piano fall before it hits the sidewalk?

b. Suppose the cat intended to smash his nemesis, the mouse, on the sidewalk below. But the mouse pulls a trampoline out of his pocket and the piano bounces back toward the cat at an initial velocity of 100 feet per second.

The height h of the piano t seconds after it hits the trampoline can be found using the formula $h = 100t - 16t^2$. Graph this equation, and use your graph to estimate how long it takes the piano to return to the fourth-story window.

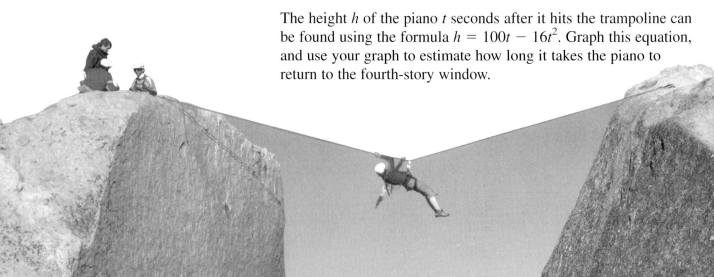

14. Desmond wants to construct a large picture frame using a strip of wood 20 feet long.

a. Let w represent the width of the frame. Write an equation that gives the height of the frame h in terms of the width w.

b. Express the area enclosed by the frame in terms of the width.

c. What dimensions would give Desmond a frame that encloses the greatest possible area?

d. What dimensions would give Desmond a frame that encloses the least possible area?

e. If the enclosed area is to be approximately 15 square feet, determine the dimensions of the frame to the nearest tenth of a foot.

15. Two objects are projected straight upward from the ground at exactly the same moment. One object is released at an initial velocity of 20 feet per second. The other is released at 30 feet per second. The two equations that relate the heights of the objects to time t are $y_1 = 20t - 16t^2$ and $y_2 = 30t - 16t^2$. Here are graphs of these two equations.

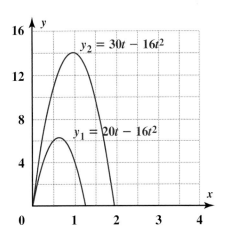

a. Write two equations you could solve to find the times at which the objects hit the ground.

b. These two objects hit the ground at different times. Use a calculator to find the exact difference between times.

Describe how
graphing an
equation can help
you estimate its
solutions.

16. Graphs of $y_1 = 4 - (x - 2)^2$ and $y_2 = (x + 2)^2 - 4$ are shown below.

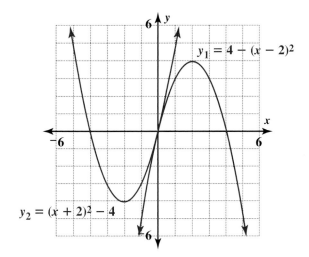

a. The maximum value of y_1 is 4. Explain how you can identify the maximum value of a parabola using a table of values.

b. The minimum value of y_2 is $^-4$. Explain how you can identify the minimum value of a parabola using a table of values.

c. For each equation, make a table of x and y values for x values between $^-3$ and 3. Without graphing, determine whether each equation has a maximum or a minimum y value and tell what it is.

  i. $y = 21 - 4x - x^2$

  ii. $y = (x - 2)^2 - 1$

  iii. $y = 8 + 2x - x^2$

Mixed Review

Find an equation of the line passing through the given points.

17. $(9, \; ^-0.5)$ and $(^-1, 4.5)$ **18.** $(1, \; ^-2)$ and $(^-4, \; ^-3)$

Recall that for linear equations, first differences are constant; and that for quadratic equations, second differences are constant. Determine whether the relationship in each table could be linear, quadratic, or neither.

19.

x	y
$^-3$	$^-4$
$^-2$	1
$^-1$	4
0	5
1	4
2	1

20.

x	y
$^-3$	7
$^-2$	2
$^-1$	$^-1$
0	$^-2$
1	$^-2$
2	$^-1$

21.

x	y
$^-3$	0
$^-2$	$^-0.5$
$^-1$	$^-1$
0	$^-1.5$
1	$^-2$
2	$^-2.5$

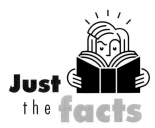

22. Physical Science Falling objects fall faster and faster, or *accelerate,* because of the force of gravity. Acceleration due to gravity is represented by *g.* Near Earth's surface, *g* has a value of about 9.806 meters per second squared, or 9.806 m/s^2.

As objects move away from Earth's surface, the force of gravity lessens—so the value of *g* falls. The table shows the approximate value of *g* for various heights above Earth's surface.

Height, *h* (m)	Value of *g* (m/s^2)
0	9.806
1,000	9.803
4,000	9.794
8,000	9.782
16,000	9.757
32,000	9.71

a. Graph the data on axes like those shown below.

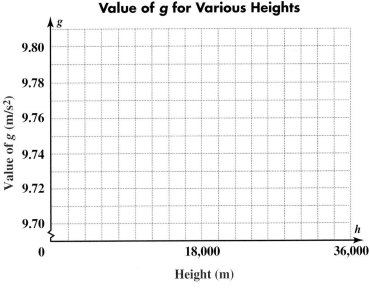

Value of g for Various Heights

b. Do the data appear to be approximately linear?

c. Draw a line that fits the data as well as possible, and find an equation of your line.

d. Use your equation or graph to predict the value of *g* at a height of 1,000,000 m.

Geometry Find the volume and the surface area of each solid.

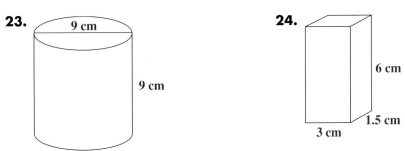

23. 9 cm, 9 cm

24. 6 cm, 1.5 cm, 3 cm

Solving Systems of Equations

You can think of an equation as a condition that a variable or variables must satisfy. For example, the equation $x + 2y = 6$ states that the sum of x and $2y$ must equal 6. An infinite number of pairs of values satisfy this condition, such as $(0, 3)$, $(2, 2)$, and $(^-2.3, 4.15)$.

When there is a second equation, or condition, that the same variables must satisfy, sometimes only one number pair satisfies *both* equations. For example, the only solution satisfying both $x + 2y = 6$ and $x = y$ is $(2, 2)$.

In the activity that follows, you and your classmates will make "human graphs" to find pairs of values that satisfy two equations.

Explore

Select a team of nine students to make the first graph. The team should follow these rules:

- Line up along the x-axis, from $^-4$ to 4.

- Multiply the number you are standing on by 2, and subtract 1.

- When your teacher says "Go!" walk forward or backward to the y value equal to your result from the previous step.

Describe the resulting "graph." What is its equation? Have two students explain why their coordinates are solutions of this equation.

You will next make two graphs on the same grid. Select two teams of nine students. Team 1 should follow their instructions and then stay on their points while Team 2 follows their instructions.

Team 1: Graph of $y = 2x$

- Line up along the x-axis.

- Multiply the number you are standing on by 2.

- When your teacher says "Go!" walk forward or backward that number of paces.

Team 2: Graph of $y = ^-x + 3$

- Line up along the x-axis.

- Multiply the number you are standing on by $^-1$, and add 3.

- When your teacher says "Go!" walk forward or backward that number of paces.

Are some students trying to stand on the same points? If a shared point is on both graphs, it should be a solution of both equations. Check that this is true.

Now choose two new teams, and make new human graphs to find ordered pairs that are solutions of both $y = 3x$ and $y = x + 4$.

Investigation ▶1▶ Graphing Systems of Equations

VOCABULARY
system of equations

MATERIALS
graph paper
graphing calculator

In the Explore activity, you found one (x, y) pair that was the solution of two equations. A group of two or more equations is called a **system of equations.** A solution of a system of equations is a set of values that makes all the equations true.

In Problem Set A, you will concentrate on the method of graphing to solve systems of equations.

Problem Set A

1. Consider these five equations.

    i. $2x + 3 = 7$

    ii. $x^2 + 5x + 6 = 0$

    iii. $(x - 2)^2 = 0$

    iv. $2x + y = 7$

    v. $3x - y = 3$

  a. Equations i, ii, and iii each involve one variable. How many solutions does each equation have? Find the solutions using any method you choose.

  b. Equations iv and v each involve two variables. How many solutions does each equation have? Can you list them? Explain why or why not.

  c. Consider Equations iv and v. Could you graph the solutions of $2x + y = 7$? Could you graph the solutions of $3x - y = 3$? Explain.

  d. Draw graphs of $2x + y = 7$ and $3x - y = 3$. At what point or points do they meet?

  e. What values of x and y make both $2x + y = 7$ and $3x - y = 3$ true? That is, what is the solution of this system of equations? How do these values relate to your answer to Part d?

In Problems 2 and 3, solve the system of linear equations using this method:

- Rewrite each equation so y is alone on one side.

- Graph both equations on your calculator.

- If the lines meet, estimate the x- and y-coordinates of their intersection. This may not be an exact solution, but it will give you an estimate of the solution.

- Check that these values satisfy both equations.

2. $2y - x = 20$ and $2x = 5 - y$ **3.** $y = 4$ and $x + y = {}^-1$

4. Consider these two equations.

$$x + y = 5 \qquad 2x = 12 - 2y$$

a. Graph both equations on one set of axes.

b. Do the graphs appear to intersect?

c. How could you verify your answer to Part b by looking at the equations?

d. Does this system of equations have a solution? How do you know?

5. Make up your own system of equations for which you think there will be no solution.

6. Consider these three equations.

$$4x + 2y = 7 \qquad y = {}^-2x + 5 \qquad 3x = 5y - 4$$

a. Select two that will form a system of equations that has a solution.

b. Select two that will form a system with no solution.

7. Consider this system of equations.

$$2x - y = 4 \qquad 5x - 2.5y = 10$$

a. The slope of the line for each equation is 2. How many solutions do you think this system will have?

b. Without using a calculator, draw graphs of these equations.

c. What do you notice about the two graphs?

d. How many solutions do you think this system has?

e. Why do you think this system of equations is different from the system in Problem 4?

8. Ben is twice as old as his younger brother Alex. Four years ago, Ben was four times as old as Alex.

a. Write two equations that relate Alex's and Ben's ages.

b. Find the ages of Ben and Alex using a graph.

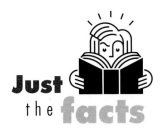

Problem Set B

It is possible to find solutions of systems of equations that are not linear.

1. Use a graph to solve this system of equations. Rewrite the equations before graphing them, if necessary. How many solutions can you find?

$$y - x - 1 = 0 \qquad y = x^2 - 3x + 4$$

2. Consider the equation $x + 1 = x^2 - 3x + 4$.

a. Follow these steps to solve this equation using a table:

- Enter two equations, $y_1 = x + 1$ and $y_2 = x^2 - 3x + 4$, into your calculator.

- Use the TableSet and Table features to create tables of the two equations.

- Examine the table for values of x for which y_1 and y_2 are equal. Create new tables, using smaller increments as necessary.

b. How do the solutions of this equation relate to what you found in Problem 1? Explain why the solutions are related in this way.

c. Now use a table or graph to solve $x^2 - 4x + 3 = 0$. How do the solutions of this equation compare to your result for Part a? Explain.

Share & Summarize

1. Describe how to estimate a solution of a system of two equations using a graph.

2. Describe how you could create a system of linear equations with no solution.

Investigation Problems Involving Systems of Equations

Sometimes the solution to a practical problem requires setting up and solving a system of equations. In this investigation you will use graphs to find solutions to such problems.

> **MATERIALS**
>
> *graph paper*

Explore

Two bookstores each fill catalog orders. Each charges the standard price for the books, but they have different shipping and handling charges. Gaslight Bookstore charges a shipping fee of $3 per order plus $1 per book, while Crimescene Bookstore charges $5 per order and $0.50 per book.

For each bookstore, write an equation that describes how the shipping cost y is related to the number of books x. Then graph both equations on the same axes.

Under what circumstances would you order from Gaslight? When would you order from Crimescene? Are there circumstances when it makes no difference from which store you buy?

What other (perhaps nonmathematical) things might you consider when making this choice?

> **MATERIALS**
>
> *graph paper*

Problem Set C

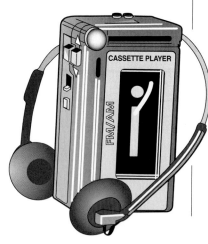

At Shikara's Music Emporium, cassette tapes sell for $10 and compact disks sell for $15.

1. Kai's family bought 16 items. List some possible values for the number of cassette tapes t and compact disks d they bought.

2. Write an equation to describe the total number of items bought by Kai's family.

3. Plot points on a grid to show all the pairs (t, d) that satisfy the equation in Problem 2. Remember, t and d must be whole numbers.

4. Suppose that instead of telling you how many items they bought, Kai told you his family spent $220. List some possible values for the number of cassette tapes t and the number of compact disks d that have a total price of $220.

5. Write an equation to describe the total number of dollars Kai's family spent on tapes and disks.

6. On the same grid, plot all the pairs of whole-number values that satisfy the equation in Problem 5. You may want to use a different color or symbol for these points.

7. Use your graph to find the possible values for *t* and *d* if both statements in Problems 1 and 4 are true.

MATERIALS

graph paper

Remember

If two items *x* and *y* are proportional, they are related by a linear equation $y = mx$, where *m* is a constant.

Problem Set D

The owner of a hat business has *fixed expenses* of \$3,000 each week. She has these expenses regardless of how many hats she makes.

The remaining expenses are *variable costs,* such as materials and labor. These costs are proportional to the number of hats made. For example, in a week in which 500 hats were made and sold, her variable costs were \$7,500. In a week in which 1,000 hats were made, they were \$15,000.

All the hats were sold to department stores at \$20 each.

1. Write an equation relating the income *i* to the number of hats sold *n*.

2. Consider the hatmaker's costs.

a. Write an expression for the *variable costs* to produce *n* hats. That is, if the shop made *n* hats in a week, what would the variable costs be?

b. Write an equation relating the total cost per week, *c,* and the number of hats made and sold that week, *n*. Remember that the total cost includes both the variable costs and the fixed expenses.

3. The difference between cost and income is the owner's profit. If the hatmaker has no sales, she will still have to pay the fixed costs, and therefore will have a large loss. If she has many sales, she will easily cover her fixed costs and make a good profit.

a. In one particular week, the hatmaker made and sold 500 hats. Did she make a profit or a suffer a loss? How much?

b. In another week, she made and sold 1,000 hats. Did she make a profit or suffer a loss? How much?

4. Somewhere between no sales and many sales is a number of sales called the *break-even point*. This is the point at which costs and income are equal, so there is no profit and no loss.

a. On one set of axes, graph the equations for income and costs. Label each graph with its equation. You will need to read amounts from your graph, so use an appropriate scale and be as accurate as you can.

b. How can you use your graph to estimate the break-even point?

c. From your graph, estimate the number of hats the hatmaker needs to make and sell each week to break even. Calculate the costs and income for that number of items to check your estimate. Improve your estimate if necessary.

d. Use your graph to estimate the number of hats she needs to sell each week to make a profit of $1,000 per week. Calculate the costs and income for that number of items to check your estimate. Improve your estimate if necessary.

e. Use the equations you wrote for Problems 1 and 2 to write an equation that gives the value of *n* for which costs and income are equal. Solve your equation to find the break-even point.

Share & Summarize

1. Describe a situation that would require solving a system of equations to find the answer.

2. What is the connection between a graph describing income and expenses for a particular business, and the break-even point for that business?

3. What is the connection between a system of equations describing income and expenses for a particular business, and the break-even point for that business?

Investigation ▶3 Solving Systems of Equations by Substitution

You can find or estimate the solution of a system of equations by graphing the equations and finding the coordinates of the points where the graphs intersect. You can also solve systems by working with the equations algebraically. To see how this works, consider the situation described below.

Ana's mother took her shopping for new socks.

To find the cost of each type of socks, you can write and solve a system of equations. Let x represent the price of the plain socks, and let y represent the price of the designer socks.

The designer socks cost $1 more than twice the cost of the plain socks.

$$y = 2x + 1$$

The price for 3 pairs of plain socks and 1 pair of designer socks is $11.

$$3x + y = 11$$

To determine the cost of the socks, you need to find an (x, y) pair that is a solution of both $y = 2x + 1$ and $3x + y = 11$. That is, you need to solve this system of equations:

$$y = 2x + 1 \qquad 3x + y = 11$$

Think & Discuss

Graphs of $y = 2x + 1$ and $3x + y = 11$ are shown at right.

* Which graph is which?

* Use the graph to solve the system of equations. How much does each type of sock cost?

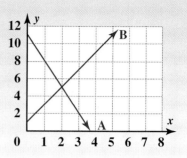

You can also solve this system algebraically. The first equation tells you that y must equal $2x + 1$. Since the value of y must be the same in *both* equations, you can *substitute* $2x + 1$ for y in the second equation. This gives you a linear equation with only one variable, which you already know how to solve.

$$3x + y = 11 \implies 3x + (2x + 1) = 11$$
$$5x + 1 = 11$$
$$5x = 10$$
$$x = 2$$

So $x = 2$, and since $y = 2x + 1$, you know that $y = 5$. That means the plain socks cost \$2 and the designer socks cost \$5. Check this solution by substituting it into *both* equations.

VOCABULARY
substitution

This algebraic method for finding solutions is called **substitution** because it involves *substituting* an expression from one equation for a variable in another equation.

Problem Set E

Use substitution to solve each system of equations. Check each solution by substituting it into both original equations.

1. $a = 3 - b$
$4b + a = 15$

2. $x = 2 - y$
$8y + x = 16$

3. Evan suggested solving Problem 1 by substituting $3 - b$ for a in the *first* equation. Will Evan's method work? Explain why or why not.

As you will discover next, sometimes you must rewrite one of the equations in a system before you can determine what expression to substitute.

Problem Set F

1. Consider this system of equations.

$$x + y = 8 \qquad 4x - y = 7$$

  a. Rewrite one of the equations to get y by itself on one side.

  b. Solve the system by substitution. Check your solution by substituting it into both of the original equations.

2. To solve the system in Problem 1, you could have rewritten the first equation to get x alone on one side. Solve the system this way. Do you get the same answer?

3. Carinne scored 23 points in last night's basketball game. Altogether, 10 of her shots went in; some were 2-point shots and the others were 3-point shots. Let a stand for the number of 2-point shots and b the number of 3-point shots.

  a. Write an equation for the total number of Carinne's shots that went in.

  b. Write an equation for the total number of points Carinne scored.

  c. How many 2-point shots did Carinne make? How many 3-point shots did she make?

4. Consider this system of equations.

$$y - 2x = 3 \qquad 2y + 5x = 27$$

  a. Graph the equations on your calculator, and estimate the point of intersection.

  b. Solve the system of equations by using substitution. Compare your solution with the approximation you found in Part a.

  c. What advantage does solving linear equations by substitution have over solving them by graphing?

Share & Summarize

Describe the steps for solving a system of two equations by substitution.

Investigation Solving Systems of Equations by Elimination

Solving systems of equations by *elimination* is sometimes simpler than solving by substitution. The elimination method uses the idea of doing the same thing to both sides of an equation.

Think & Discuss

Consider this system of equations.

$$5x + 4 = 13 + 3y$$
$$2x = 3y$$

To solve the system, Gabriela first tried the following:

$$
\begin{array}{rcl}
5x + 4 & = & 13 + 3y \\
-\,2x & & -\,3y \\
\hline
3x + 4 & = & 13
\end{array}
$$

Gabriela said she was doing the same thing to both sides of the first equation. Is this true? Explain.

In a way, what Gabriela did above was to subtract the second equation from the first. By doing this, which variable did Gabriela *eliminate*?

How do you think eliminating this variable will help her find the solution?

If $3x + 4 = 13$, we know that $x = 3$. Substituting this value into either of the original equations gives $y = 2$. Therefore, the solution of this system is $(3, 2)$.

VOCABULARY
elimination

By subtracting or adding equations, you can sometimes eliminate one variable and solve for the other. This process is called the method of **elimination.**

Sometimes you will need to rewrite one or both equations before you can eliminate one of the variables. The Example on the next page demonstrates this technique.

Solve the system consisting of Equations A and B.

① The coefficient of y in Equation A is $^-4$, so multiply both sides of Equation B by 2, so that ^-2y becomes ^-4y. Rewrite the two equations.

② Now subtract the new Equation C from Equation A, as Gabriela did in the Think & Discuss. Since $^-4y - (^-4y) = 0$, the variable y has been eliminated from the equations.

This leaves just one equation with x as a variable. Solve for x, which gives the equation $x = 20$.

③ Now substitute that value into either of the original equations to find the value of y.

$$7x - 4y = 100 \qquad [A]$$
$$3x - 2y = 40 \qquad [B]$$

$$① \; 7x - 4y = 100 \qquad [A]$$
$$6x - 4y = 80 \qquad [C]$$

$$② \; 7x - 6x = 100 - 80$$
$$x = 20$$

③ Substituting $x = 20$ into [B]:
$$3(20) - 2y = 40$$
$$-2y = -20$$
$$y = 10$$

Check in Equation [A] and Equation [B].
So, the solution to the system is (20, 10).

Just the facts

People in many professions use systems of equations in their work. Makers of feed for farm animals might determine the amounts of several ingredients for a particular feed, for example, to ensure it contains certain proportions of protein, fiber, and fat.

Think & Discuss

In the Example, x could have been chosen as the variable to eliminate, but it seemed easier to eliminate y. Why might this be so?

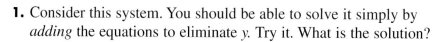

Problem Set G

1. Consider this system. You should be able to solve it simply by *adding* the equations to eliminate *y*. Try it. What is the solution?

$$9x - 2y = 3$$
$$3x + 2y = 9$$

Confirm that the equations in each pair are equivalent. Explain what you must do to both sides of the first equation to obtain the second equation.

2. $3m + 2 = 13.5$
   $12m + 8 = 54$

3. $x + 4y = 2$
   $2x = {}^-8y + 4$

4. $7d + 1 = 4p$
   $21d = 12p - 3$

The first task in using the elimination method is deciding how to eliminate one of the variables. The trick is to write equations—equivalent to those you started with—for which the coefficients of one of the variables are the same or opposites.

5. Consider these four systems of equations.

   i. $x + 2y = 9$ [A]
       $3x + y = 7$ [B]

   ii. $7x - y = 4$ [C]
       $2x + 3y = 19$ [D]

   iii. $35x - 6y = 1$ [E]
        $7x + 3y = 10$ [F]

   iv. $5x + 3y = 42$ [G]
       $2x + 8y = 78$ [H]

   a. Look at System i. You could eliminate *x* by replacing Equation A with an equivalent equation that contains the term $3x$ or ^-3x. What would you do to Equation A? What would you get?

   b. For System i, you could choose to eliminate *y* instead of *x*. How could you do this?

   c. Look at System ii. Which variable would you eliminate? Which equation would you rewrite? What would you do then?

   d. For System iii, state which variable you would eliminate and how you would do it.

   e. For System iv, you will need to write equivalents for both equations. Which variable would you choose to eliminate, and how would you do it?

6. Solve Systems i, iii, and iv using elimination. Check that each solution fits both equations.

Remember

Equivalent equations have the same solutions. You can always change an equation into an equivalent one by doing the same thing to both sides, as long as you don't multiply by 0.

Substitution and elimination work for any system of linear equations. However, some problems are easier to solve using one method than the other.

MATERIALS
graph paper

Problem Set **H**

Solve each system of equations two ways, first by *substitution* and then by *elimination*. In each case, compare the two approaches for ease, speed, and likelihood of mistakes. For each system of equations, do you think one method is better? Explain.

1. $3x = y + 7$
$5x = 9y + 41$

2. $5x - 3y = 10$
$15x + 6y = 30$

3. Rima is puzzled. She has tried to solve this system of equations by substitution and elimination, but neither method seems to work.

$$x + y = 4$$
$$3x + 3y = 11$$

a. Try to solve the equations by making a graph. Explain why Rima is having trouble.

b. Try to solve the equations by substitution. What happens?

c. Now solve the equations by elimination. What happens?

d. Rima discovered that she had copied down the equations incorrectly. This is the correct system.

$$x + y = 4$$
$$3x + 3y = 12$$

Solve this system of equations by substitution, elimination, or graphing.

Share & Summarize

1. Make up a system of linear equations that is easy to solve using substitution. Explain why you think substitution is a good method.

2. Make up a system of linear equations that is easy to solve using elimination. Explain why you think elimination is appropriate.

3. Make up a system of linear equations that does not have a solution. Explain why there is no solution.

Using a Spreadsheet to Solve Problems with Two Variables

MATERIALS

computer with spreadsheet software (1 per group)

Jeans Universe is having a grand opening sale.

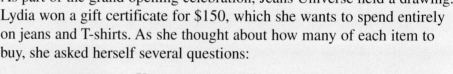

Jeans Universe

All jeans just $24.95 per pair!

T-shirts just $11.95 each!

GRAND OPENING CELEBRATION!

As part of the grand opening celebration, Jeans Universe held a drawing. Lydia won a gift certificate for $150, which she wants to spend entirely on jeans and T-shirts. As she thought about how many of each item to buy, she asked herself several questions:

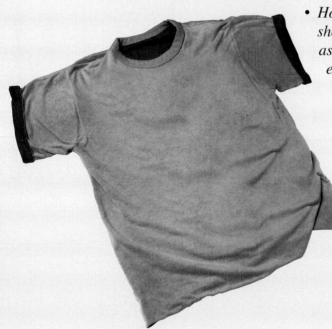

- *How many pairs of jeans and how many T-shirts should I buy to spend as much of my gift certificate as possible? What if I want to buy at least one of each?*

- *If I want to buy the same number of jeans as T-shirts, how many of each can I buy? What if I want to buy two T-shirts for every pair of jeans? Two pairs of jeans for every T-shirt?*

- *My mother said, "I really don't think you need more than five new T-shirts." How many pairs of jeans and how many T-shirts can I buy without going over this limit?*

Setting Up a Spreadsheet

Lydia had a hard time answering her questions because there are so many possible combinations. She set up the spreadsheet below to help keep track of the possibilities.

When Lydia completes her spreadsheet, it will show the costs of various combinations of jeans and T-shirts. For example, Cell F7 will show the cost for three pairs of jeans and four T-shirts.

	A	B	C	D	E	F	G	H	I
1					Jeans	24.95			
2			0	1	2	3	4	5	6
3		0							
4		1							
5		2							
6		3							
7		4							
8	T-shirts	5							
9	11.95	6							
10		7							
11		8							
12		9							
13		10							
14		11							
15		12							

1. To help fill the spreadsheet, Lydia decided she needed an equation showing how the total cost c depends on the number of pairs of jeans j and the number of T-shirts t she buys. What equation should she use?

Lydia then entered formulas in Cells C3–I3.

	A	B	C	D	E	F	G	H	I
1					Jeans	24.95			
2			0	1	2	3	4	5	6
3		0	=24.95*0+11.95*B3	=24.95*1+11.95*B3	=24.95*2+11.95*B3	=24.95*3+11.95*B3	=24.95*4+11.95*B3	=24.95*5+11.95*B3	=24.95*6+11.95*B3

2. Explain what the formulas in Cells C3–I3 do.

3. Set up your spreadsheet like Lydia's. Use the currency format for the numbers in Rows 3–15 of Columns C–I. What values appear in Cells C3–I3?

Now select the seven cells with formulas, Cells C3–I3, and use the Fill Down command to copy the formulas down to Row 15. You should now have values in all the cells of your table.

4. Look at the formulas in the cells in Column D.

    a. What parts of the formula change as you move down the column from cell to cell? What parts of the formula stay the same?

    b. What do the formulas in Column D do?

Using the Spreadsheet

You can use the values in your spreadsheet to answer Lydia's questions.

5. What combination of T-shirts and jeans would allow Lydia to spend as much of her $150 gift certificate as possible? How much would she spend?

6. What combination of T-shirts and jeans would allow Lydia to spend as much of her $150 gift certificate as possible, if she wants to buy at least one of each item? How much would she spend?

7. What combination should she buy if she wants the same number of T-shirts as jeans? How much would she spend?

8. What combination should she buy if she wants twice as many T-shirts as jeans? How much would she spend?

9. What combination should she buy if she wants twice as many jeans as T-shirts? How much would she spend?

10. If she wants to buy no more than five T-shirts, what combination should she buy? How much would she spend?

Try It Again

What if, instead of jeans and T-shirts, Lydia decides to buy two different items, such as shorts and turtlenecks or tennis shoes and socks? Instead of creating a separate spreadsheet for each pair of items, Lydia changed her spreadsheet so it would work for any pair of prices.

Notice that, in Lydia's original spreadsheet, the value in Cell F1 is the cost of a pair of jeans and the value in Cell A9 is the cost of a T-shirt. Lydia used this information to type new formulas into Cells C3–I3. She also changed the label "Jeans" to "Item 1" and the label "T-shirts" to "Item 2."

	A	B	C	D	E	F	G	H	I
1					Item 1	24.95			
2			0	1	2	3	4	5	6
3		0	=$F$1*0+ $A$9*B3	=$F$1*1+ $A$9*B3	=$F$1*2+ $A$9*B3	=$F$1*3+ $A$9*B3	=$F$1*4+ $A$9*B3	=$F$1*5+ $A$9*B3	=$F$1*6+ $A$9*B3
4		1							
5		2							
6		3							
7		4							
8	Item 2	5							
9	11.95	6							
10		7							
11		8							
12		9							
13		10							
14		11							
15		12							

In place of 24.95 and 11.95, the new formulas use $F$1 and $A$9. The $ symbol tells the spreadsheet to use the value in the indicated cell as a *constant.*

- $F$1 tells the spreadsheet to use the value in Cell F1, 24.95, as a constant.

- $A$9 tells the spreadsheet to use the value in Cell A9, 11.95, as a constant.

When a formula with a constant is copied from one cell to another, the spreadsheet does not "update" the constant to reflect a new row and column.

Set up your spreadsheet in this new way.

- Enter the new formulas shown in Cells C3–I3.

- Select Cells C3–I3, and fill all seven columns down to Row 15.

11. Look at how the formulas change as you move down Column D.

  a. What parts of the formula change as you move from cell to cell? What parts of the formula stay the same?

  b. What do the formulas in Column D do?

By simply changing the prices in Cells F1 and A9, Lydia can now easily find prices for combinations of any two items.

Suppose shorts cost $19.95 and turtlenecks cost $13.95. Update your spreadsheet to show prices for combinations of these two items.

  • Type "Shorts" in Cell E1 and "Turtlenecks" in Cell A8.

  • Type the price of the shorts, $19.95, in Cell F1 and the price of the turtlenecks, $13.95, in Cell A8.

12. Kai won a $75 gift certificate. He wants to spend as much of it as possible on shorts and turtlenecks. What should he buy? How much will he spend?

13. Tamika wants to buy hats and socks with her $30 gift certificate. Hats cost $6.75; socks are $2.95 a pair. Alter your spreadsheet to show prices for combinations of hats and socks.

How many of each item should Tamika buy to spend as much of her gift certificate as possible? How much will she spend?

14. Work with your partner to invent a problem situation that involves combinations of two items with different prices, and write some questions about that situation. Use your spreadsheet to answer your questions.

What Did You Learn?

15. Imagine you have a part-time job working at a bookstore. You earn $6.25 per hour now, but expect a raise soon. Design a spreadsheet you could use to keep track of the hours you work and the money you earn each day of a week. Your spreadsheet should have these characteristics:

  • It should allow you to enter the number of hours you work each day, Monday through Sunday.

  • It should automatically compute the dollars you earn each day, as well as the total for the week.

  • It should use a constant for your hourly rate of pay, so you can easily update it each time you get a raise.

Just the facts

The federal hourly minimum wage was established in 1938 at 25¢ per hour. Sixty-five years later, the minimum wage had risen about $5, to $5.15 per hour.

On Your Own Exercises

Practice & Apply

1. There are six ways to pair these four equations.

  i. $y = 2x + 4$ **ii.** $y + 2x = {}^-4$

  iii. $x = 4 - \frac{y}{2}$ **iv.** $2y - 4x = 10$

 a. Predict which pairs of equations do not have a common solution.

 b. Verify your results for Part a by carefully graphing both equations in each pair you selected. Explain how the graphs do or do not verify your prediction.

 c. Predict which pairs, if any, have a common solution.

 d. Verify your results for Part c by graphing both equations in each pair you selected. Explain how the graphs do or do not verify your prediction.

 e. Use your graphs to find a common solution of each pair of equations you listed in Part c.

2. Cheryl and Felipe each have some keys on their key chains. Cheryl said, "I have twice as many as you have." Felipe said, "If you gave me four, we would have the same number."

 a. Write two equations that relate how many keys each has.

 b. How many keys does Cheryl have? How many does Felipe have?

3. A group of friends enters a restaurant. No table is large enough to seat the entire group, so the friends agree to sit at several separate tables. They want to sit in groups of 5, but there aren't enough tables: 4 people wouldn't have a place to sit. Someone suggests they sit in groups of 6, which would fill all the tables, with 2 extra seats at one table.

Answer these questions to find how many people are in the group and how many tables are available.

 a. Write a system of two equations to describe the situation.

 b. Solve your system of equations. How many friends are in the group, and how many tables are available? Check your work.

Rockefeller Center,
New York City

4. Physical Science A bicyclist riding from Boston to New York City maintains a steady speed of 18 miles per hour. An hour after the bicyclist started out, a car left from the same point, traveling along the same road at an average speed of 50 miles per hour.

  a. If the car and the bicyclist travel in the same direction, how far from Boston will each be an hour and a half after the bicyclist left town? Who will be in front?

  b. After 2 hours, who will be in front?

  c. On the same set of axes, draw distance-time graphs for the car and the bicycle. Put distance, in miles from Boston, on the horizontal axis. Put time, in hours from the time the cyclist left Boston, on the vertical axis.

  d. Use your graphs to determine the approximate distance from Boston when the car overtook the bicyclist.

  e. Approximately how long had the cyclist been riding when the car caught up?

  f. Write an equation relating the bicyclist's distance from Boston in miles to the time in hours after the bicyclist left. Using the same variables, write another equation relating the car's distance from Boston to the time after the bicyclist left.

  g. If you were to find the pair of values that fits both of your equations, what would those values represent?

5. Solve one of these systems of equations by drawing a graph. Solve the other by substitution.

  a. $f + g = 20$
  $\quad 3f + g = 28$

  b. $y - x = 3$
  $\quad 2y + 3x = 16$

6. The sum of two numbers is 31. One of the numbers is 9 more than the other.

  a. Write a system of two equations relating the numbers.

  b. Solve the system to find the numbers.

7. Economics Andre bought 11 books at a used book sale. Some cost 25¢ each; the others cost 35¢ each. Andre spent $3.15.

  a. Write a system of two equations to describe this situation, and solve it by substitution.

  b. How many books did Andre buy for each price?

Solve the systems of equations in Exercises 8–11 by elimination, and check your solutions. Give the following information:

- which variable you eliminated
- whether you added or subtracted equations
- the solution

8. $x + y = 12$
$x - y = 6$

9. $3p + 2q = 13$
$3p - 2q = {}^-5$

10. $5a + 4b = 59$
$5a - 2b = 23$

11. $9s + 2t = 3$
$4s + 2t = 8$

Solve the systems of equations in Exercises 12–15 by elimination, and check your solutions. Give the following information:

- which equation or equations you rewrote
- how you rewrote each equation
- whether you added or subtracted equations
- the solution

12. $3m + n = 7$ [A]
$m + 2n = 9$ [B]

13. $6x + y = {}^-54$ [C]
$2x - 5y = {}^-50$ [D]

14. $2a + 5b = 12$ [E]
$3a + 2b = 7$ [F]

15. $y = \frac{3}{4}x - 4$ [G]
$4y = 2x + 3$ [H]

Connect & Extend

16. Economics A manager of a rock group wants to estimate, based on past experience, how many tickets will be sold in advance of the next concert and how many tickets will be sold at the door on the night of the concert.

At a recent concert, the 1,000-seat hall was full. Tickets bought in advance cost $30, tickets sold at the door cost $40, and total ticket sales were $38,000.

a. Write a system of two equations to represent this information.

b. On one set of axes, draw graphs for the equations.

c. Use your graphs to estimate the number of advance sales and the number of door sales made that night.

d. Check that your estimates fit the conditions by substituting them into both equations.

17. Three tourists left a hotel one morning and headed in the same direction. Tyrone, the walker, arose early and set off at a steady pace. Manuela, the cyclist, slept in and left 2 hours later. Kevin caught the bus half an hour after Manuela had left. These graphs show their distances over time.

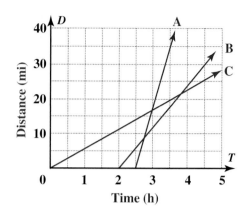

a. Match each graph with the tourist it represents.

b. When and where did Manuela pass Tyrone? When and where did Kevin pass Manuela? When and where did Kevin pass Tyrone?

c. From the graphs, estimate the speed at which each tourist traveled.

d. Write equations for the three lines.

e. Which system of equations would you solve to determine when and where Manuela passed Tyrone? Check your answer by solving the system and comparing the solution with your answer to Part b.

18. Make up a situation involving two variables that can be expressed in terms of two linear equations. Write and solve the equations to give an answer to the problem.

19. Solve this system of quadratic equations by drawing a graph.

$$y = x^2 \qquad y = 4 - 3x^2$$

20. History Solve this problem, which was posed by Mahavira, an Indian mathematician who lived around 850 A.D.

The mixed price of 9 citrons and 7 fragrant wood apples is 107; again, the mixed price of 7 citrons and 9 fragrant wood apples is 101. O you Arithmetician, tell me quickly the price of a citron and of a wood apple here, having distinctly separated those prices well.

21. Rachel and Cuong are going to the fair. Rachel has $13.50 to spend and calculates that she will spend all her money if she takes four rides and plays three games. Cuong has $15.50 and calculates that he will spend all his money if he takes two rides and plays five games.

a. Choose variables and write equations to describe this situation.

b. Use substitution to find the costs to go on a ride and to play a game.

In your
own
words

You've used three methods for solving systems of equations in this lesson: graphing, substitution, and elimination. Explain how you decide which solution method to use when solving a particular system of equations.

22. Economics For a school concert, a small printing business charges $8 for printing 120 tickets and $17 for printing 300 tickets. Both of these total costs include a fixed cost and a unit cost per ticket. That is, the cost for n tickets is $c = A + Bn$, where A is the business's fixed charge per order and B is the charge per ticket.

   a. Using the fact that it costs $8 to have 120 tickets printed, write an equation in which A and B are the only unknown quantities.

   b. Write a second equation using the cost of $17 for 300 tickets.

   c. Use the substitution method to find the business's fixed and variable charges.

23. Evan and his younger brother Keenan are comparing their collections of CDs. Set up two equations and solve them by elimination to find how many CDs each has.

24. Economics The Perez and Searle families went to the movies. Admission prices were $9 for adults and $5 for children, and came to a total of $62 for the 10 people who went.

   a. Write two equations to represent this situation.

   b. Solve your system using elimination to determine how many adults and how many children went to the movies.

25. You can use a system of equations to find an equation of a line when you know two points on the line. This exercise will help you find an equation for the line that passes through the points $(1, ^-2)$ and $(3, 4)$.

   a. An equation of a line can always be written in $y = mx + b$ form. Substitute $(1, ^-2)$ in $y = mx + b$ to find an equation using m and b.

   b. Substitute the point $(3, 4)$ in $y = mx + b$ to find a second equation.

   c. Find the common solution of the two equations from Parts a and b. Use them to write an equation of the line.

   d. Use the same technique to find an equation of the line passing through the points $(^-1, 5)$ and $(3, ^-3)$. Check your answer by verifying that both points satisfy your equation.

The *x*-intercept is the *x* value at which a line crosses the *x*-axis. Find an equation of the line with the given *x*-intercept and slope.

26. *x*-intercept 5, slope $^-2$　　**27.** *x*-intercept $^-2.5$, slope 0.5

For each table, tell whether the relationship between *x* and *y* could be linear, quadratic, or an inverse variation, and write an equation for the relationship.

28.

x	0.5	2	4	5	20
y	5	1.25	0.625	0.5	0.125

29.

x	0.5	2	4	5	20
y	$^-1.25$	2.5	7.5	10	47.5

30.

x	1	2	3	4	5
y	0.25	1	2.25	4	6.25

Geometry Find the volume of each cylindrical water tank.

31. a tank with height 9 m and circumference 19 m

32. a tank with circumference 34 m and a height equal to twice its radius

Geometry Tell whether the figures in each pair are congruent, similar, both, or neither.

33.

34.
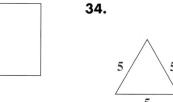

35.

36.
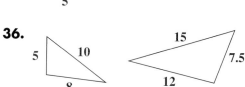

37.
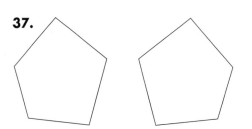

38.
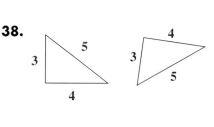

Remember
The circumference of a circle is 2πr, where r is the radius.

Remember
Congruent figures must be exactly the same shape and size. In similar figures, corresponding sides have lengths that share a common ratio, and corresponding angles are congruent.

Chapter Summary

VOCABULARY
elimination
inequality
substitution
**system of
equations**

You began this chapter by reviewing methods for solving equations in preparation for solving inequalities. Graphing was introduced as a way to understand inequalities and to solve equations. You learned to use calculator graphs and tables to estimate solutions of equations.

You also discovered that a *solution* of a system of equations is a set of values that satisfies all the equations in the system. You found solutions of systems by graphing and by using the algebraic methods of substitution and elimination.

MATERIALS
graph paper
graphing calculator

Strategies and Applications

The questions in this section will help you review and apply the important ideas and strategies developed in this chapter.

Using algebraic methods to solve equations

1. Explain how to solve an equation by backtracking. Illustrate your explanation with an example.

2. Explain how to solve an equation by doing the same thing to both sides. Illustrate your explanation with an example.

3. The sum of three consecutive numbers is 57. Write and solve an equation to find the three numbers.

Understanding and solving inequalities

4. List all the whole numbers that make both of these inequalities true.

$$t < 8 \qquad 2 < t \leq 10$$

5. Write the following inequality in words, and give all the whole-number values for x that make it a true statement.

$$8 < x \leq 11$$

6. A small video store has determined that its monthly profit must be *at least* $1,300 in order to stay in business. Movies rent for $2 each. Business expenses, including rent, for one month are $800.

Use the formula *total sales − expenses = profit* to write and solve an inequality that shows how many movies, *m,* must be rented each month for the store to stay in business.

7. A banquet hall puts on dinners for large groups. The cost to rent the hall for one night is $1,500 plus $40 per person.

Write and solve an inequality that answers this question: *Rod and Masako are planning to have their wedding reception in this banquet hall. How many people can they invite if their total budget for renting the hall is $7,500?*

Graphing inequalities

8. Consider these two statements.

$$x > 3 \text{ or } x < 5 \qquad\qquad x > 3 \text{ and } x < 5$$

a. Does the set of values that satisfies "$x > 3$ or $x < 5$" include 3 and 5? Draw a number-line graph to illustrate your answer.

b. Does the set of values that satisfies "$x > 3$ and $x < 5$" include 3 and 5? Use a number-line graph to illustrate your answer.

9. Explain the steps involved in graphing an inequality with two variables. Give an example to illustrate your steps.

Using graphs to estimate solutions of equations

10. Explain how you can use the Trace and Zoom features on your calculator to help estimate solutions of equations.

11. Use this graph for the height of an object over time to estimate, to the nearest tenth, the time at which $h = 5$.

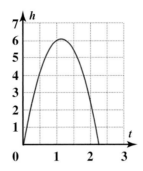

12. Draw a graph of $y = 4x - 0.5x^2$. Find a value of c for which $4x - 0.5x^2 = c$ has only one solution.

Using tables to estimate solutions of equations

13. This table of values is for the equation $y = x^2 + x$. Between which two values of x would you expect to find a solution for $x^2 + x = \frac{11}{4}$?

x	y
1	2
1.1	2.31
1.2	2.64
1.3	2.99
1.4	3.36
1.5	3.75
1.6	4.16

14. The equation $0.5x^2 - x - 1 = 3 - x^2$ has two solutions. In this table, $y_1 = 0.5x^2 - x - 1$ and $y_2 = 3 - x^2$. What information does this table give you about the solutions?

x	y_1	y_2
$^-3$	6.5	$^-6$
$^-2$	3	$^-1$
$^-1$	0.5	2
0	$^-1$	3
1	$^-1.5$	2
2	$^-1$	$^-1$
3	0.5	$^-6$

Solving systems of equations graphically and algebraically

15. Use a graph to find solutions of this system of equations. If necessary, rewrite the equations before graphing them.

$$y - 1 = {}^{-}x \qquad y - x^2 = {}^{-}1$$

16. Two competing building-supply stores have different prices for structural brick. Store A charges $0.75 per brick and $15 for delivery. Store B charges $1.05 per brick but delivery is free.

a. For each building-supply store, write an equation to represent the total cost for bricks, C, related to the number of bricks bought, b. Then graph both equations on the same set of axes.

b. Under what circumstances would you order structural bricks from Store A? From Store B? Are there circumstances when it makes no difference from which store you buy?

17. Solve this system of equations using either substitution or elimination. State which method you used, and explain why you chose it.

$$2x - 2y = 5 \qquad y = {}^{-}x - 3$$

Demonstrating Skills

18. Solve this equation by backtracking. Check your solution.

$$\tfrac{1}{3}\left(3 + \tfrac{x}{2}\right) - 1 = 4$$

19. Solve this equation by doing the same thing to both sides. Check your solution.

$$1 - 2p = 2 + 2p$$

Solve each inequality.

20. $3(j + 1) - 2(2 - j) \leq 9$

21. $6(k - 5) - 2 \leq 10$

22. $^-2(b + 1) > ^-5$

23. On a number line, graph all x values for which $x \leq ^-1$ or $x > 2$.

24. Graph the solution of this inequality on a number line.

$$\tfrac{t}{2} + 2 > 3$$

Graph each inequality.

25. $y \geq x - 2$

26. $y < ^-2x + 1$

27. Use a calculator and tables of values, in the interval from $^-3$ to 7, to estimate both solutions of this equation to the nearest hundredth.

$$0.5x(0.5x - 2) = 2$$

28. Use a table of values to solve this equation to the nearest hundredth. If there is more than one solution, find and check both.

$$2x^2 - x = 2x + 1$$

29. Use substitution to solve this system of equations. Check your solution.

$$x = y + 8$$
$$3y + 1 = 2x$$

30. Use elimination to solve this system of equations. Check your solution.

$$4x - 10y = 2$$
$$3x + 5y = 9$$

Solve each system of equations. Check your solutions.

31. $^-x + 1 = y$
$\quad y - x = 2$

32. $3x = 3y + 1$
$\quad x = 1 - y$

33. $3.5y - 1.5x = 6$
$\quad ^-3y + x = 3$

CHAPTER 5

Transformational Geometry

Real-Life Math

A History of Symmetry Symmetry is a type of balance created by repeating a basic shape of design in a regular pattern. Many cultures have used symmetry in their clothing, pottery, and artwork throughout history. Perhaps one of the greatest uses of symmetry in art can be found in the work of the Dutch artist M.C. Escher.

Think About It Can you find a basic design that is repeated in the Escher work shown above?

Family Letter

Dear Student and Family Members,

   Our next chapter in mathematics is about *transformational geometry*. We will learn about four basic transformations that can be applied to two-dimensional objects: reflections (flips), rotations, translations (slides from one place to another), and dilations (enlargements and size reductions). Using these transformations, we can move an object, reposition it, or place it on top of another object of the same shape. We will learn to recognize these types of transformations, describe them, and create symmetric designs using them.

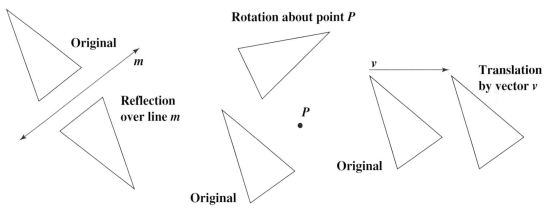

   Some of the most interesting border patterns, wallpaper designs, and quilt patterns are created by combinations of transformations.

Vocabulary
Along the way, we'll be learning about several new vocabulary terms.

dilation	reflection over a line	scale factor
image	reflection symmetry	transformation
line of reflection	rotation	translation
line of symmetry	rotation symmetry	vector
perpendicular bisector	scale drawing	

What can you do at home?

   In class, your student will be creating designs by performing transformations. You might ask your student to show you some of these. Also, be on the lookout, for patterns or figures with symmetry and arrangements of dilated figures on buildings, in wallpaper and quilts, and other places in your home and neighborhood. It's really quite surprising to see how many examples exist!

Reflection

In this chapter, you will learn about four kinds of transformations. To *transform* means to change. A **transformation** is a way to take a figure and create a new figure that is similar or congruent to the original.

One helpful way to begin thinking about transformations is through symmetry. *Symmetry* is a form of balance in figures and objects. When studying different cultures, archaeologists and anthropologists often look at the symmetry of designs found on pottery and other crafts. Chemists and physicists study how symmetry in the arrangements of atoms and molecules is related to their functions. Artists often use symmetry in their creations.

Maricopa jar by Mary Juan, ca. 1940s. Courtesy California Academy of Sciences, Elkus Collection, catalog #370-1820.

Explore

Follow these directions to create a symmetric "snowflake" design.

- Start with a square piece of paper. Fold it in half in one direction and then in half in the other direction. This will make a smaller square.

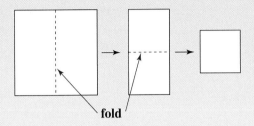

- Along each edge of the folded square, draw a simple shape—something like these.

- Cut out the shapes you drew, and unfold the square.

Describe the ways your snowflake design looks "balanced," or symmetric.

Investigation ▶ 1 ▶ Lines of Symmetry

VOCABULARY
line symmetry
line of symmetry
reflection
  symmetry

There are different types of symmetry. One type is **reflection symmetry,** also called **line symmetry.** With this type of symmetry, a line can be drawn between two halves of a figure or between two copies of a figure. The line is like a mirror between them, with one half of the image identical to the other half, but flipped. The line is called a **line of symmetry.**

The dashed lines below are lines of symmetry.

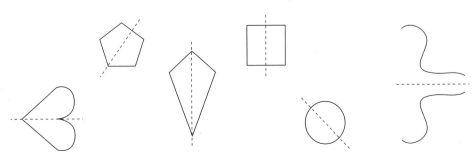

The dashed lines below are *not* lines of symmetry. Though they do cut the figures in half, they don't create "mirror image" halves.

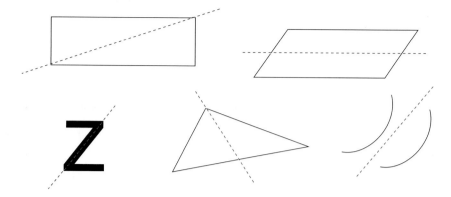

Héctor and Dante are talking about how they find lines of symmetry.

M A T E R I A L S
GeoMirror

Problem Set A

For each figure, find all the lines of symmetry you can, and record how many lines of symmetry the figure has. Then sketch the figure and show the lines of symmetry.

1. This is an equilateral triangle. (All three sides are the same length.)

2. This is a scalene triangle. (All three sides are different lengths.)

3. These are two overlapping polygons.

4. This is a rectangle.

5. This is a circle.

6. This is a parallelogram.

Share & Summarize

Find the lines of symmetry in this square. Then sketch the square, using dashed lines to show the lines of symmetry. How many lines of symmetry does a square have?

Investigation 2 ▶ Reflection

VOCABULARY
**reflection over
 a line**

You can create figures with reflection symmetry using a transformation: a **reflection over a line.** Suppose you have a figure and a line.

Imagine that the line is a line of symmetry and that the curve is only half of the figure. The other half is a *reflection* of the curve.

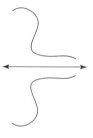

VOCABULARY
**line of reflection
image**

The reflection of a figure is its mirror image. The line acts like a mirror and is called a **line of reflection.** The result of a reflection, or of any transformation, is called an **image.**

Each point in the original figure has an image. In a reflection, the image of a point P is the point that matches with point P when you look through a GeoMirror or when you fold along the line of reflection.

The image of point P is called point P', pronounced "P prime." In the drawing below, points P' and Q' are the images of points P and Q.

As with real faces, this mask is not exactly symmetric.

Tamika used her GeoMirror to reflect Figure S over the line.

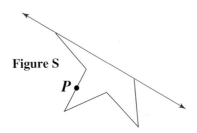

Figure S

When I put the GeoMirror on the line I can see the reflection of Figure S.

But I can also see what I'm drawing on the other side!

I'll draw the lines over the reflection. I can even see where Point P' goes.

Add the labels, and I'm done!

MATERIALS
• GeoMirror
• protractor
• metric ruler

Problem Set B

Problems 1–3 show a figure and a line of reflection. Carefully copy each picture. Then, using a GeoMirror, reflect the figure over the line to create the image. Mark the image of point *A* and name it point *A'*. Make your drawings as accurately as you can.

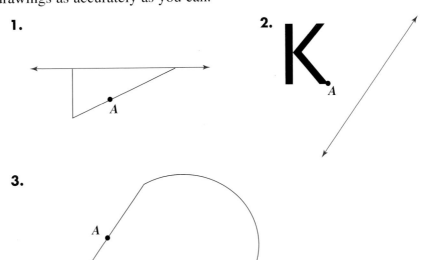

1.

2.

3.

4. When you reflect a figure over a line, is the image you create *congruent* to the original (the same size and shape), *similar* to the original (the same shape but possibly a different size), or *neither*? Explain how you know.

5. For each drawing you made in Problems 1–3, connect points A and A' with a straight line. Measure the angle between each line of reflection and Segment AA' $(\overline{AA'})$. What are the measures?

6. For each drawing you made in Problems 1–3, measure these lengths:

  a. the length of the segment between point A and the line of reflection

  b. the length of the segment between point A' and the line of reflection

7. Make a conjecture about the relationship between the line of reflection and the segment connecting a point to its image.

V O C A B U L A R Y
perpendicular bisector

Lines called *perpendicular bisectors* are particularly useful in working with reflections. A **perpendicular bisector** of a segment has two important characteristics.

  • It meets the segment at its midpoint (it *bisects* the segment).

  • It is perpendicular to the segment.

This line is perpendicular to $\overline{AA'}$ but is not a bisector of it.

This line bisects $\overline{AA'}$ but is not perpendicular to it.

This line is a perpendicular bisector of $\overline{AA'}$.

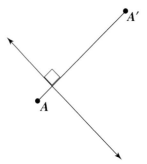

not a perpendicular bisector

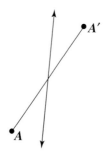

not a perpendicular bisector

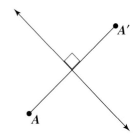

perpendicular bisector

Problem Set C

1. Copy each segment, and draw a perpendicular bisector of it.

a.

b.

2. Explain how you found the perpendicular bisectors in Problem 1.

3. How many perpendicular bisectors does a segment have?

Share & Summarize

Describe the relationship between a line of reflection and the segment joining a point to its reflected image. Use the idea of a perpendicular bisector in your description.

Investigation ▶3 The Perpendicular Bisector Method

If you don't have a GeoMirror handy, you can still reflect a point over a line using the *perpendicular bisector method*. If point A' is the image of point A in a figure with a line of symmetry, the line of symmetry is the perpendicular bisector of the segment connecting the points.

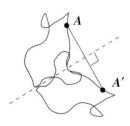

To reflect a point B over a line by the perpendicular bisector method, draw a segment from point B, perpendicular to the line.

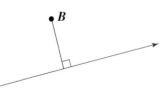

Continue the segment past the line, until you have doubled its length. At the other endpoint of the segment, mark the image point, point B'.

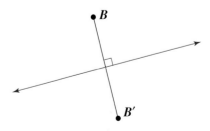

Think & Discuss

Suppose you want to reflect △*ABC* over line *m*.

Which points would you reflect? Why?

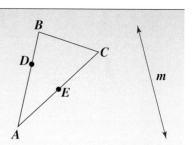

MATERIALS

- tracing paper
- ruler
- protractor
- GeoMirror (optional)

Problem Set D

For each problem, copy the figure and the line. Reflect the figure over the line using the perpendicular bisector method. Check your work using a GeoMirror or by folding.

1.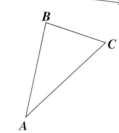

2.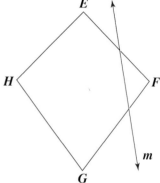

3.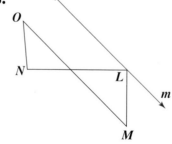

Share & Summarize

Here is a triangle and a line. Explain how you would use the perpendicular bisector method to reflect the triangle over the line.

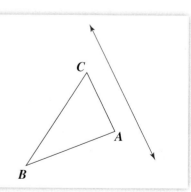

On Your Own Exercises

Practice  **Apply**

Remember

An *isosceles trapezoid* is a trapezoid with two opposite, nonparallel sides that are the same length.

In Exercises 1–4, copy the figure and draw all the lines of symmetry you can. Check the lines you drew by folding the paper.

1. This is an isosceles triangle.

2. This is an isosceles trapezoid.

3. This is an ellipse.

4. This is a trapezoid.

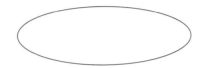

Exercises 5 and 6 show an original figure and a line of reflection.

- Copy the picture.
- Reflect the figure over the line, and draw the image.
- Mark the image of point *A* and name it point *A'*.

5.

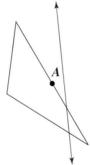

6.

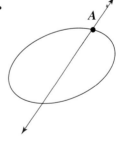

7. This picture shows two lines of reflection.

- Copy the picture.
- Reflect the figure over one line to create an image.
- Now reflect both figures— the original and the image—over the other line.

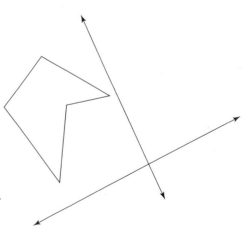

Copy each segment, and draw its perpendicular bisector.

8.

9.

10. Copy this picture.

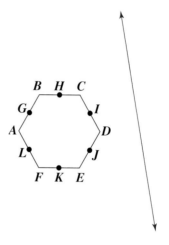

    a. Reflect the figure over the line using the perpendicular bisector method.

    b. What is the minimum number of points you have to reflect by this method to be able to draw the whole image figure? Explain.

11. Copy this picture.

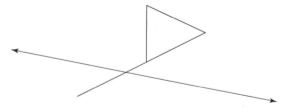

    a. Reflect the figure over the line using the perpendicular bisector method.

    b. What is the minimum number of points you have to reflect by this method to be able to draw the whole image figure? Explain.

Connect & Extend

12. Create a paper snowflake with four lines of symmetry. Check the lines of symmetry by folding.

13. Similar to lines of symmetry for two-dimensional figures, three-dimensional objects can have *planes* of symmetry. For example, a cube has nine planes of symmetry, including these three:

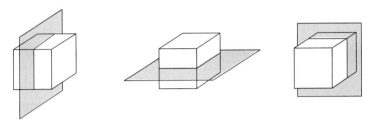

a. How many planes of symmetry does a regular square pyramid have? Describe or sketch them.

b. How many planes of symmetry does a sphere have? Describe or sketch them.

c. How many planes of symmetry does a right hexagonal prism have? Describe or sketch them.

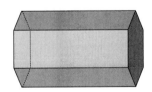

How many planes of symmetry does this treasure chest have?

14. Sports Bianca enjoys playing billiards. At the end of one close game she played, the only balls on the table were the 8 ball and the cue ball. It was Bianca's turn. To win, all she had to do was hit the 8 ball into a pocket of her choosing.

Bianca hit the cue ball, which collided with the 8 ball. This sent the 8 ball in the direction shown, sinking it into the correct pocket— and Bianca won the game.

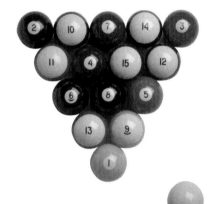

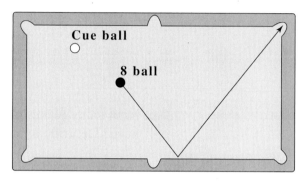

a. Was the path the ball traveled symmetric in some way? If so, find the line of symmetry.

b. Bianca said the angles made by the side of the table and the 8 ball's path, coming and going, are always equal to each other. Is this true in this case? How do you know?

15. Kai said that to reflect this star over the line, he needs to reflect only five points.

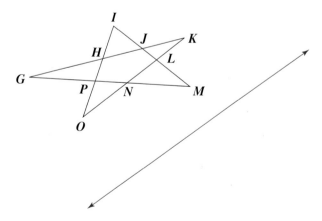

In your
own
words

Find a picture, design, or object in your home that has at least one line of symmetry. Describe or sketch it, and indicate each line of symmetry.

Is Kai correct? If so, explain how he could draw the image. If not, explain why he couldn't do it.

16. List three natural or manufactured things that have lines of symmetry.

17. Copy this picture.

   a. Reflect the circle over the line using the perpendicular bisector method. How many points did you reflect?

   b. Check your work by folding. Do the figures match exactly?

   c. What is the minimum number of points you must reflect by this method in order to reflect the whole figure? Explain.

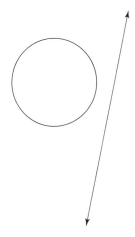

Mixed Review

Evaluate each expression for $a = 2$ and $b = 3$.

   18. 4^a **19.** $\left(\frac{1}{9}\right)^a$ **20.** $\left(9 \cdot \frac{1}{b}\right)^b$ **21.** $\left(\frac{b}{a} \cdot \frac{a}{b}\right)^a$

Write an equation of a line that is parallel to the given line.

   22. $y = 7x - 6$ **23.** $2x = 4y$ **24.** $x = {}^-2$

25. Geometry There are 360° in a full circle. Without using a protractor, match each angle with one of the angle measurements below.

      30°   270°   100°   50°   220°   80°

a.   **b.**    **c.**

d.   **e.**   **f.**

26. The table shows the way a certain patch of fungus might spread. The fungus initially covered an area of 12 mm².

   a. Write an equation for the area A covered by the fungus after w weeks.

   b. During which week will the fungus have spread to cover an area of at least 3,000 mm²?

Week	Area (mm²)
0	12
1	24
2	48
3	96
4	192

In Lesson 5.1, you looked closely at reflection symmetry, or line symmetry. These designs show another kind of symmetry.

 Z

Think & Discuss

Examine the figures above. In what way does it seem reasonable to say that these figures have symmetry?

Investigation 1 ► Rotation Symmetry

At the beginning of Lesson 5.1, you examined a paper snowflake for symmetry. You probably noticed that it had reflection symmetry, but you might not have realized your creation had other symmetry as well.

Problem Set A

MATERIALS

- scissors
- tracing paper
- pin

Follow these directions to create another paper snowflake.

1. Start with a square piece of paper.

    a. Fold it in half three times as shown.

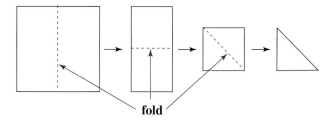

fold

b. On each side of the folded triangle, draw some shapes. For example, you might draw shapes like these:

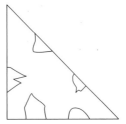

c. Cut out along the lines you drew, and then unfold the paper. The design above made this snowflake:

Now place your paper snowflake over a sheet of tracing paper. Copy the design by tracing around the edges, including the holes.

2. Pin the centers of your snowflake and your tracing together. *Rotate* (turn) the snowflake about its center until the design on the tracing paper coincides with the design on the snowflake. Did you need to turn the snowflake all the way around?

3. Rotate the snowflake again, until the designs match once more. Have you returned the paper to the position it was in before Problem 2? If not, rotate it again, until the designs match. How many times do you have to rotate the snowflake before you have turned it all the way around?

4. How many *degrees* must you rotate the snowflake each time to get the designs to match?

Remember
A full turn has 360°.

V O C A B U L A R Y
rotation symmetry

A figure has **rotation symmetry** if you can copy the figure, rotate the copy about a centerpoint *without turning it all the way around,* and find a place where the copy exactly matches the original. The *angle of rotation*—the smallest angle you need to turn the copy for it to match with the original—must be less than 360°.

Problem Set B

Each of these figures has rotation symmetry. Find the angle of rotation for each figure.

1.

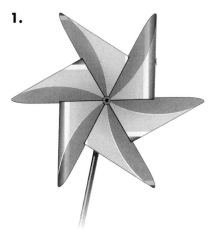

2.

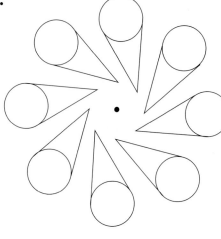

3.

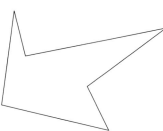

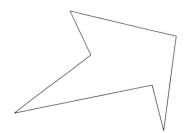

Share & Summarize

Look again at the figures in Problem Set B. Make a conjecture about the relationship between the angle of rotation and the number of identical elements in the figure.

Investigation Rotation as a Transformation

To create your own design with rotation symmetry, you need three things: (1) a figure, called a *basic design element;* (2) a center of rotation; (3) an angle of rotation.

By convention, angles of rotation assume a figure is rotated *counterclockwise.* To indicate a clockwise rotation, use a negative sign.

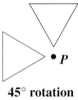

45° rotation about point *P*

‾45° rotation about point *P*

MATERIALS

- tracing paper
- protractor
- pin

Problem Set C

You will now create a design by rotating the figure below. The figure is your basic design element, and the point is the center of rotation. For this design, you will use an angle of rotation of 60°.

1. Begin by copying this picture, including point *P* and the reference line.

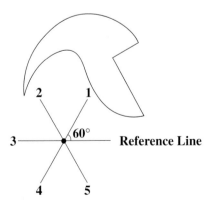

Reference Line

P

2. Draw a new segment with point *P* as one endpoint and forming a 60° angle with the reference line. Label this Segment 1. Then draw and label four more segments from point *P,* each forming a 60° angle with the previous segment, as shown here.

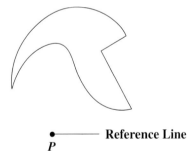

3. Place a sheet of tracing paper over your figure. Pin the papers together through the center of rotation. Trace the figure, including the reference line, but *don't* trace Segments 1–5.

4. Now rotate your tracing until the reference line on the tracing is directly over Segment 1. Trace the original figure again. Your tracing should now look like this.

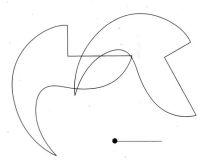

5. Rotate the tracing until the reference line on the tracing is directly over Segment 2. Trace the original figure again.

6. Repeat the process, rotating to place the reference line over the next segment and tracing the figure. Do this until the reference line on the tracing is back on the original reference line.

V O C A B U L A R Y
rotation

In Lesson 5.1, you learned about the reflection transformation. You have now used the second transformation, **rotation.** Each time you turned the basic design element and traced it, you performed a rotation.

M A T E R I A L S
• tracing paper
• protractor

Problem Set **D**

Use the given basic design element, center of rotation, and angle of rotation to create a design with rotation symmetry.

1. angle of rotation: 120°

2. angle of rotation: 90°

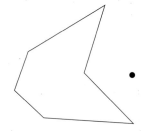

3. Do your completed designs have lines of symmetry? If so, where?

Share & Summarize

1. When you reflect a figure over a line once, you create a design with reflection symmetry. If you rotate a figure about a point once, does that always create a design with rotation symmetry?

2. Look at the designs you created in Problem Set D. Is there a pattern between the number of identical elements in the finished designs and the angle of rotation? If so, what is it?

Investigation 3 ▶ The Angle of Rotation

In Investigation 2, you used the *rotation* transformation to create figures with rotation symmetry. Now you will look more closely at an important part of these transformations: the *angle of rotation.*

M A T E R I A L S

• tracing paper
• ruler
• protractor
• colored pencils (optional)

Problem Set E

$\triangle A'C'D'$ is the image of $\triangle ACD$ when it is rotated a certain angle about point *P*. Carefully make a copy of the entire picture.

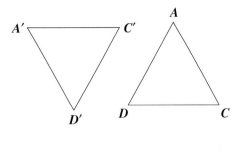

•*P*

1. Consider point *A* and its image, point *A'*.

   a. Draw $\overline{AP}$ and $\overline{A'P}$. You may want to use a different color for these segments than you used to make your drawing.

   b. Measure the two segments you just drew. What do you find?

   c. Measure $\angle APA'$.

2. Now consider point *C* and its image, point *C'*.

   a. Add $\overline{CP}$ and $\overline{C'P}$ to your picture. Again, you may want to use a different color.

   b. Measure these two segments. What do you find?

   c. Measure $\angle CPC'$.

3. Finally, consider point *D* and its image, point *D'*.

   a. Add $\overline{DP}$ and $\overline{D'P}$ to your picture.

   b. Measure these two segments. What do you find?

   c. Measure $\angle DPD'$.

4. Compare the three angles you measured. What do you notice?

The recycle logo, with its ever-revolving arrows, has rotation symmetry. (Can you determine the angle of rotation?)

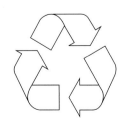

Problem Set F

1. Sketch this picture. Use what you discovered in Problem Set E to rotate the figure about point *P*, using 140° as the angle of rotation.

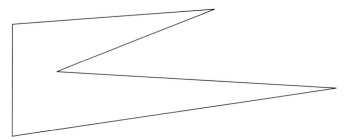

•*P*

2. Sketch this picture. Draw a copy of the figure rotated about point *Q*, using ⁻80° as the angle of rotation. Remember: A negative angle of rotation means to rotate clockwise.

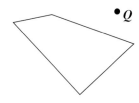

•*Q*

3. When you rotate a figure, is the image *congruent* to the original, *similar* to the original, or *neither*? Explain how you know.

Share & Summarize

Suppose you have a protractor, a ruler, and a pencil, but no other paper. You are asked to perform a rotation of the segment at right, using the point as the center of rotation.

•

1. What additional information do you need?

2. Suppose you are given the information you need. How would you perform the rotation of the segment?

On Your Own Exercises

1. Copy this figure onto tracing paper. Then rotate the tracing until the copy matches with the original.

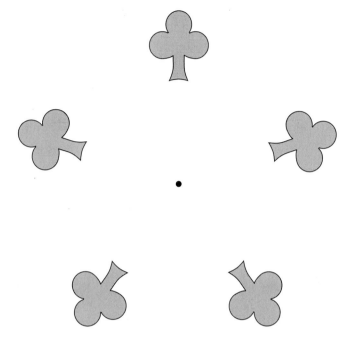

a. How many times do you have to rotate the tracing, matching the tracing to the original, to return it to its starting position?

b. What is the angle of rotation?

c. Describe how the angle of rotation is related to the number of identical elements.

2. Copy this figure onto tracing paper. Then rotate the tracing until the copy matches with the original.

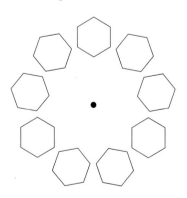

a. How many times do you have to rotate the tracing, matching the tracing to the original, to return it to its starting position?

b. What is the angle of rotation?

c. Describe how the angle of rotation is related to the number of identical elements.

Just the facts

Three letters of the alphabet that have rotation symmetry are shown below. Can you find the others?

S   O   N

3. Use this as a basic design element.

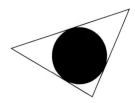

    a. Choose a center of rotation, and make a design with a 72° angle of rotation. Use tracing paper if you need it.

    b. Make another design using this basic design element and a 72° angle of rotation, using a different center of rotation.

4. Copy $\overline{AB}$ and point *P.*
Create a design with rotation
symmetry by rotating the
segment 60° about point *P*
several times.

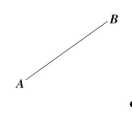

5. Copy △*ABC*, and rotate it ⁻60° about point *Q.* Remember: A negative angle of rotation means to rotate clockwise.

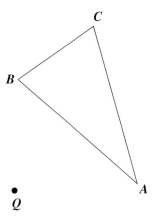

Do your work for Exercises 6 and 7 without using tracing paper.

    6. Draw a quadrilateral *ABCD.* Choose a point to be the center of rotation, and rotate your quadrilateral 80° about that point. Use prime notation to label the image vertices. For example, the image of vertex *A* will be vertex *A′*.

    7. Draw a pentagon *EFGHI.* Choose a point to be the center of rotation, and rotate your pentagon ⁻130° about the center. Use prime notation to label the image vertices. For example, the image of vertex *E* will be vertex *E′*.

Connect & Extend

8. Make a snowflake as you did in Problem Set A (or use the one you've already made). Unfold it completely.

   a. How many lines of symmetry does your snowflake have?

   b. What is the relationship between your lines of folding and the lines of symmetry?

   c. Fold your snowflake again. Choose one design that you cut out, and make a sketch of it. Unfold the snowflake and look for that design, using your sketch for comparison. How many copies of it can you find? Where are they?

   d. What is the angle of rotation for your snowflake?

9. Three-dimensional objects can have rotation symmetry. A three-dimensional object with rotation symmetry has an *axis of rotation* instead of a center of rotation.

While a two-dimensional figure can have only one center of rotation, a three-dimensional object can have more than one axis of rotation. This rectangular prism has three axes of rotation.

   a. The prism below has bases that are equilateral triangles. How many axes of rotation does this triangular prism have? Explain where they are. (You may use a diagram to explain, if necessary.)

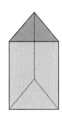

Remember
A prism is named for the shape of its bases.

   b. For each axis of rotation you found, what is the angle of rotation?

10. Create your own design with rotation symmetry.

   a. What is the angle of rotation?

   b. How many identical elements are there in your design?

   c. Does your design have reflection symmetry? If so, how many lines of reflection does it have?

11. List at least four natural or manufactured objects or figures that have rotation symmetry. Which of your examples also have line symmetry?

In your **own words**

Describe how rotation and reflection are similar and how they are different. Give examples.

12. If the angle of rotation for a figure with rotation symmetry is an integer, it is also a factor of 360. Consider what might happen if you tried to create a figure using an angle measure that isn't a factor of 360, such as 135°.

Choose a point A and a center of rotation. Rotate Point A 135°, and rotate the image 135°. Keep rotating the images until you return to the original point. (When you perform the rotations, you will pass the original point, because you have made one full turn.)

  a. How many full circles did you make?

  b. How many copies of the point do you have in your drawing?

  c. There is an angle of rotation smaller than 135° that you could have used to create this same design. What is its measure?

  d. Now find the *greatest common factor* (GCF) of 135 and 360.

  e. Divide 135 and 360 by your answer to Part d.

  f. Compare your answers for Parts a–c to your answers for Parts d and e. What do you notice?

  g. Suppose you created a figure by rotating a basic design element 80° each time. What angle of rotation will the final design have? Test your answer by rotating a single point.

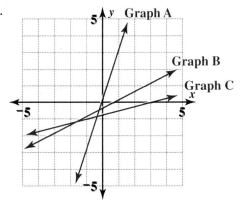

Remember

The *greatest common factor* is the greatest number that will exactly divide into two or more integers.

Mixed Review

Find the value of c in each equation.

13. $\sqrt[c]{8} = 2$ **14.** $\sqrt[c]{81} = 3$ **15.** $\sqrt[4]{1} = c$

16. Match each equation to a graph.

  a. $x - \frac{3}{5} = 2y$

  b. $3y - 4x - 1 = 8x - y$

  c. $x + 5 = 4(y + 2)$

17. Write an equation to represent the value of A in terms of t.

t	0	1	2	3	4
A	9	27	81	243	729

Translation and Combining Transformations

VOCABULARY
translation

You will now explore a third kind of transformation: a **translation.** You can think of translating a figure as moving it a specific distance in a specific direction. Unlike rotations and reflections, the image of a translation and the original figure have the same *orientation*—the top is still the top, and the bottom is still the bottom. Some people call a translation a *slide* or a *glide.*

MATERIALS
ruler

Think & Discuss

Look at this series of figures.

If the original figure is the one on the left, how far was it translated at each stage, and in what direction?

If the original figure is the one on the right, how far was it translated at each stage, and in what direction?

Investigation Translation

VOCABULARY
vector

To describe a translation, you need to give both a distance and a direction. *Vectors* can be used to describe translations. A **vector** is a line segment with an arrowhead. The length of the segment tells how far to translate, and the arrowhead gives the direction. The two vectors below, for example, indicate translations of the same length but different directions.

Translate this hexagon using the given vector.

First trace the hexagon and the vector. Extend the vector with a dotted line. This line will help you keep the figure's orientation.

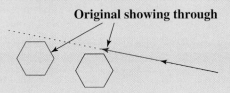

Now slide the *original* in the direction indicated by the vector until the *head* of the traced vector touches the *tail* of the original vector. Keeping the original vector under the dotted line on the tracing, trace the original hexagon and vector again.

Original showing through

You can repeat the process. When you slide the original hexagon and vector, position the original vector so that its tail touches the head of the second vector.

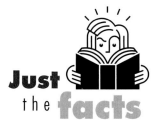

Just the facts

If you could continue this pattern forever, in both directions, the result would have translation symmetry.

Problem Set A

Translate the figure using the given vector to create a design with four copies of the figure.

1.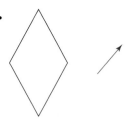

2.

3. Now draw your own figure and vector of translation. Use tracing paper to create a design with at least three copies of your figure.

4. Are a figure and its translated images *congruent, similar,* or *neither*? Explain.

Problem Set B

You will now examine what happens in a translation more closely. Triangle *ABC* has been translated by the given vector to create an image, △*A′B′C′*.

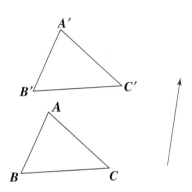

1. Trace the picture. On your copy, connect point *A* to its image, point *A′*. Also connect points *B* and *C* to their images.

2. Measure the lengths of $\overline{AA'}$, $\overline{BB'}$, and $\overline{CC'}$. Compare these lengths to the length of the vector.

3. Imagine extending each of $\overline{AA'}$, $\overline{BB'}$, and $\overline{CC'}$ to form lines. Suppose you also extended the vector to form a line. What would be true about these four lines?

4. Suppose you have a single point and a vector. How can you translate the point by the vector without using tracing paper?

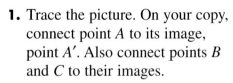

Just the facts

Each of the Japanese border designs below involves translation symmetry.

Share & Summarize

Explain what a *translation* is in your own words. Be sure to include what information is needed to perform one.

Investigation 2 ▶ Combining Transformations

What happens when you combine two transformations? For example, suppose you reflect a figure and then reflect its image.

MATERIALS

- tracing paper
- GeoMirror (optional)
- protractor
- ruler

Problem Set C

Consider what happens when you reflect over two lines that intersect. Copy each picture. Reflect the figure over line *l*, and then reflect the image over line *m*. You might want to use a GeoMirror to do the reflections.

1.

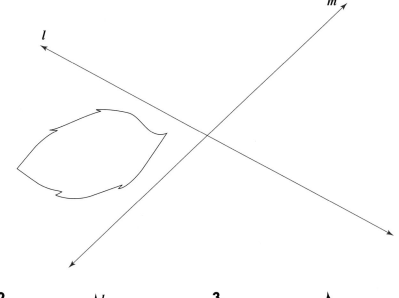

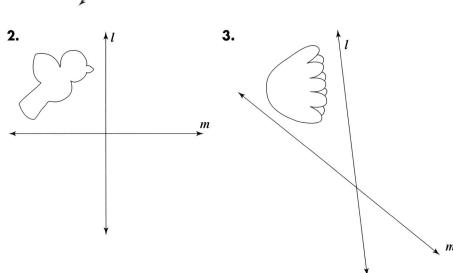

4. The final images in Problems 1–3 could have been created using only one transformation of the original figure. Describe that single transformation.

You have looked at combinations of two reflections. There are many other ways to combine transformations. Now you will explore combining a reflection and a translation.

MATERIALS

tracing paper

Problem Set D

1. Vector *c* is parallel to the line of reflection, line *m*. Trace △*FGH*, vector *c*, and line *m*.

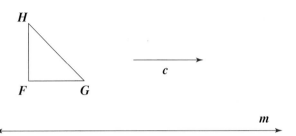

   a. Reflect △*FGH* over line *m*.

   b. Translate the image from Part a by vector *c* to get a second image.

2. Now try reversing the order of the transformations.

   a. Translate △*FGH* from Problem 1 by vector *c*.

   b. Reflect the image you created in Part a over line *m*.

3. What do you notice about the relationship between the final images in Problems 1 and 2?

The combination of a reflection over a line and a translation by a vector parallel to that line is called a *glide reflection*. Footprints in the sand, leaves on a stem, and patterns on wallpaper borders can show glide reflection.

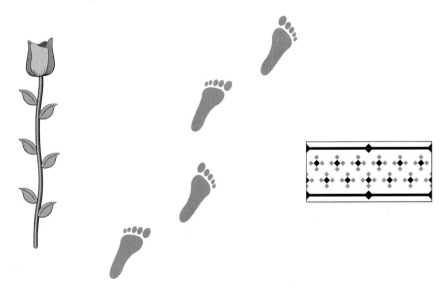

Lab Investigation ▶ Making Tessellations

The Dutch artist M. C. Escher (1898–1972) often translated, rotated, or reflected a figure or a collection of figures to create fascinating images.

Just the facts

The mathematical themes of symmetry, filling the plane, and approaching infinity are prominent in many of Escher's most famous works.

Symmetry Drawing E71 by M. C. Escher. © 1999 Cordon Art–Baarn–Holland. All rights reserved.

A design using one or more shapes to cover the plane without any gaps or overlaps is called a *tessellation*. Many shapes will *tessellate*. One of the easiest shapes to tessellate is a square.

You can use a square and the techniques of reflection, rotation, and translation to create your own tessellation artwork.

M A T E R I A L S

• stiff paper squares
• scissors
• tape
• posterboard
• markers or crayons

Try It Out

1. Start with a paper square. On one side of the square, draw a shape. The shape should be a single piece that starts and ends on the same side. Here are some examples.

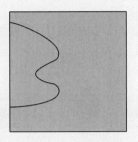 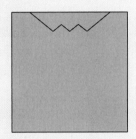 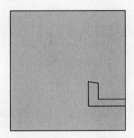

Carefully cut out the shape you drew, keeping it in a single piece.

2. You now have some options for moving your shape.

• You could translate it to the other side of the square.

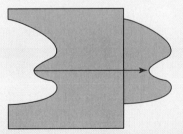

• You could translate it to the other side of the square, as above, and then reflect it over a line through the center of the square and parallel to the direction you translated.

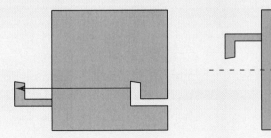

- You could rotate it 90° about one of the vertices adjacent to your shape. The shape below has been translated about the upper-left corner of the square.

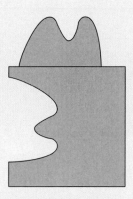

Choose one of these three options, and tape the shape to the square at the place where you moved it.

3. Now you can create a tessellation. Place your figure on a large sheet of paper or posterboard and trace around it. Then decide how to move your figure (by translation, rotation, or some combination of transformations) so that the cutout on the tracing fits together with the attached piece on your figure. Trace the figure again.

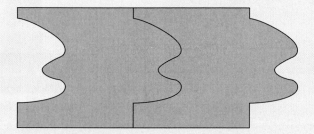

4. Repeat the process several times, filling your paper or posterboard with your shape. You have a tessellation!

Try It Again

To create more interesting tessellations, you can cut shapes from two sides of a square. After creating the tessellation, you can color each figure to look like birds, fish, people—or just about anything else you can imagine.

5. Start with a new square. On one side of the square, draw a shape. Cut the shape out and translate, translate and reflect, or rotate it as described in Step 2. Reattach the shape to the square.

6. You have now used two of the four sides of your square. Choose one of the other sides and draw a new shape. Cut the shape out and translate, translate and reflect, or rotate it. You will have to choose your rule based on which side of the square is left to attach the shape to.

For example, here is a translation of one shape, and then a translation and reflection of another shape.

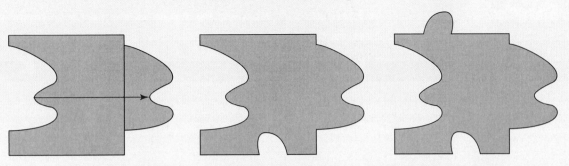

7. Trace your figure onto a large sheet of paper or posterboard.

   a. Decide how to move the shape so that the *first* cutout on the tracing fits correctly with the attached piece. Trace the figure again.

   b. Now move the shape so that the *second* cutout fits correctly, and trace it.

   c. Continue the process until you have created a tessellation. Decorate the figures however you like.

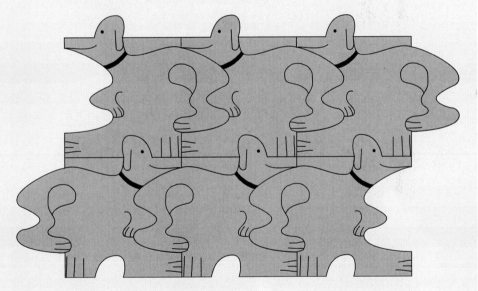

What Have You Learned?

8. Suppose you *translate* a cutout to create a figure. How would you move the figure to make a tessellation?

9. Suppose you *translate and then reflect* a cutout to create a figure. How would you move the figure to make a tessellation?

10. Suppose you *rotate* a cutout to create a figure. How would you move the figure to make a tessellation?

On Your Own Exercises

Practice & Apply

In Exercises 1–3, translate the figure by the vector to create a design with three elements.

1.

2.

3.

4. Suppose someone asked you to reflect this figure over the line, and then to reflect the image over the same line. Describe a simpler way to find the final image.

5. Carefully copy this picture, including lines *l* and *m*, which are perpendicular.

   a. Reflect the figure over line *l* and its image over line *m*.

   b. Describe the single transformation that would give you the same final image. Give as much detail as you can.

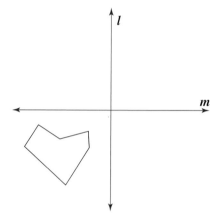

6. Carefully copy this picture.

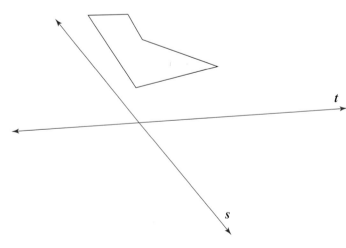

   a. Reflect the figure over line *t* and reflect its image over line *s*.

   b. Describe the single transformation (reflection, rotation, or translation) that would give you the same final image. Give as much detail as you can.

impactmath.com/self_check_quiz

7. Carefully copy this picture.

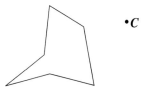

a. Rotate the figure about point C through an angle of 30°. Then rotate the image about point C through an angle of 40°.

b. What single transformation of the original figure would give the same final image as in Part a?

c. Now rotate the original figure about point C through an angle of 30°, and rotate the image about point C through an angle of ⁻40°.

d. What single transformation of the original figure would give the same final image as in Part c?

8. Carefully copy this picture, including the line of reflection and the vector. Follow the directions to perform a glide reflection.

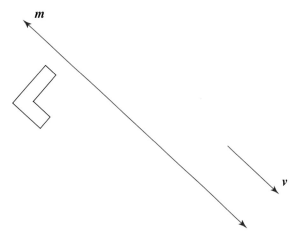

a. Reflect the figure over line m.

b. To complete the glide reflection, translate the image by vector v.

c. Take the final image from Part b and reflect it over line m. Then translate that figure by vector v.

d. Your image in Part c can be created by a single transformation of the original. Give as much information as possible about that transformation.

Connect & Extend

In Exercises 9 and 10, an original figure (colored orange) has been translated. Find the length and direction of translation. Record it as a vector on your paper.

9.

10.

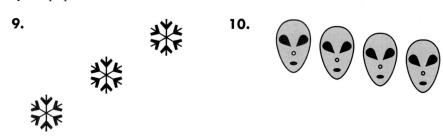

11. Tala and Evan were talking about this figure.

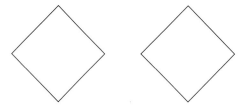

Tala said it was created using a reflection, but Evan thought it was created using a translation.

a. Could Tala be right? Try to find a line of reflection that would work.

b. Could Evan be right? Try to find a vector of translation that would work.

c. Here is a pair of figures that could have been created by either translation or reflection. Find the line of reflection and a vector of translation that would work.

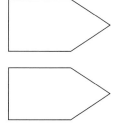

d. Create a design that could be created by translation or reflection.

e. A figure must have a particular kind of symmetry to look the same reflected or translated. Look back over the basic elements in this exercise. What kind of symmetry is necessary?

12. In Parts a–c, an equation is given. The three equations are graphed at right. Draw what each graph would look like translated by the given vector.

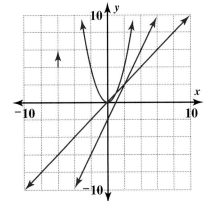

a. $y = x$

b. $y = x^2$

c. $y = 2x - 2$

d. For Parts a–c, write the equation of the new graph you created.

13. Suppose you reflected this figure over line p and reflected its image over line r.

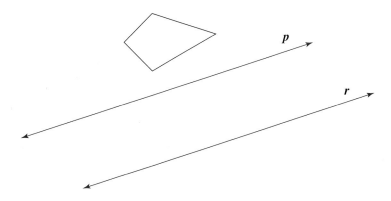

What single transformation—a *reflection*, a *rotation*, or a *translation*—would give you the same final image?

14. Sports Jesse wants to impress her friends with her skill at billiards. She clears the table of all the balls except the cue ball.

When a ball *banks*, or bounces, off a side of the table, the angle it makes as it travels away from the side is the same as the angle it made when it approached the side.

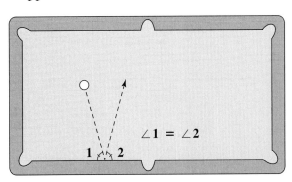

$\angle 1 = \angle 2$

a. The lines along which the ball travels are related by reflection symmetry. Use the figure on page 325 to find the line of symmetry.

b. Jesse challenged her friend Marcus to sink the cue ball into a side pocket after banking it off the sides of the table exactly twice. Marcus made the following shot, which failed.

Carefully copy the table below, and find the path that the ball traveled. Marcus tried to put the ball in the marked side pocket.

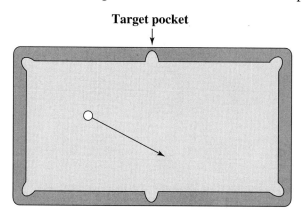

Target pocket

c. Jesse used her understanding of reflections to estimate a shot that would work. Complete these steps to find the shot Marcus should have used.

- On another sheet of paper, copy the table and ball from Part b.

- Imagine that the inner edge of the right side of the table is a line of reflection. Find the image of the target side pocket when you reflect over that line.

- Now imagine that the inner edge of the bottom side of the table is a line of reflection. Reflect the image from the previous step over that line.

- Draw a line connecting this final image to the cue ball. This line shows the direction in which the shot should be made.

- Verify the shot by finding the path the ball would follow.

15. Carefully copy this picture.

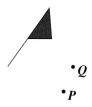

• *Q*

• *P*

a. Rotate the flag 50° about point *P* and then rotate the image ⁻30° about point *Q*.

b. Extend the flagpoles of the original flag and the final image until the two segments intersect. What is the measure of the angle between them?

c. Now rotate the original flag through the angle measure you gave in Part b, using the point of intersection you created in Part b.

d. Now rotate the original flag once more. Use the same angle measure from Part b, but this time use *any point* you haven't used as a center of rotation. (For example, use one of the vertices of the flag.)

e. Compare the images in Parts c and d, and the final image in Part a. What do you notice about their orientations?

16. Challenge Carefully copy this picture.

a. Rotate the flag 90° about point P. Then rotate the image $^-90°$ about point Q. What single type of transformation of the original flag would give the same final image?

b. Make a new copy of the picture. Rotate the flag 90° about point P, and then rotate the image 90° about point Q. What single transformation of the original flag would give the same final image?

c. Now rotate the original flag 50° about point P, and then rotate the image $^-30°$ about point Q. (If you completed Exercise 15, you can refer to your rotation in that exercise.) Extend the flagpoles of the original flag and the final image, until the two segments intersect. What is the measure of the angle between them?

d. Add the two angles of rotation in Part a. Then do the same for the pairs of angles in Parts b and c.

e. How are these sums connected to your answers to Parts a, b, and c?

Mixed Review

Find the value of m in each equation.

17. $\sqrt[m]{512} = 8$

18. $\sqrt[3]{1,331} = m$

19. $\sqrt[3]{m} = 7$

20. $\sqrt[4]{m^4} = 10$

Simplify.

21. $\sqrt{50}$ **22.** $\sqrt{150}$ **23.** $\sqrt{162}$

24. $\sqrt{210{,}000}$ **25.** $\sqrt{147}$ **26.** $\sqrt{448}$

27. $2\sqrt{72}$ **28.** $0.1\sqrt{68}$ **29.** $^{-}1.1\sqrt{171}$

30. Recall that you can use the distance formula to find the length of a segment, where (x_1, y_1) and (x_2, y_2) are the coordinates of the endpoints.

$$\text{distance} = \sqrt{(x_2 - x_1)^2 + (y_2 - y_1)^2}$$

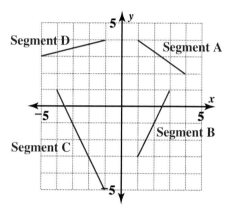

Use the distance formula to find the length of each segment.

a. Segment A

b. Segment B

c. Segment C

d. Segment D

31. Devon said, "I thought of a number and subtracted 5. I multiplied the answer by 6 and divided that result by $\frac{1}{2}$. The answer was 36. What was my number?"

32. Ian said, "I thought of a number, doubled it, and then added 10. I multiplied the answer by $^{-}0.5$ and divided that result by 2. The answer was $^{-}3.5$. What was my number?"

33. A particular type of sunflower grows to an average height of 7 to 10 feet. Write an inequality to express the average height h of this type of sunflower in inches.

34. Boxed-In sells cube-shaped shipping boxes with edge lengths ranging from 12 centimeters to 1.5 meters. Write an inequality to represent the range of possible volumes v, in cubic meters, for these boxes.

35. Ninety-six square tiles cover a 6-foot-by-12-foot area.

    a. Write an equation that would help you find the side length of a tile, s.

    b. What is the side length of one of the tiles, in inches?

5.4 Dilation

VOCABULARY
dilation

You have studied three transformations that create congruent figures. The fourth transformation, **dilation** (which is sometimes called *scaling*), creates figures that are similar, but not necessarily congruent, to the original.

You might recall that matching, or *corresponding,* sides of similar polygons like the triangles below are in proportion. That is, the side lengths share a common ratio. Corresponding angles of similar figures are congruent.

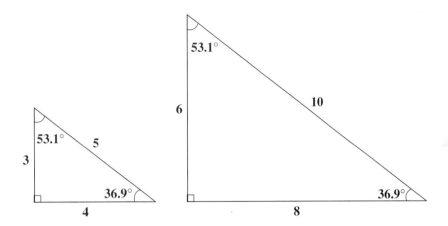

Remember
Congruent angles have the same measure.

In this lesson, you will learn two ways to create similar figures by dilating.

Explore

MATERIALS
- graph paper
- ruler
- protractor

Copy the outline of a cat's head onto a coordinate grid.

- Find the coordinates of each point.

- Multiply the coordinates of each point by 2, creating points A' through H'. In other words, if the coordinates of point A are (x, y), the coordinates of point A' are $(2x, 2y)$.

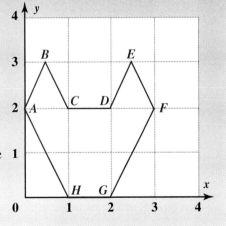

- Plot these new points on the same grid, and connect them in the same order.

You have just *scaled* or *dilated* the original cat's head. Is the new, dilated figure similar to the original? How do you know?

Investigation ▶ 1 ▶ Scale Drawings

VOCABULARY

scale drawing
scale factor

When you dilated the cat's head, you were using the *coordinate method* to create a similar figure. You can use this method—which involves multiplying coordinates of vertices by a number—to make *scale drawings* of figures on a coordinate grid. A **scale drawing** is a drawing that is similar to some original figure. The **scale factor** is the ratio between corresponding side lengths of the similar figures.

Every pair of similar figures of different sizes has *two* scale factors associated with it. One describes the dilation from the small figure to the large figure; the other describes the dilation from the large figure to the small figure.

For example, the lengths in your scale drawing of the cat's head are all twice the corresponding lengths in the original figure, so the scale factor from the small figure to the large figure is 2. And since the lengths in the small figure are $\frac{1}{2}$ the lengths in the large figure, you can also say that the scale factor from large to small is $\frac{1}{2}$.

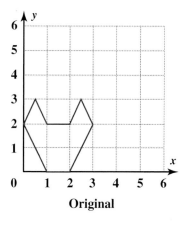

Original

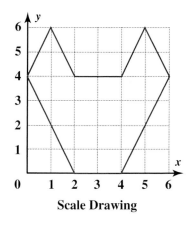

Scale Drawing

Problem Set A

1. Dilate the figure below by a scale factor of $\frac{1}{3}$.

2. Dilate the figure below by a scale factor of 3.

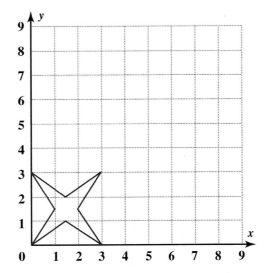

Some computer drawing and animation programs use a technique similar to the coordinate method to create scale drawings. The software treats the screen as a coordinate plane and calculates the placement of points. Slide projectors also make similar figures, but they work differently.

Slide projectors use a bright bulb and a series of lenses to project enlarged images onto a screen. The most important purpose of the lenses is to focus the light so that it comes from a single point. The focused light passes through the slide and onto the screen, spreading out to create a larger image.

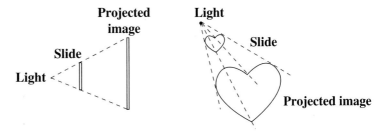

You can make scale drawings using this concept by applying a technique called the *projection method.*

Here's how to use the projection method to make a figure similar to polygon *ABCDE* with sides that are half as long—that is, to make a figure that is dilated by $\frac{1}{2}$.

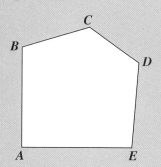

- First choose any point. Although the point can be on the polygon, you may find it easier to work with a point that is inside or outside of the figure, like point *F*.

  After you have chosen the point, draw segments from it to every vertex of the polygon.

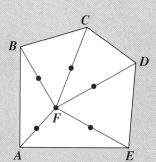

- To *halve* the polygon, find the *midpoints* of segments *FA* through *FE*.

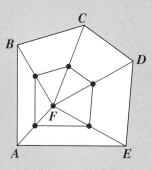

- Finally, connect the midpoints, in order, to form the new polygon.

A point that helps you to make a similar figure, such as point *F,* is called a *projection point.*

Remember
The *midpoint of a segment* is the point that lies halfway between the endpoints of the segment.

* tracing paper
* ruler

Problem Set **B**

1. Use the projection method to dilate this polygon by a factor of $\frac{1}{2}$.

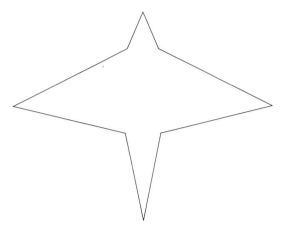

2. Below is a figure of a house.

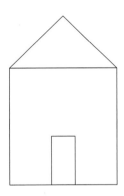

   a. Use the projection method to create a similar figure smaller than this original figure.

   b. Now modify the projection method to enlarge the original figure by a scale factor of 2.

Share & Summarize

You will now dilate this figure using both methods you have learned.

1. Dilate the figure by $\frac{1}{2}$ using the coordinate method.

2. Dilate the figure by $\frac{1}{2}$ using the projection method. Use the origin as the projection point.

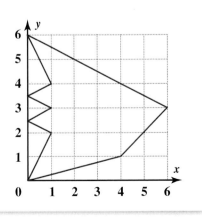

On Your Own Exercises

1. Santiago drew a picture using line segments on a coordinate grid. He then multiplied the coordinates of all the endpoints by 1.5, plotted the resulting points on a new grid, and connected them to form a new picture.

   a. One segment in Santiago's original drawing was 2 in. long. How long was the corresponding segment in the new drawing?

   b. One segment in the new drawing was 2 in. long. How long was the corresponding segment in Santiago's original drawing?

2. Imagine using the given scale factor on some Figure X to create a similar Figure Y. What scale factor would you use on Figure Y to create another figure the same size as Figure X?

   a. $\frac{1}{3}$ **b.** 5 **c.** 1

3. Consider how to reduce a figure by a factor other than $\frac{1}{2}$.

   a. Look back at the Example on page 332. Which step involves $\frac{1}{2}$ of a distance or length?

   b. How might you change that step to create a figure with sides $\frac{1}{3}$ as long as the original?

   c. Try it. Use the projection method to dilate this polygon by a factor of $\frac{1}{3}$.

impactmath.com/self_check_quiz

4. Dilate this figure by a factor of 2 using the coordinate method.

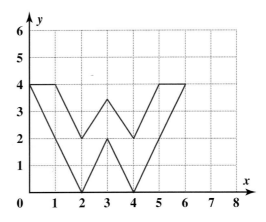

5. Use the projection method to scale this figure.

a. Dilate the figure by a factor of $\frac{1}{4}$.

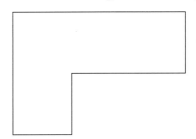

b. Dilate the figure you created in Part a by a factor of 4.

c. How does the figure you created in Part b compare to the original?

6. Consider this figure.

a. Shrink the figure using a scale factor of $\frac{1}{3}$.

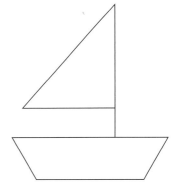

b. Now enlarge the figure from your answer for Part a using a scale factor of 6.

c. Compare your answer for Part b with the original figure. What single scale factor would change the original figure to this new one?

7. Draw a figure of your own using line segments on a coordinate grid. Enlarge your figure using the coordinate method. Tell the scale factor you used. Verify the scale factor from the original figure to the enlargement by checking at least two corresponding pairs of line segments.

8. Brian dilated the figure on the left to get the figure on the right.

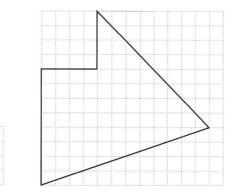

a. If Brian had dilated the figure in one step using only one scale factor, what scale factor did he use?

b. If Brian had dilated the figure in two steps, using two dilate factors, what pair of scale factors might he have used? List three possibilities.

9. What would you get if you used the coordinate method to dilate a picture by a factor of 0?

10. **Fine Arts** *Perspective drawings* look three-dimensional. The projection method for making scale drawings is related to a method for making perspective drawings.

On your own paper, follow the steps below to make a perspective drawing of a box. Use a pencil.

a. Start by drawing a rectangle. This will be the front of your box.

b. Choose a point outside your rectangle. This point is called the *vanishing point* for your drawing. Connect each vertex to that point, and then find the midpoint of each connecting segment.

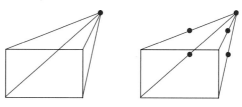

Just the facts

This technique of drawing in perspective was not known until the Renaissance. The Italian architect Brunelleschi, who lived about 600 years ago, invented it.

In your own words

Draw a simple picture, and make a dilated copy of it. Explain your method, being sure to say the scale factor you used. Then show how someone else could test that your two pictures are similar.

Remember
There are 5,280 feet in a mile.

In this painting, the tops and the bottoms of the trees form two lines that will intersect at a point called the *vanishing point*.

c. Connect the four midpoints you found in Part b to each other, in order. This gives you the back of the box. Then erase the lines connecting them to the vanishing point.

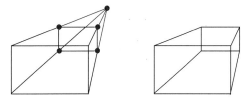

d. To make the box clearer, erase the lines that should be hidden on the back of the box, or make them dashed.

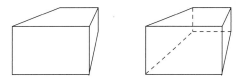

e. Follow the same steps to make a perspective drawing of a triangular prism. That is, start with a triangle (instead of a rectangle) and follow Part a–d.

f. In this method of three-dimensional drawing, at what step do you create a pair of similar figures? Explain.

11. Mile Square Park in Fountain Valley, California, is exactly 1 mile on each side. The city officials want to create a park map, showing visitors where playgrounds, drinking fountains, restrooms, and paths are located. They want the map to fit on a standard ($8\frac{1}{2}$ in. by 11 in.) sheet of paper.

Because the officials want the map to be easy to read, it should be as large as possible—that is, the scale factor from the map to the park should be as small as possible. If you consider only whole numbers for scale factors, what is the smallest possible scale factor they could use?

12. Technology Many photocopy machines reduce and enlarge figures automatically. However, copy machines often have only a limited number of scale factors. Suppose you are using a photocopy machine that has three settings for reducing or enlarging: 50%, 150%, and 200%.

a. Suppose you want to reduce a picture using a scale factor of $\frac{1}{4}$. How would you do it?

b. Suppose you want to enlarge a picture using a scale factor of 3. How would you do it?

c. How can you reduce a picture to 75% its original size?

Find the value of t in each equation.

13. $t^4 = 81$ **14.** $t^5 = 32$ **15.** $3^t = 729$ **16.** $4^t = 1,024$

17. Life Science In 1999, the world's tallest tree was an ancient redwood that stands in Montgomery Woods State Reserve in Northern California. The tree is 367.5 feet tall.

  a. The tree is estimated to be 600 to 800 years old. How many inches per year did the tree grow, on average, during its lifetime?

  b. A nearby redwood is 363.4 feet tall. What percentage of the tallest tree's height is this tree?

Write an equation to represent the value of y in terms of x.

18.

x	y
0	1.2
1	2.4
2	4.8
3	9.6
4	19.2

19.

x	y
1	21
2	63
3	189
4	567
5	1,701

20. Statistics Consider this set of data.

14.5   15.6   18.1   16.2   15.9

  a. Find the mean and the median of this data set.

  b. What two values can you add to the data set so that the median remains the same but the mean is higher?

  c. What two values can you add to the *original* data set so that the mean remains the same but the median is higher?

Geometry Write an expression for the volume of each cylinder.

21.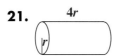

22.

Remember
The volume of a cylinder is the area of its base multiplied by its height.

5.5

Coordinates and Transformations

You already know how to use coordinates to graph relationships. Coordinates are also helpful for describing a position of a geometric figure and for performing transformations. In this lesson, you will learn how rules for coordinates can create reflections and rotations.

Explore

Copy the triangle below onto graph paper.

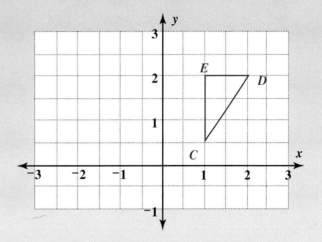

Transform each vertex of the triangle using this rule: $(x, y) \rightarrow (^-x, y)$. This rule says two things:

- The *x*-coordinate of the image is the opposite of the original *x*-coordinate.

- The *y*-coordinate of the image is the same as the original *y*-coordinate.

Describe how the image is related to the original figure: is it a *reflection*, a *rotation*, or a *translation*? Give as much information about the transformation—such as the line of reflection or the vector of translation—as possible.

Investigation Reflection and Rotation with Coordinates

Looking at figures on coordinate axes can help you think about reflections and rotations.

MATERIALS

- graph paper
- GeoMirror (optional)
- protractor (optional)
- tracing paper (optional)

Problem Set A

Problems 1–3 each show a figure on a coordinate plane and a rule to perform on the coordinates. For each problem, do Parts a–d to create an image of the figure.

    a. Explain the rule in words.

    b. Find the coordinates of each vertex, and copy the figure onto graph paper.

    c. Perform the given rule on the coordinates of each vertex to find the image vertices.

    d. On the same set of axes, plot each image point. Connect them in order.

1. Rule: $(x, y) \rightarrow (x, {}^-y)$

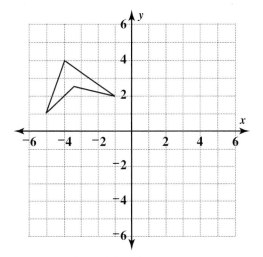

Tiny microcircuits such as this one replaced whole networks of thousands of transistors and other electrical components and made the personal computer possible. What kinds of symmetry can you see in this microcircuit?

2. Rule: $(x, y) \rightarrow (y, x)$

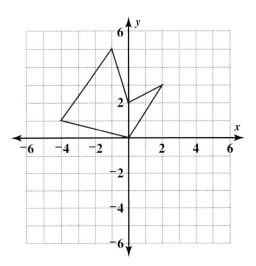

3. Rule: $(x, y) \rightarrow (^-y, x)$

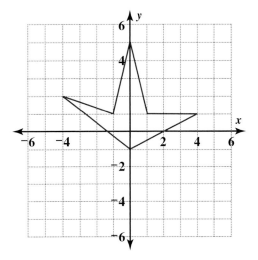

4. Compare the original figures and the images in Problems 1–3.

   a. For which problems can you get the image by *reflecting* the original figure? For each of these, give the line of reflection. You might want to use your GeoMirror to check.

   b. For which problems can you get the image by *rotating* the original figure? For each of these, give the center and the angle of rotation. You might want to use tracing paper and a protractor to check.

Problem Set **B**

In this problem set, you will find a rule for a given transformation.

1. Triangle ABC has vertices A (2, 5), B ($^-$2, 4), and C (0, 3). A certain transformation of these vertices gives A' ($^-$5, 2), B' ($^-$4, $^-$2), and C' ($^-$3, 0).

   a. Draw the two triangles on a grid.

   b. Is this a reflection or a rotation? If it's a reflection, give the line of reflection. If it's a rotation, give the center and angle of rotation.

   c. What is the image of (x, y) under this transformation?

2. Consider a reflection over the line $y = {}^-x$.

   a. Reflect the point (1, 0) over this line. What are the coordinates of the image?

   b. Reflect the point (0, 1) over the line. What are the coordinates of the image?

   c. If you reflect a point (x, y) over the line, what will the coordinates of its image be? Test your answer by reflecting $\triangle ABC$ from Problem 1 over the line and comparing the image's coordinates to the original figure's coordinates.

Share & Summarize

1. If you reflect the point ($^-$2, 4.3) over the y-axis, what will the coordinates of its image be?

2. If you rotate the point (1, 3) about the origin with a 90° angle of rotation, what will the coordinates of its image be?

3. Explain how you knew that the transformation in Problem 1 of Problem Set B was a rotation.

4. Explain how you wrote the rule for the reflection in Problem 2 of Problem Set B.

Investigation ▶2▶ Translation with Coordinates

In Investigation 1, you saw that you can perform a rule on the coordinates of a figure to rotate or reflect that figure in the plane. In a similar way, some rules will produce a translation of a figure.

M A T E R I A L S

graph paper

Problem Set C

Problems 1–3 each show a figure on a coordinate plane and a rule to perform on the coordinates. For each problem, follow these steps to create an image of the figure.

- Find the coordinates of each vertex, and copy the figure onto graph paper.
- Perform the given rule on the coordinates of each vertex to find the image vertices.
- On the same set of axes, plot each image point. Connect them in order.

1. Rule: $(x, y) \rightarrow (x + 2, y)$. That is, to get the image point, add 2 to the x-coordinate and leave the y-coordinate the same.

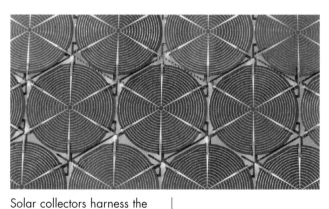

Solar collectors harness the sun's energy for such uses as heating homes and swimming pools. What kinds of symmetry can you see in this solar collector?

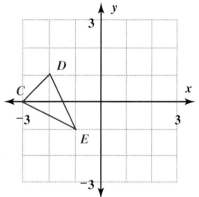

2. Rule: $(x, y) \rightarrow (x, y + 3)$

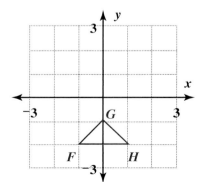

3. Rule: $(x, y) \rightarrow (x - 2, y - 2)$

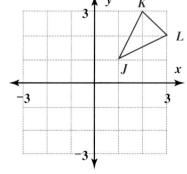

Think & Discuss

All the rules you have seen for translations have the same form: add some number to the *x*-coordinate, and add some number to the *y*-coordinate. The numbers added might be positive, negative, or 0.

What kind of numbers would you add to each coordinate of a point to move the point

- straight down (not to the left or right)?

- up and to the left?

Problem Set D

You will now practice writing rules to give a desired translation.

1. Here is a quadrilateral on a coordinate plane.

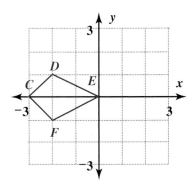

a. Copy the quadrilateral onto graph paper. Then draw an image of the quadrilateral that has been translated 2 units to the right and 1 unit down.

b. What rule performed on the coordinates would create this translation?

2. Here is a triangle on a coordinate plane.

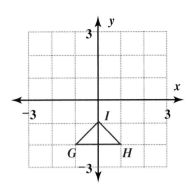

a. Copy the triangle onto graph paper. Then draw an image of the triangle that has been translated 2 units to the right and 1 unit up.

b. Write a coordinate rule to describe the translation in Part a.

c. Translate the *image* 2 units up and 3 units to the left.

d. Write a coordinate rule to describe the translation in Part c.

e. Compare the image you created in Part c with the original figure. Is there a single translation that would create that image from the original? If so, describe the translation, and explain how it relates to the two separate translations you performed.

Share & Summarize

1. Write a coordinate rule for a translation that moves a point 5 units to the right and 7 units down.

2. Make up your own rule for translating a figure. Write it as a coordinate rule, and show how it works on a copy of the figure below.

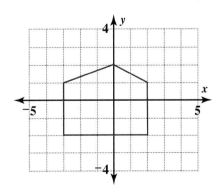

On Your Own Exercises

Practice & Apply

For Exercises 1 and 2, complete Parts a–e.

a. Explain the rule in words.

b. Find the coordinates of each vertex, and copy the figure onto graph paper.

c. Perform the given rule on the coordinates of each vertex to find the image vertices.

d. On the same set of axes, plot each image point. Connect them in order.

e. Compare the image to the original: is it a reflection, a rotation, a translation, or some other transformation?

1. Rule: $(x, y) \rightarrow (^-x, ^-y)$

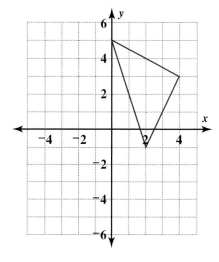

2. Rule: $(x, y) \rightarrow (^-x, y)$

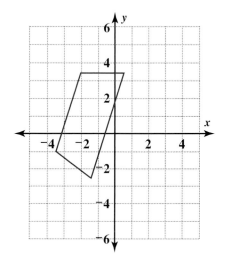

3. A rule has been applied to the original quadrilateral *ABCD* to create the image quadrilateral *A′B′C′D′*.

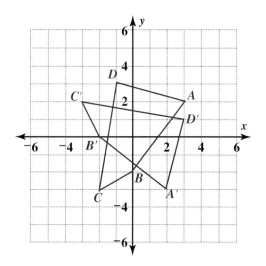

impactmath.com/self_check_quiz

a. Copy the table, and add the coordinates of each point. Remember, point A' is the image of point A, point B' is the image of point B, and so on.

Points	Original Coordinates	Image Coordinates
A, A'		
B, B'		
C, C'		
D, D'		

b. Write the rule that creates the image coordinates from the original coordinates.

c. Does the rule produce a reflection, a rotation, a translation, or some other transformation?

For Exercises 4 and 5, complete Parts a–d.

a. Find the coordinates of each vertex, and copy the figure onto graph paper.

b. Perform the given rule on the coordinates of each vertex to find the image vertices.

c. On the same set of axes, plot each image point. Connect them in order.

d. Compare the image to the original: is it a reflection, a rotation, a translation, or some other transformation?

4. Rule: $(x, y) \rightarrow (x + 2, y - 3)$ **5.** Rule: $(x, y) \rightarrow (x - 1, y - 1)$

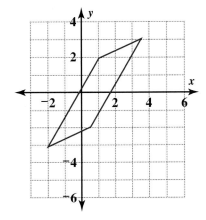   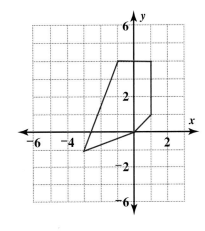

6. Suppose you translate a figure using this rule:

$$(x, y) \rightarrow (x + 2, y - 3)$$

You then translate the image using this rule:

$$(x, y) \rightarrow (x - 1, y - 1)$$

Where is the final image in relation to the original figure?

Connect & Extend

7. In Lessons 5.1 and 5.2, you may have conjectured that reflections and rotations produce figures congruent to the original figures. If you have a rule that produces a reflection or a rotation, you can use the distance formula to check that segments stay the same length.

a. Copy the grid and the segment below. Then use the rule $(x, y) \rightarrow (x, {}^{-}y)$ to create a new segment reflected over the x-axis.

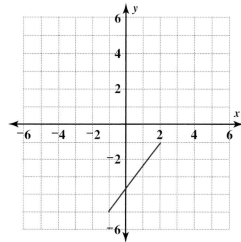

b. Use the distance formula to find the length of the original segment.

c. Use the distance formula to find the length of the image segment. Is it the same as the original?

d. Think about creating reflections using a GeoMirror or folding. Explain why the length of a segment would not change when you reflect it over *any* line.

Remember

To find the length of a segment, use the distance formula:

$$d = \sqrt{(x_2 - x_1)^2 + (y_2 - y_1)^2}$$

8. Consider what happens when you reflect a linear graph.

a. Graph the line $y = 2.5x + 4$.

b. On the same axes, draw the image of this line after reflection over the x-axis.

c. Write an equation of the new line.

d. On the same axes, draw the image of the original line after reflection over the y-axis.

e. Write an equation of the new line.

f. What do you notice about the two image lines you drew?

g. Do your equations in Parts c and e support your observation in Part f? Explain.

9. Consider what happens when you rotate a linear graph 180°.

a. Graph the line $y = 2x + 4$.

b. On the same grid, draw the image of the line under a 180° rotation centered at the origin.

c. What do you notice about the image line and the original line?

d. Write an equation of the new line.

e. Does your equation in Part d support your observation in Part c? Explain.

10. Draw a graph of $y = x^3$. Then draw the image of this line under a 180° rotation about the origin. What do you notice?

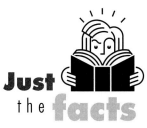

Just the facts

When you look into a flat mirror, light rays from you are reflected right back toward you—creating the image with which you are so familiar.

Remember
To find the length of segment, use the distance formula:

$$d = \sqrt{(x_2 - x_1)^2 + (y_2 - y_1)^2}$$

11. Here is another rule to perform on coordinates:

$$(x, y) \rightarrow (x + 0, y + 0)$$

That is, add 0 to both the x-coordinate and the y-coordinate. This is called the *identity transformation*.

a. Explain what the rule does.

b. Why do you think this transformation has the name *identity*?

c. The identity transformation is written above like a translation. What rotation would have the same result? That is, what angle of rotation could you use, and what center of rotation?

d. Is there a single reflection that would have the same result as the identity transformation? If so, draw a triangle and the appropriate line of reflection.

e. Is there a scaling that would have the same result as the identity transformation? If so, by what number would you multiply the coordinates?

12. In Lesson 5.3, you may have conjectured that translations produce figures congruent to the original figures. If you have a rule that produces a translation, you can use the distance formula to check that segments stay the same length.

a. Copy the grid and the segment at right. Then use the rule $(x, y) \rightarrow (x + 3, y + 1)$ to create a new segment translated to the right 3 units and up 1 unit.

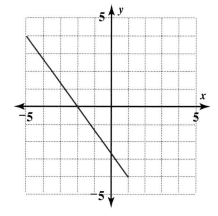

b. Use the distance formula to find the length of the original segment.

c. Now find the length of the image segment. Is it the same as the original?

d. Think about creating translations using tracing paper. Explain why the length of a segment would not change when you translate by *any* vector.

13. Copy this figure.

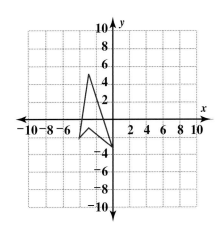

a. Use the rule $(x, y) \rightarrow (^-x, y - 3)$ to create an image of the quadrilateral.

b. Describe the transformation the rule performed. Is it a rotation, a reflection, a translation, a scaling, or a glide reflection?

Mixed Review

Tell whether the relationship in each table could be linear.

14.

x	0	1	2	3	4
y	2.2	0	$^-2.2$	$^-4.4$	$^-6.6$

15.

x	0	1	2	3	4
y	$^-2$	2	$^-4$	4	$^-6$

16.

x	0	1	2	3	4
y	4.1	13.1	22.1	31.1	40.1

17. Devon has 100 meters of fishing line. She cuts the line in half, stores one strand in her tackle box, and cuts the second strand in half again. She continues to make cuts in this manner.

a. Copy and complete the table for this situation.

Cut Number	0	1	2	3	4	5
Line Length (m)	100					

b. What kind of relationship is this?

18. Statistics For her fourteenth birthday, Fran's grandfather is giving her his comic book collection. Her grandfather explains that he has organized the comic books by year of publication into six categories often used by collectors. Fran counted the number of comic books in each category.

Category	Years Covered	Number of Comic Books
Pre-Golden Age	1896–1937	21
Golden Age	1938–1945	117
Post-Golden Age	1946–1949	32
Pre-Silver Age	1950–1955	93
Silver Age	1956–1969	67
Post-Silver Age	1970–present	23

a. Make a pie chart to represent the number of comic books in each category. Write the percentage of each section on the chart, rounded to the nearest 0.1%.

b. The collection contains 32 comic books for the four years in the Post-Golden Age, an average of 8 comic books per year. Which category contains the most comic books per year? Which contains the least? For each of these, what is the average number of comic books per year?

c. Do the percentages from your chart total 100%? If not, why do you think they don't?

Chapter Summary

In this chapter, you studied transformational geometry. You learned how to perform four transformations: *reflection, rotation, translation,* and *dilation.* You also learned how to recognize designs with both reflection and rotation symmetry.

Three of the translations—reflection, rotation, and translation—produce images that are congruent to the original figures. Dilation produces images that are similar but not necessarily congruent.

Strategies and Applications

The questions in this section will help you review and apply the important ideas and strategies developed in this chapter.

Recognizing reflection and rotation symmetry

Each of the figures in Questions 1–3 has reflection symmetry, rotation symmetry, or both. Copy each figure, and then do Parts a–c.

a. Determine the type or types of symmetry the figure has.

b. Indicate the line or lines of symmetry, if any, and the center of rotation, if any.

c. If the figure has rotation symmetry, determine the angle of rotation.

1.

2.

3.

Performing reflections

Three of the methods for reflecting figures that you learned are folding, using a GeoMirror, and using perpendicular bisectors.

4. Choose one of these three methods, and explain in your own words how to reflect a figure using that method.

5. A particular rule for reflecting a figure on a coordinate grid changes an original coordinate (x, y) to the image coordinate $(^-y, ^-x)$. What is the line of reflection for this rule?

Performing rotations

Two of the methods for rotating figures that you learned are using tracing paper and using a protractor and a ruler.

6. Choose one of these two methods, and explain in your own words how to use it to rotate a figure about a point with a given angle of rotation.

7. A particular rule for rotating a figure on a coordinate grid changes an original coordinate (x, y) to the image coordinate $(^-y, x)$. What are the center and angle of rotation for this rule?

Performing translations

8. Explain in your own words how to translate a figure by a given vector.

9. A particular rule for translating a figure on a coordinate grid changes an original coordinate (x, y) to the image coordinate $(x + 4, y - 3)$. On a coordinate grid, show the translation vector for this rule.

Performing dilations

In this chapter, you learned two methods for dilating figures: the coordinate method and the projection method.

10. Choose one of these two methods. Explain in your own words how to use that method to dilate a figure by a given scale factor.

11. A polygon has a vertex at the point (8, 6). A dilated version of the polygon has a corresponding vertex at the point (12, 9). What scale factor was used to dilate the original polygon?

Combining transformations

12. Figure Z is the image of Figure A.

a. Find a way to transform Figure A into Figure Z using *two* transformations.

b. Does the order in which you perform your transformations matter?

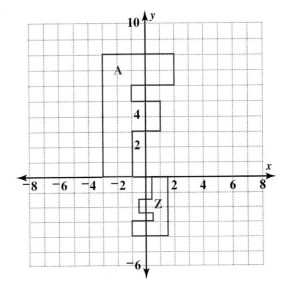

Demonstrating Skills

13. Copy this picture. Reflect the figure across the line using a method other than the one you described in Question 4.

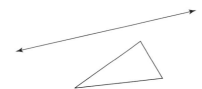

14. Copy this picture. Rotate the figure around point *P* with a $-40°$ angle of rotation. Use the method you did not describe in Question 6.

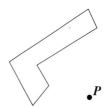

15. Copy this picture. Translate the figure by the given vector.

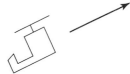

16. Copy this figure onto graph paper. Enlarge the figure by a scale factor of 3. Use the method you did not describe in Question 10.

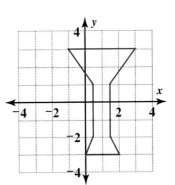

CHAPTER 6

Working with Expressions

Real-Life Math

Heavy Lifting Do you think you could lift a 10,000-pound elephant? If you could find a lever long and strong enough, you could! Using a lever allows you to apply more force to an object than you could with your bare hands.

Suppose the elephant is 4 feet from the fulcrum of the lever. Then, the amount of force F you would need to apply is $F = \dfrac{40,000}{d}$, where d is the distance from the fulcrum to where you apply the force.

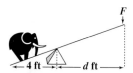

If your lever was as long as the world's tallest tree (about 368 feet), you would need to apply only about 110 pounds of force to the end of the lever to lift the elephant!

Think About It In the equation $F = \dfrac{40,000}{d}$, what do you think the number 40,000 represents?

Family Letter

Dear Student and Family Members,

Our class is about to begin a new topic of study in mathematics, one in which students create, work with, and simplify *algebraic expressions*. Much of the work of this chapter may feel familiar to you if you recall your own study of algebra.

Algebra is one of the most powerful tools of mathematics. Part of its usefulness—and even beauty—is the way algebra allows you to state and solve many problems with little effort. Here is an example:

> The cost of a movie is $4.50 for each adult plus half of the adult price for each of four children, less your $3 coupon. How much will you pay? The answer is $[4.50x + 2.25(4) − 3]$, where x is the number of adults.

In this chapter, we will begin by learning to multiply expressions like $x(x + 5)$ and $(x + 1)(x + 5)$ by referring to a *geometric model*. For example, to find the area of the large rectangle below, you can multiply the length by the width, or $x(x + 3)$. However, we can also find the area of the square, $x \cdot x$ or x^2, and add the area of the small rectangle, $x \cdot 3$ or $3x$, to get $x^2 + 3x$. This shows that $x(x + 3) = x^2 + 3x$.

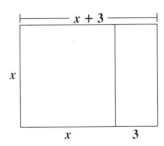

Vocabulary Along the way, we'll be learning about these new vocabulary terms:

binomial **expanding** **like terms**

What can you do at home?

Throughout this chapter, students will work with concrete situations and geometric models so that they can develop a sense of *why* their calculations with symbols work the way that they do. You can encourage a deeper understanding by asking your student to explain his or her work to you, using both geometric models and symbols.

6.1

Rearranging Algebraic Expressions

Ideas from geometry can sometimes shed light on certain concepts in algebra. In this investigation, you will look at a geometric model involving rectangles to help you work with and simplify algebraic expressions.

Think & Discuss

This rectangle can be thought of as a square with a strip added to one side. The large rectangle's width is 1 unit longer than its height.

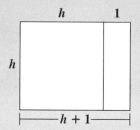

If you cut the large rectangle apart, you get a square and a small rectangle with the dimensions shown below.

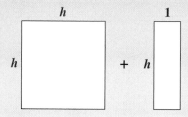

- What is the area of the square?

- What is the area of the small rectangle?

- Using the expressions you wrote above, write an expression for the area of the large rectangle.

- What does this tell you about $h(h + 1)$ and $h^2 + h$? Why?

Rectangle diagrams, like those above, are geometric models that can help you think about how the distributive property works. That is, they can help you understand why $a(b + c) = ab + ac$.

V O C A B U L A R Y

expanding

Using the distributive property to multiply the factors a and $(b + c)$ is called **expanding** the expression. For example, to expand the expression $2(x + 1)$, multiply 2 and $(x + 1)$ to get $2x + 2$. The expanded version of $h(h + 1)$ is $h^2 + h$.

Investigation ▶**1**▶ Using Geometric Models

In this investigation, you will use rectangle models to represent algebraic expressions.

Problem Set **A**

1. One of the rectangles in this diagram has an area of $x(x + 3)$.

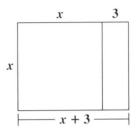

a. Copy the diagram, and indicate the rectangle that has area $x(x + 3)$.

b. Use your diagram to expand the expression $x(x + 3)$. Explain what you did.

2. Dante and Héctor are making another rectangle diagram.

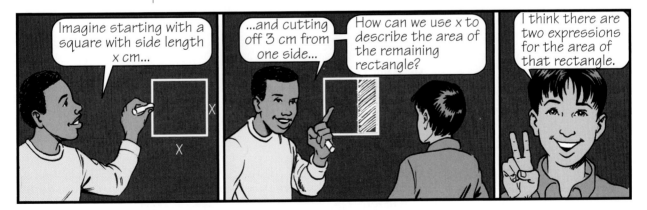

a. Answer Dante's question by writing an expression.

b. Is Héctor right? Explain your thinking.

3. In this diagram, the rectangle that is shaded has been removed from the square.

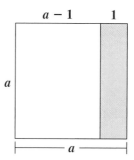

Explain two ways you could use the diagram to find an expression for the area of the unshaded rectangle. Give your expressions.

4. Use the distributive property to rewrite each expression. Then draw a rectangle diagram that shows why the two expressions are equivalent. Use shading to indicate when a region's area is being removed.

a. $b(b + 4)$

b. $m(m - 6)$

You will now explore more complex combinations of rectangles and the algebraic expressions they represent.

Problem Set B

1. Start with a square that has side length x. Create a large rectangle by adding another square of the same size and a 1-unit strip.

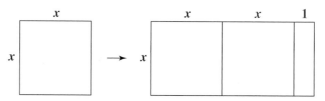

a. What is the height of the large rectangle?

b. Write a simplified expression for the width of the large rectangle.

c. Use the dimensions from Parts a and b to write an expression for the area of the large rectangle.

d. The large rectangle is composed of two squares and a smaller rectangle. Write an expression for the area of each of these parts. Then use the areas to write an expression for the area of the large rectangle, simplifying it if necessary.

e. Your two expressions for the area of the large rectangle, from Parts c and d, are equivalent. Write an equation that states this, and then use the distributive property to verify your equation.

2. Start with a square with side length x, add another square of the same size, and *remove* a strip with width 1.

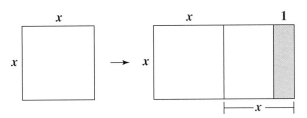

Use the diagram to help explain why $x(2x - 1) = 2x^2 - x$.

3. Now start with a square with side length x and make a rectangle with sides $2x$ and $x + 1$.

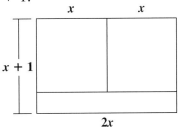

a. Use the distributive property to expand $2x(x + 1)$.

b. Use the diagram to explain why your expansion is equivalent to the original expression.

4. Draw a rectangle diagram that models each expression. Use your diagram to help you write the expression in a different form.

a. $2a(a - 1)$

b. $b^2 - 3b$

c. $c^2 + 2c$

5. Expand each expression.

a. $3a(a + 4)$

b. $2m(3m - 2)$

c. $4x(3 + 2x)$

Share & Summarize

According to the distributive property, $a(b + c) = ab + ac$. Use a rectangle model to explain why the distributive property makes sense.

Investigation 2 ▶ Simplifying Expressions

In Investigation 1, you learned about expanding expressions by removing parentheses. To make expanded expressions easier to use, you will sometimes want to shorten, or simplify, them.

Think & Discuss

This floor plan is for a living room. Imagine that you want to buy new carpeting for the room. All dimensions are in feet.

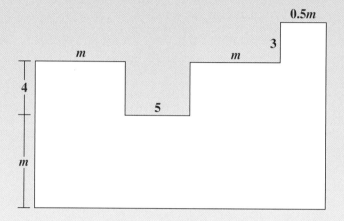

• Write an expression for the area of the floor.

• Tamika wrote this expression for the area of the floor:

$$m^2 + 4m + 5m + m^2 + 4m + 3.5m + 0.5m^2$$

Is her expression correct?

• Evaluate Tamika's expression for $m = 6$.

Tamika's area expression has several terms, but you can write an equivalent expression that is easier to work with.

For example, consider this expression:

$$k + 4k^2 + 3 - 2k^3 + 2k - 16 - 6k^4 + 3k^2 + 7k^3 + 19k^8$$

Two of the terms are k and $2k$. You can reason that their sum is $3k$ even though you don't know what k stands for. A number plus twice that number is three times that number, no matter what the number is.

VOCABULARY
like terms

The parts k and $2k$ are called *like terms*. **Like terms** have the same variable raised to the same power; they can be added or subtracted and then written as a single term. For example, in the expression above,

$$4k^2 + 3k^2 = 7k^2 \qquad \text{and} \qquad {}^-2k^3 + 7k^3 = 5k^3$$

Similarly, 3 and $^-16$ are like terms because they are both constants (terms with no variable) and can be combined to give $^-13$. Since the terms $^-6k^4$ and $19k^8$ are unlike each other and unlike the other terms, they stand alone.

You can rewrite the expression more simply as

$$19k^8 - 6k^4 + 5k^3 + 7k^2 + 3k - 13$$

Notice that, in the above expression, the terms are ordered by the exponent on the variable k.

Problem Set C

1. Which of these expressions are equivalent to the expression $p + 2p - p + 6 - 3 + 2p$?

$3p \qquad 7p \qquad 4p + 3 \qquad 6p + 3 \qquad 2p + 3 + 2p \qquad 4p - 3$

2. Which of these expressions are equivalent to the expression $y(2y + 3) - 5 + 2y - 2 + 3y^2 + 7$?

$10 \qquad 2y^2 + 3y^2 + 5y \qquad 10y^2 \qquad 12y + 14 \qquad 5y^2 + 5y$

Evan tried to simplify the expressions below but made some errors. For each, tell whether the simpler expression is correct. If it's not, identify Evan's mistake and write the correct expression.

3. $x + x + 7 = 2x + 7$

4. $m^2 + m^2 - 4 = m^4 - 4$

5. $2 + b + b^2 = 2 + b^3$

6. $3 - b^2 + b(b + 2b) = 2b^2 + 3$

7. Copy the expression you wrote for the area of the floor in the Think & Discuss on page 362.

    a. Write the expression as simply as you can.

    b. How many square feet of carpet do you need if m is 6 ft?

    c. In the Think & Discuss, you evaluated Tamika's expression for $m = 6$. Which was easier, evaluating Tamika's expression or your expression from Part a?

8. Lana and Keenan simplified $5a^2 + 10 - 4a^2 - 5 + 3a^2$ to $3a^2 + 5$. Lana checked the answer by substituting 0 for a. She found that both expressions equal 5 when a is 0, and she concluded that they are equivalent.

Keenan asked, "But what happens when we substitute 2?" Using $a = 2$, he found that the first expression equals 21 and that the simplified expression equals 17.

    a. Did Lana and Keenan simplify correctly? Explain.

    b. Lana and Keenan tested the equivalence of the expressions by substituting the same value into each expression to see whether the results were equal. Do you think this test should work? What did you learn from the results of their tests?

M A T E R I A L S

graphing calculator

Remember

Subtracting a number is equivalent to adding its opposite; for example, $3 - 5 = 3 + {}^{-}5$.

Problem Set **D**

Write each expression as simply as you can.

    1. $3(x + 1) + 7(2 - x) - 10(2x - 0.5)$

    2. $3a + 2(a - 6) + \frac{1}{2}(8 - 4a)$

    3. $3y + 9 - (2y - 9) - y$

    4. $(x^2 - 7) - 2(1 - x + x^2)$

In this addition chart, the expression in each white cell is the sum of the first expressions in that row and column. For example, the sum of a^2 and $a(a - 1)$ is $a^2 + a(a - 1) = a^2 + a^2 - a = 2a^2 - a$.

+	a	$a(a - 1)$
$a - 1$	$2a - 1$	$a^2 - 1$
a^2	$a^2 + a$	$2a^2 - a$

5. Copy and complete this chart by finding the missing expressions.

+		$2a(a - 5)$	$a(2a + 1)$
		$2a^2 - 9a$	
$a(a + 1)$	$a^2 + 2a + 3$		
			$3a^2 + 5a$

6. By completing the expression below, create an expression that simplifies to $4x - 3$.

$$3(x^2 + x - 2) + 2(\underline{\hspace{3cm}})$$

7. Lucita and Mikayla are discussing this equation.

$$y = 2x^2 + 5x + 4 - 2(x^2 + 1) - 3x$$

Mikayla says it is a quadratic equation because it has x^2 terms. Lucita graphed the equation and thinks it is not quadratic.

a. Graph the equation. Based on the graph, what kind of relationship does this equation appear to represent: linear, quadratic, cubic, reciprocal, exponential, or something else?

b. How can you tell for certain what type of relationship this equation represents?

Share & Summarize

Try to stump your partner! Write an expression that simplifies to one of these three expressions.

$$3x - 1 \qquad 5x + 2 \qquad 3x^3 - 7x - 2$$

Include at least five terms in your expression, let no more than two terms be single numbers, and include some terms with variables raised to a power. When you are done, swap with your partner, and figure out which expression above is equivalent to your partner's.

Lab
Investigation ▶ Making the Cut

In this H-shape, a and b are positive numbers and the angles are all right angles.

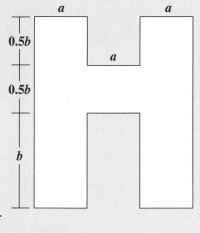

Analyze the H-Shape

1. On graph paper, draw an H-shape in which $a = 1$ cm and $b = 2$ cm.

2. Find the perimeter and the area of your H-shape.

3. Write an algebraic expression for the perimeter of your H-shape in terms of a and b.

4. Write an expression for the area of your H-shape in terms of a and b.

5. Use your expression from Question 3 to calculate the perimeter of your H-shape when $a = 1$ cm and $b = 2$ cm. Does it agree with your answer to Question 2?

6. Now use your expression from Question 4 to calculate the area of your H-shape when $a = 1$ cm and $b = 2$ cm. Does it agree with your answer to Question 2?

Transform the H-Shape

7. Create a new figure—different from the H-shape—with an area the same as the H-shape shown above but a different perimeter.

   a. Draw your figure.

   b. Write an algebraic expression for your figure's area. Is it equivalent to the expression for the area of the H-shape from Question 4?

   c. Write an expression for your figure's perimeter.

   d. The expression for the perimeter of the figure you drew probably looks very different from the perimeter expression for the H-shape. Try to find values of a and b for which the perimeters of the two figures are the same. If you find such values, does that mean the perimeters of the general shapes are equivalent? Explain.

8. Now consider a figure that has the same perimeter as the H-shape but a different area.

 a. Draw such a figure.

 b. Write an expression for your figure's perimeter. Is it equivalent to the expression for the perimeter of the H-shape?

 c. Write an expression for your figure's area.

 d. Are there any values of a and b for which the areas of the two figures are the same? (To answer this question, you might want to write and try to solve an equation.)

Make a Rectangle

Copy the H-shape from page 366, cut it into pieces, and rearrange the pieces to form a rectangle. Keep track of the lengths of the sides of your pieces in terms of a and b. You will need this information later.

 9. Draw the rectangle you formed from the H-shape, and label the lengths of the sides.

Make a Prediction

 10. Without doing any calculations, think about how the perimeter of the original H-shape compares to the perimeter of your rectangle. Are they the same or different?

 11. Without doing any calculations, think about how the area of the original H-shape compares to the area of your rectangle. Are they the same or different?

Check Your Prediction

 12. Write an expression for the perimeter of your rectangle in terms of a and b. Check your prediction from Question 10. Are there specific values of a and b that would make the perimeters the same? Different?

 13. Write an expression for the area of your rectangle in terms of a and b. Check your prediction from Question 11. Are there specific values of a and b that would make the areas the same? Different?

What Did You Learn?

 14. Jenny thinks that if you increase the perimeter of a figure, the area must also increase. Write a letter to her explaining whether she is correct and why. You may want to include examples or illustrations.

Practice & Apply

1. Start with a square with side length x cm. Imagine extending the length of one side to 7 cm.

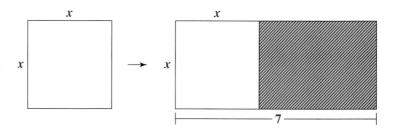

   a. What is the area of the original square? What is the area of the new, large rectangle?

   b. Use the areas you found to write an expression for the area of the striped rectangle.

   c. Now write expressions for the length and width of the striped rectangle.

   d. Use the dimensions from Part c to write an expression for the area of the striped rectangle.

   e. What do your answers to Parts b and d suggest about the expansion of $x(7 - x)$?

2. Use the distributive property to expand $3x(x - 2)$. Then draw a rectangle diagram and use it to help explain why the two expressions are equivalent.

Use the distributive property to expand each expression.

3. $3z(z + 1)$

4. $\frac{1}{2}x(x - 2)$

5. $t(2 - t)$

Draw a rectangle diagram to match each expression, and use it to write the expression in factored form.

6. $2x^2 + x$

7. $2x^2 - x$

impactmath.com/self_check_quiz

In your
own
words

Describe some of the steps you take when you simplify an algebraic expression, and explain how you know when an expression is simplified as much as possible.

8. This diagram shows the grassy area between two buildings and the rectangular walkway through the middle of the area. The length of the grassy area is four times its width. The edges of the walkway are 2.5 meters from the sides of the rectangle.

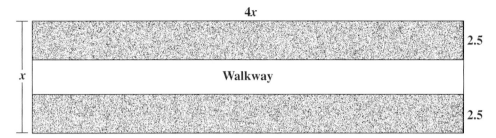

$4x$

2.5

x Walkway

2.5

Write two expressions for the area of the walkway, one in factored form and one in expanded form.

Write each expression as simply as you can.

9. $2a(a + 2) - 4a + 3$

10. $n(n + 1) - n$

11. $x(3 - 2x) + 2(x^2 - 4)$

12. $p(3p - 4) - 2(3 - 5p)$

13. $n(n^2 - 1) - n(1 - n)$

14. $q - 3q - 4q(1 - q)$

15. $2(c - 3) + c(c - 2)$

16. $2(c + 3) - c(c - 2)$

Simplify each expression and then tell whether it is linear, quadratic, cubic, or none of these.

17. $p(p + 1) - \frac{2}{p} + 2 + p^2 - \left(1 - \frac{2}{p}\right) - 1$

18. $w(1 - w) + 2w\left(\frac{1}{w}\right) - 2w - (1 - w^2)$

19. $6x - 2(1 + x) + 2\left(\frac{1}{x} - 1\right) - (4x - 1)$

20. Complete this expression to create an expression that simplifies to $x + 2$.

$$-4(x + x^2 - 1) + \underline{\hspace{3cm}}$$

21. Sort these expressions into groups of equivalent expressions.

    a. $5(x^4 - 1) - 10 - 2x^4 + 5 - 5x^2 + 2x^4 + 2x^2 + 7$

    b. $3x^5 + 2x^4 + 3x^2$

    c. $5x^5 + 2(x^4 - x^5) - 10 - 2x + 3x^2 + 2x + 3 + 7$

    d. $^-3x^2 - 3 + 5x^4$

    e. $x + 4 + 5(x + x^2) - 8 + 2x - 10x + 5 - 2x - 5x^2$

    f. $3(x^5 + x^2) + 2(x^4 + 3x) - 6x$

For each expression, copy the diagram and shade an area that matches the expression.

22. $(2x)^2$

23. $2x^2$

24. $x(2x + 1)$

25. $2x + 2$

26. $(2x + 2)^2$

27. $x(2x + 2)$

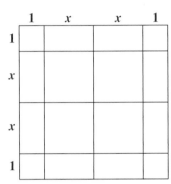

Expand each expression.

28. $x(a - b + c)$ **29.** $\frac{k}{7}(21a - 0.7)$

30. $\frac{x}{3}\left(\frac{a}{2} - \frac{b}{3} + \frac{c}{4}\right)$ **31.** $4w(4w - 2x - 1)$

32. This diagram shows a drawer and the surrounding cabinet. The drawer is 2 inches wider than it is tall, and there is a 1-inch gap between the drawer and the outside of the cabinet on all sides. The length of both the drawer and the cabinet is y.

Write two equivalent expressions for the volume of the drawer, one in factored form and one in expanded form.

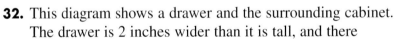

33. Ben and Lucita are discussing algebraic expressions.

a. After the 1-cm strip is removed, there are two ways to fold the remaining piece of paper. For each possibility, write an expression for the area of the final piece (after the paper is folded and then cut in half). Are the expressions equivalent? Explain.

b. Ben posed a new problem: *Imagine a square with side length x. Fold the square in half, cut it, and throw away one half. Now remove a 1-cm strip from the remaining half. Write an expression for the area of the remaining piece.*

Lucita said there are two ways to interpret Ben's instructions. Find both ways, and write an expression for each. Are the two expressions equivalent? Explain.

34. A construction worker made a stack of bricks 16 layers high. Each layer consists of three bricks arranged in the pattern shown at left below. Each brick is twice as long as it is wide and has a thickness $1\frac{1}{2}$ inches less than its width.

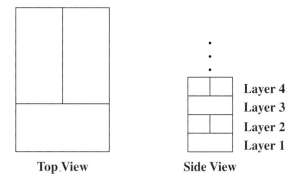

Top View   **Side View**

a. How many bricks are in the 16-layer stack?

b. Write an expression for volume of the stack, using w for the width of each brick.

35. Find an equation for the line through the points (3, 2) and (8, $^-$5).

Make a rough sketch showing the general shape and location of the graph of each equation.

36. $y = x^2 + 3$ **37.** $y = \frac{2}{x}$ **38.** $y = x^3 - 1$

Simplify each expression as much as possible.

39. $\sqrt{34}$ **40.** $\sqrt{99x^4}$ **41.** $^-\sqrt{60b}$

Use the distributive property to rewrite each expression without parentheses.

42. $^-(3n - 4)$ **43.** $3p(4 - p)$ **44.** $^-k(^-k - k)$

45. Probability The game of backgammon involves two players, one with black markers and one with white. Players can remove one of their opponent's markers from the board by landing on a space occupied by a single opposing marker. The opponent can return the marker to the board on one of six spaces, if any of the six holds no more than 1 of the first player's markers.

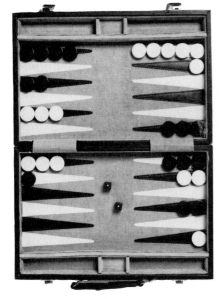

For example, on the board below, the white marker can enter if a 5 or a 3 is rolled. If a 3 is rolled, the black marker in that space is removed.

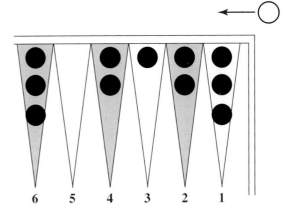

On each turn, a player rolls two standard dice.

a. What is the probability that the white marker can enter the board on this turn? That is, what is the probability of rolling either a 5 or a 3 on either of two dice?

b. What is the probability that the white marker will send out a black marker when it enters? That is, what is the probability of rolling a 3 on either die?

c. What is the probability that the white marker *cannot* enter the board on this turn?

6.2

Expanding Products of Binomials

In Lesson 6.1, you examined rectangle diagrams consisting of a square with a rectangular strip added to or taken away from one side. What if you were to add or remove *two* rectangular strips?

Think & Discuss

Start with a square with side length *m* cm. Add a 3-cm strip to one side.

Now add a 1-cm strip to an adjacent side of the new rectangle.

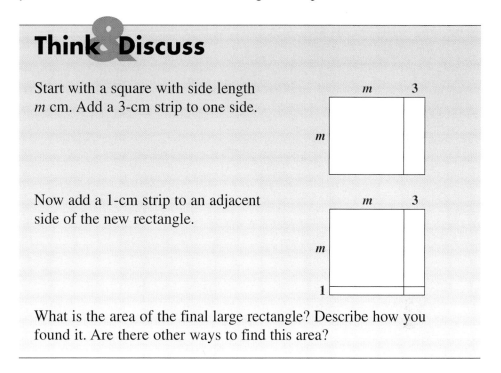

What is the area of the final large rectangle? Describe how you found it. Are there other ways to find this area?

VOCABULARY
binomial

In Lesson 6.1, you used the distributive property to multiply expressions in the forms $m(m + a)$ and $m(m - a)$. That is, you found the product of a number or variable and a binomial. A **binomial** is the sum or difference of two unlike terms.

The expression $x + 5$ is a binomial because it is the sum of two unlike terms. Similarly, $x^2 - 7$ is a binomial; it is the difference of two terms that can't be combined into one term.

Expressions such as $x^2 + x - 1$ and x^2 have more than or fewer than two terms, so they are not binomials. The expression $x + 2x$ is not a binomial either: its terms are like terms, and the expression is equivalent to $3x$.

The area of the final large rectangle in the Think & Discuss is the product of *two* binomials: $m + 3$ and $m + 1$. In this lesson, you will learn how to multiply two binomials.

Investigation ▶ 1 ▶ Using Geometric Models to Multiply Binomials

The geometric model you used to think about multiplying a term and a binomial can be adapted for multiplying two binomials.

Problem Set A

1. Look at this rectangle.

   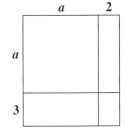

   a. Write two expressions, one for the length of the large rectangle and one for the width.

   b. Use your expressions to write an expression for the area of the large rectangle.

   c. Use the diagram to expand your expression for the area of the large rectangle. That is, write the area of the large rectangle without using parentheses.

2. Arturo wanted to expand $(x + 4)(x + 3)$. He drew a rectangle $x + 4$ units wide and $x + 3$ units high.

   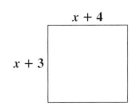

   He then drew lines to break the rectangle into four parts.

   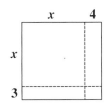

   Arturo then wrote: Area of large rectangle $= (x + 4)(x + 3)$.

   He then found the area of each of the four smaller parts and used them to write another expression for the area of the large rectangle. Finally, he simplified his expression by combining like terms. What was his final expression?

3. For Parts a–c, do the following:

 • Draw a rectangle diagram to model the product.

 • Use your diagram to help expand the expression.

 a. $(m + 7)(m + 2)$

 b. $(w + 2)^2$

 c. $(2n + 3)(n + 1)$

4. A certain rectangle has area $y^2 + 6y + 3y + 18$.

 a. Draw a rectangle diagram that models this expression.

 b. Use your diagram to help you rewrite the area expression as a product of two binomials.

5. Challenge Another rectangle has area $y^2 + 5y + 6$.

 a. Draw a rectangle diagram that models this expression.

 b. Use your diagram to help you rewrite the area expression as a product of two binomials.

Share & Summarize

Ben thinks that $(n + 3)(n + 5) = n^2 + 15$. Show him why he is incorrect. Include a rectangle diagram that models the correct expansion of $(n + 3)(n + 5)$.

Just the facts

At right is a geometric model of carbon-60. This molecule is composed of 60 interlinked carbon atoms arranged in 12 pentagons and 20 hexagons. Because of its structural similarity to the geodesic dome, designed by U.S. architect R. Buckminster Fuller, it was named *buckminsterfullerene.*

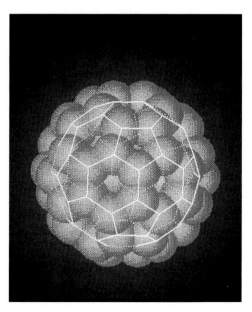

Investigation 2 ▶ Multiplying Binomials That Involve Addition

Rectangle models can help you understand how the distributive property works. For example, the expression $m(m + 3)$ is expanded below with the distributive property and a rectangle diagram.

$$m(m + 3) = m \cdot m + m \cdot 3$$
$$= m^2 + 3m$$

You can also use the distributive property and rectangle diagrams to expand such expressions as $(m + 2)(m + 3)$. Just think of $m + 2$ in the same way you thought about the first variable m in the expression $m(m + 3)$ above. That is, multiply $m + 2$ by each term in $m + 3$:

$$(m + 2)(m + 3) = (m + 2) \cdot m + (m + 2) \cdot 3$$
$$= m(m + 2) + 3(m + 2)$$

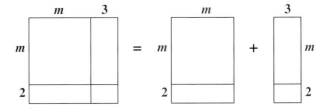

The distributive property can then be used to simplify each term. Start by simplifying the first term, $m(m + 2)$:

$$m(m + 2) = m \cdot m + m \cdot 2$$
$$= m^2 + 2m$$

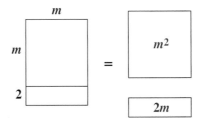

And then simplify the second term, $3(m + 2)$:

$$3(m + 2) = 3 \cdot m + 3 \cdot 2$$
$$= 3m + 6$$

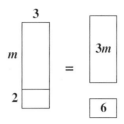

Finally, put everything together and combine like terms:

$$(m + 2)(m + 3) = m(m + 2) + 3(m + 2)$$
$$= m^2 + 2m + 3m + 6$$
$$= m^2 + 5m + 6$$

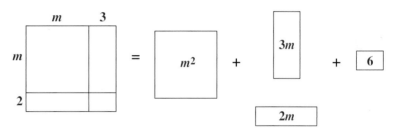

Problem Set B

For each problem, do the following:

- Expand the expression using the distributive property.
- Draw a rectangle diagram to model the expression, and check your expansion.

1. $(x + 3)(x + 4)$

2. $(k + 5)^2$

3. $(x + a)(x + b)$

Problem Set C

1. Complete the table by substituting values for x into this equation.

$$y = x(x + 1) - 3(x + 1)(x + 2) + 3(x + 2)(x + 3) - (x + 3)(x + 4)$$

x	0	1	2	3	4	5
y						

2. Make a conjecture about the value of y for other values of x. Test your conjecture by trying more numbers, including some decimals and negative numbers. Revise your conjecture if necessary.

3. Prove It! Use your knowledge of expanding binomial products to show that your conjecture is true.

Share & Summarize

Use the distributive property to expand $(a + b)(c + d)$. Draw a rectangle diagram, and explain how it shows your expansion is correct.

Investigation ▶3 Multiplying Binomials That Involve Subtraction

You have used rectangle models to think about multiplying binomials involving addition—expressions of the form $(a + b)(c + d)$. Now you will learn to expand products of binomials that involve subtraction.

EXAMPLE

Here's one way to create a rectangle diagram to represent $(d - 1)(d + 3)$.

First draw a square with side length d.

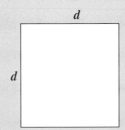

Then subtract a 1-cm strip from one side, and add a 3-cm strip to the adjacent side.

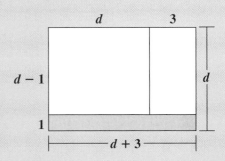

The unshaded rectangle that remains has an area of $(d - 1)(d + 3)$.

Think & Discuss

Expand the expression $(d - 1)(d + 3)$ using either the distributive property or the diagram in the Example on page 378. Describe how you found the answer.

Problem Set D

For each expression below, do the following:

- Expand the expression using the distributive property.
- Draw a rectangle diagram that represents the expression. Use shading to indicate areas that are being removed.
- Use your diagram to check that your expansion is correct.

1. $(b - 2)(b + 3)$

2. $(a + 1)(a - 4)$

3. $(2 + e)(3 - e)$

4. A certain rectangle has area $y^2 + 5y - 2y - 10$.

   a. Draw a rectangle diagram to represent this expression.

   b. Write the area of this rectangle as a product of two binomials.

You will now use the distributive property to expand products of two binomials that both involve subtraction.

Problem Set E

Expand each expression using the distributive property, and then combine like terms. Your final answer should have no parentheses and no like terms.

   1. $(x - 4)(x - 5)$

   2. $(R - 2)^2$

   3. $(2 - f)(3 - f)$

   4. $(a - 2b)(3a - b)$

You will now apply what you have learned about expanding products of binomials to analyze some number tricks.

Problem Set F

Remember

A *counterexample* is an example for which a conjecture does not work.

Lydia thinks she's found some number tricks.

Prove It! In Problems 1 and 2, determine whether Lydia's trick really works. If it does, prove it. If not, give a counterexample.

1. Lydia said, "Take any four consecutive integers. Multiply the least number and the greatest number, and then multiply the remaining two numbers. If you subtract the first product from the second, you will always get 2.

   For example, for the integers 3, 4, 5, and 6, the product of the least and greatest numbers, 3 and 6, is 18. The product of the remaining two numbers, 4 and 5, is 20. The difference between these products is 2." (Hint: If the least integer is x, what are the other three?)

2. Lydia said, "Take any three consecutive integers and multiply them. Their product is divisible by 4. For example, the product of 4, 5, and 6 is 120, which is divisible by 4."

3. Here's another number trick Lydia proposed: "Take any two consecutive even integers, multiply them, and add 1. The result is always a perfect square. For example, 4 and 6 multiply to give 24; add 1 to get 25, which is a perfect square."

   a. Since the numbers are both even, they have 2 as a factor. Suppose the lesser number is $2x$. What is the greater number?

   b. Using $2x$ and the expression you wrote for Part a, find the resulting expression, which Lydia claims is always a perfect square.

   c. Assume the result in Part b is a perfect square. Try to draw a rectangle diagram showing the binomial that can be squared to get that product.

   d. What binomial is being squared in Part c?

   e. Have you proved that Lydia's trick always works? Explain.

Share & Summarize

Use the distributive property to expand $(a - b)(c + d)$. Then draw a rectangle diagram to show why your expansion is correct.

Investigation Shortcuts for Multiplying Binomials

You have been using the distributive property and rectangle diagrams to think about how to multiply two binomials. Have you noticed any patterns in your computations? In this investigation, you will look at patterns that can help you multiply binomials quickly and efficiently.

Here's how Mikayla multiplied $(2x + 7)(x - 3)$.

Problem Set G

Use Mikayla's method to multiply each pair of binomials.

1. $(y + 6)(y - 3)$

2. $(p + 4)(p + 3)$

3. $(t - 11)(t - 3)$

4. $(2x + 1)(3x + 2)$

5. $(2n + 3)(2n - 3)$

Think & Discuss

Why does Mikayla's method work?

Here's how Tamika approaches these problems.

Problem Set H

Use Tamika's method to multiply each pair of binomials.

1. $(y + 7)(y - 4)$

2. $(p + 1)(p - 5)$

3. $(t - 4)(t - 4)$

4. $(2x - 1)(x - 2)$

5. $(3n + 2)(2n - 3)$

Think & Discuss

Why does Tamika's method work?

In the next problem set, you will apply what you have learned about expanding binomials.

Problem Set I

For each equation, do Parts a and b.

    a. Decide whether the equation is true for all values of x, for some but not all values of x, or for no values of x.

    b. Explain how you know your answer is correct. If the equation is true for some but not all values of x, indicate which values make it true.

1. $(x - 2)(x - 3) = 0$

2. $(x + 3)(x + 2) = x^2 + 5x + 6$

3. $(x - b)^2 = x^2 - 2xb + b^2$

4. $(x + 3)(x - 1) = x^2 + 2x + 3$

5. $(x - 3)(x + 3) = x^2 - 6x - 9$

6. Brian thinks he has found some number patterns on a calendar. He says his patterns work for any 2-by-2 square on a calendar, such as the three shown here. Decide whether each of his patterns works, and justify your answers.

JULY						
				1	2	3
4	5	6	7	8	9	10
11	12	13	14	15	16	17
18	19	20	21	22	23	24
25	26	27	28	29	30	31

    a. Find the product of each diagonal. Their positive difference is always 7. For example, for the square containing 2, 3, 9, and 10, the products of the diagonals are $2 \cdot 10 = 20$ and $3 \cdot 9 = 27$. Their difference is $27 - 20 = 7$.

    b. Find the product of each column. Their positive difference is always 12. For example, for the square containing 2, 3, 9, and 10, the products of the columns are $2 \cdot 9 = 18$ and $3 \cdot 10 = 30$. Their difference is $30 - 18 = 12$.

    c. Find the product of each row. Their difference is always even. For example, for the square containing 2, 3, 9, and 10, the products of the rows are $2 \cdot 3 = 6$ and $9 \cdot 10 = 90$. Their difference is $90 - 6 = 84$, which is even.

Share & Summarize

You have seen several methods for expanding the product of two binomials.

    1. Choose the method you like best, and explain how to use it to expand $(2x + 3)(x - 1)$.

    2. Why do you like your chosen method?

On Your Own Exercises

Practice **&** **Apply**

1. This diagram shows a rectangle with area $(a + 10)(a + 2)$.

   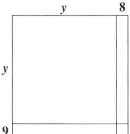

   a. Write an expression for the area of each of the four regions.

   b. Use your answer from Part a to expand the expression for the area of the large rectangle. That is, express the area without using parentheses. Simplify your answer by combining like terms.

2. This diagram shows a rectangle with area $(y + 9)(y + 8)$.

   a. Write an expression for the area of each of the four regions.

   b. Use your answer from Part a to expand the expression for the area of the large rectangle. That is, express the area without using parentheses. Simplify your answer by combining like terms.

Draw a rectangle diagram to model each product, and then use your diagram to expand the product. Simplify your answer by combining like terms.

3. $(3 + 2k)(4 + 3k)$

4. $(1 + 4x)(x + 2)$

5. A certain rectangle has area $y^2 + 5y + 2y + 10$.

   a. Draw a rectangle diagram that models this expression.

   b. Use your diagram to help you rewrite the area expression as a product of two binomials.

6. A certain rectangle has area $2y^2 + 6y + y + 3$.

   a. Draw a rectangle diagram that models this expression.

   b. Use your diagram to help you rewrite the area expression as a product of two binomials.

7. Consider the expression $(p + 3)(p + 5)$.

   a. Use the distributive property to expand the expression.

   b. Draw a rectangle diagram to model the expression, and check your expansion.

　　impactmath.com/self_check_quiz

Use the distributive property to expand each expression.

8. $(1 + 3a)(5 + 10a)$

9. $3\left(2x + \frac{1}{3}\right)(x + 2) - (x + 3)(1 + x)$

10. $\left(s + \frac{1}{4}\right)(3s + 1) - \left(\frac{1}{4} + s\right)(1 + s)$

Draw a rectangle diagram to model each product. Then expand the product using your diagram. Simplify your answer by combining like terms.

11. $(x + 3)(x - 3)$ **12.** $(p - 4)(3p + 2)$

13. Consider the expression $(h - 2)(h + 2)$.

   a. Use the distributive property to expand the expression.

   b. Draw a rectangle diagram to represent the expression. Shade areas that are being removed. Use your diagram to check that your expansion is correct.

Expand each expression using the distributive property.

14. $(x - 7)(x - 2)$ **15.** $(3 - g)(4 - g)$

16. $(4 - 2p)(4 - p)$ **17.** $(2w + 1)(w - 6)$

18. $(1 - 5q)(2 + 2q)$ **19.** $(3v - 5)(v + 1)$

20. A certain rectangle has area $y^2 - 4y + 8y - 32$.

   a. Draw a rectangle diagram to represent this expression.

   b. Use your diagram to help you rewrite the area expression as a product of two binomials.

21. Challenge A certain rectangle has area $2y^2 + 4y - 3y - 6$.

   a. Draw a rectangle diagram to represent this expression.

   b. Use your diagram to help you rewrite the area expression as a product of two binomials.

Prove It! Determine whether each number trick below works. If it does, prove it. If not, give a counterexample.

22. Take any three consecutive integers. Multiply the least and the greatest. That product is equal to the square of the middle integer minus 1.

23. Think of any two consecutive odd integers. Square both integers, and subtract the lesser result from the greater. The result is always evenly divisible by 6.

Expand each expression. Simplify your expansion if possible.

24. $(4x + 1)(4x - 1)$

25. $(r - 12)(r - 12)$

26. $(2x + 2)(x - 2)$

27. $(4x + 1)^2$

28. $(5M + 5)^2$

29. $(n + 1)^2 + (n - 1)^2$

Decide whether each equation is true for all values of x, for some but not all values of x, or for no values of x.

30. $(x + 3)(x - 4) = x^2 + 7x - 12$

31. $(2x + 1)(x - 1) = 2x^2 - x - 1$

32. $(3x + 1)(3x - 1)x = 9x^3 - x - 1$

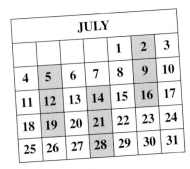

33. After working with the calendar problems in Problem Set I, Chapa wrote one: "Choose a block of three dates in a column (so they're all the same day of the week). Square the middle date, and subtract the product of the first and the last dates. The result is always 49."

Does Chapa's number trick always work? If so, show why. If not, give a counterexample.

Connect & Extend

34. Consider this graph.

 a. Write an expression for the area of the unshaded region.

 b. Write two expressions, one with and one without parentheses, for the area of the shaded region.

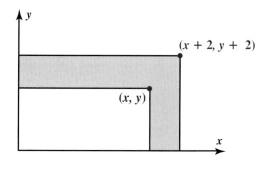

35. Write an expression for the area of the shaded region in this graph.

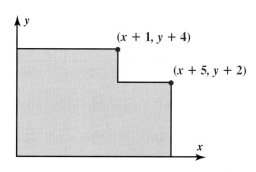

Expand each expression. Simplify your results by combining like terms.

36. $(1 + x^2)(x + y) + (1 + y)(x + y) - x^2(x + y) - xy$

37. $2(x + y) + x(3 + y)(x + 2)$

A *trinomial* is an expression with three unlike terms. Expand the following products of a trinomial and a binomial. You may find it helpful to use
a rectangle diagram.

38. $(x + y + 1)(x + 1)$

39. $(a + b + 1)(a + 2)$

40. $(x + y + 2)(x + 1)$

41. Consider the product $(2x + y)(x + 2y)$.

   a. Expand the expression using the distributive property.

   b. Draw a rectangle diagram that represents the expression. Use your
   diagram to check that your expansion is correct.

42. This block of wood has length y
and a square base with sides of
length x. A woodworker will cut
the block of wood twice, taking
off two strips from adjacent sides.
Each cut removes $1\frac{1}{8}$ inches.

   a. Write an expression for the
   volume of the wood before
   any cuts are made.

   b. Write an expression without parentheses for the volume of the
   wood after the cuts are made.

43. Consider the product $(2a + 3)(3a - 4)$.

   a. Expand and then simplify the expression.

   b. Draw a rectangle diagram to represent the expression. Use your
   diagram to check that your expansion is correct.

Expand the following products of a trinomial and a binomial.

44. $(x + y + 1)(x - 1)$

45. $(x - y - 1)(x + 1)$

46. $(x + y - 2)(x - 1)$

47. Preview Expand and simplify the expressions in Parts a–d.

    a. $(x - 3)^2$ **b.** $(x - 4)^2$

    c. $(x - 5)^2$ **d.** $(x - a)^2$

    e. Look for a pattern relating the factored and expanded forms of the expressions in Parts a–c. Describe a shortcut for expanding the square of a binomial difference.

48. Consider the expression $y^2 - 9$.

    a. Write the expression as the product of two binomials.

    b. Create a diagram to illustrate $y^2 - 9$ as a rectangular area.

Expand and then simplify each expression.

49. $(n + 2)^2 - n(n + 4)$

50. $(n + p)^2 - n(n + 2p)$

Mixed Review

Solve each equation.

51. $\frac{2x}{3} - 7 = 3$ **52.** $3K + \frac{4}{5} = \frac{1}{5}$ **53.** $3.2 - 2b = 1.1$

54. Copy this figure and vector. Translate the figure using the vector.

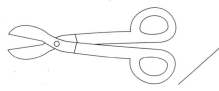

55. Graph the inequality $y > 3x + 2$.

56. Suppose r is a number between $^-1$ and 0. Order these numbers from least to greatest.

$$r \qquad r^{-3} \qquad r^2 \qquad r^3$$

Tell whether the pattern in each table can best be described by a *linear, quadratic, exponential,* or *reciprocal* relationship.

57.

x	y
$^-2$	6
$^-1$	1
0	$^-2$
1	$^-3$
2	$^-2$

58.

a	b
$^-3$	$^-4$
$^-1$	$^-12$
1	12
3	4
5	2.4

59.

s	n
$^-2$	$0.\bar{1}$
$^-1$	$0.\bar{3}$
0	1
1	3
2	9

Solve each proportion.

60. $\dfrac{8}{3} = \dfrac{x}{9}$ **61.** $\dfrac{18}{y} = \dfrac{4}{10}$ **62.** $\dfrac{9.2}{3.6} = \dfrac{2.3}{w}$

63. Economics When students first begin attending a college or university, they often receive applications for credit cards. Some students use credit cards to furnish or decorate their dormitory rooms or apartments, and quickly go into debt. Unfortunately, credit cards often have very high interest rates.

Suppose Jay charges $2,000 for electronic equipment and books using a credit card that has an interest rate of 18% per year, which is 1.5% per month.

a. If Jay doesn't pay any of the $2,000, how much interest will be added at the end of the month?

b. Jay figures he can afford to pay $100 per month on his bill. Assume he follows through with this plan and makes no more charges. (That may be a big assumption!) Copy and complete the table to show how much he still owes after 6 months. The interest added each month is based on the unpaid portion of the previous month's bill.

Month	Balance	Interest Added	Amount Paid	New Balance
1	$2,000.00	—	$100.00	$1,900.00
2	1,900.00		100.00	
3			100.00	

c. How much money has Jay paid to his credit card company?

d. By how much has Jay's original $2,000 debt decreased?

e. How much money has Jay paid in interest?

6.3 Patterns in Products of Binomials

You have learned several methods for expanding products of binomials. Some binomials have products with identifiable patterns. Recognizing these patterns will make your work easier.

Explore

Expand and simplify each product.

$(x + 1)^2 = (x + 1)(x + 1) =$

$(x + 2)^2 =$

$(x + 3)^2 =$

$(x + 4)^2 =$

Describe the pattern you see in your work. Use the pattern to predict the expansion of $(x + 10)^2$.

Check your prediction by expanding and simplifying $(x + 10)^2$.

Investigation Squaring Binomials

In this investigation, you will learn some shortcuts for expanding squares of binomials.

Problem Set A

1. Apply the pattern you discovered in the Explore to expand these squares of binomials.

   a. $(m + 9)^2$ **b.** $(m + 20)^2$ **c.** $(m + 0.1)^2$

2. Apply the pattern to predict the expansion of $(x + a)^2$. Check your answer by using the distributive property.

3. Use the pattern you discovered to explain, without calculating, why $100^2 \neq 93^2 + 7^2$. Is 100^2 greater than or less than $93^2 + 7^2$?

4. In Problem 3, you saw that $(93 + 7)^2 \neq 93^2 + 7^2$. Are there any values of x or a for which $(x + a)^2$ *does* equal $x^2 + a^2$? If so, what are they? How do you know?

You have just studied expressions of the form $(a + b)^2$. Now you will look at expressions of the form $(a - b)^2$.

Problem Set B

1. Recall that $(x - 1)^2$ can be thought of as $(x + {}^-1)^2$.

a. Use this fact, along with your findings in Problem Set A, to expand $(x - 1)^2$.

b. Check your answer to Part a by using the distributive property to expand $(x - 1)^2$.

2. Expand each expression using any method you like

a. $(m - 9)^2$ **b.** $(m - 20)^2$ **c.** $(m - 0.1)^2$

3. What is the expansion of $(x - a)^2$?

4. Use a rectangle diagram to help explain your answer to Problem 3.

You have seen by now that these two statements are true:

$$(a + b)^2 = a^2 + 2ab + b^2 \qquad (a - b)^2 = a^2 - 2ab + b^2$$

The variables a and b can represent any expressions. For example:

$$(2x + 3y)^2 = (2x)^2 + 2 \cdot 2x \cdot 3y + (3y)^2 = 4x^2 + 12xy + 9y^2$$

Problem Set C

Expand each expression.

1. $(3m + 2)^2$

2. $(2x - y)^2$

3. $(2m - 4n)^2$

4. Challenge $(g^2 - a^4)^2$

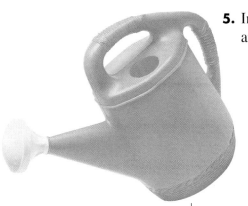

5. Imagine a square garden surrounded by a border of square tiles that are 1 unit by 1 unit. A variety of sizes are possible.

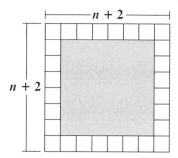

a. The garden has sides of length *n*, not including the tiles, where *n* is divisible by the length of a tile. Write an expression for the area of the ground covered by both the garden and the tiles.

b. Write an expression for the number of tiles needed for the garden in Part a.

6. **Prove It!** Evan noticed that when he found the difference of 3^2 and 4^2, the result was odd: $16 - 9 = 7$. This is also true for the difference of 6^2 and 7^2: $49 - 36 = 13$. He conjectured that the difference of squares of consecutive numbers is always odd.

Prove Evan's conjecture, if possible. If not, give a counterexample.

Share & Summarize

Bharati is confused about squaring binomials. She thinks that for any numbers *a* and *b*,

$$(a + b)^2 = a^2 + b^2 \text{ and } (a - b)^2 = a^2 - b^2$$

Write her a letter explaining why she is incorrect. Include the correct expansions of $(a + b)^2$ and $(a - b)^2$ in your letter.

Investigation ▶2 Differences of Squares

In Investigation 1, you used a shortcut to expand squares of binomials such as $(a + b)^2$ and $(a - b)^2$. In this investigation, you will find a shortcut for a different kind of product of binomials.

Problem Set D

1. Expand each expression.

a. $(x + 10)(x - 10)$ **b.** $(k + 3)(k - 3)$

c. $(S + 1)(S - 1)$ **d.** $(x + 5)(x - 5)$

e. $(2t + 5)(2t - 5)$ **f.** $(3y - 7)(3y + 7)$

2. How are the factors of the original products in Problem 1 similar?

3. How are the expansions of the products in Problem 1 similar?

4. Prove It! Expand $(x + a)(x - a)$, and show that the pattern you noticed in Problem 3 will always be true for this kind of product.

5. If the expression fits the pattern of the products in Problem 1, expand it using the pattern you described in Problem 3. If it doesn't fit the pattern, say so.

a. $(x + 20)(x - 20)$

b. $(b + 1)(b - 1)$

c. $(n - 2.5)(n - 2.5)$

d. $\left(2m - \frac{1}{2}\right)\left(m + \frac{1}{2}\right)$

e. $(J + 0.2)(J - 0.2)$

f. $(z + 25)(z - 100)$

g. $\left(2n - \frac{1}{3}\right)\left(2n + \frac{1}{3}\right)$

h. $(3 - p)(3 + p)$

6. Find two binomials with a product of $x^2 - 49$.

7. Some people call the expanded expressions you wrote in Problems 1 and 5 *differences of squares*. Explain why this name makes sense.

EXAMPLE

The shortcut you found in Problem Set D can help you do some difficult-looking computations with stunning speed!

Problem Set E

1. Show how to use Lydia's method to calculate $99 \cdot 101$ without using a calculator.

Use a difference of squares to calculate each product. Check the first few until you are confident you're doing it correctly.

2. $49 \cdot 51$ **3.** $28 \cdot 32$

4. $43 \cdot 37$ **5.** $35 \cdot 25$

6. $4.1 \cdot 3.9$ **7.** $^{-}14 \cdot 16$

For which products below do you think using a difference of squares would be a reasonable method of calculation? If it seems reasonable, find the product.

8. $41 \cdot 38$ **9.** $99 \cdot {}^{-}101$

10. $10\frac{1}{4} \cdot 9\frac{3}{4}$ **11.** $1.2 \cdot {}^{-}0.7$

12. Think about the kinds of products for which it is helpful to use differences of squares.

    a. Make up a set of three multiplication problems for which you might want to use a difference of squares. Be adventurous! Do them yourself, and record the answers.

    b. Give your problems to a partner to solve while you do your partner's set. Check that you agree on the answers and that the method works well for those problems.

13. If you combine the difference-of-squares method of fast calculation with some other mathematical tricks, you can do even more astounding computations in your head.

a. Consider the product $32 \cdot 29$.

i. Explain why $32 \cdot 29 = 31 \cdot 29 + 29$.

ii. Now use the difference-of-squares pattern to help compute the value of $31 \cdot 29$.

iii. Finally, use the product of 31 and 29 to compute $32 \cdot 29$.

b. Compute $21 \cdot 18$ in your head.

c. How did you find the answer to Part b?

Share & Summarize

1. Suppose you want to expand an expression containing two binomials multiplied together.

a. How do you know whether you can apply the shortcut you used to expand some of the products in Problem 5 of Problem Set D? Write two unexpanded products to help show what you mean.

b. Describe the shortcut you use to write the expansion.

2. You may have noticed that when you multiply two binomials, you sometimes end up with another binomial and other times with an expression with three unlike terms, or a *trinomial*.

a. Make up a product of two binomials that results in a binomial.

b. Make up a product of two binomials that results in a trinomial.

On Your Own Exercises

Practice & Apply

Expand and simplify each expression.

1. $(a + 5)^2$ **2.** $(m + 11)^2$ **3.** $(x + 2.5)^2$

4. $(t - 11)^2$ **5.** $(p - 2.5)^2$ **6.** $(2.5 - k)^2$

7. $\left(q - \frac{1}{4}\right)^2$ **8.** $(g^2 - 1)^2$ **9.** $(s^2 - y^2)^2$

10. $(3f + 2)^2$ **11.** $(3x + y)^2$ **12.** $(3m - 2n)^2$

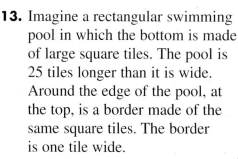

13. Imagine a rectangular swimming pool in which the bottom is made of large square tiles. The pool is 25 tiles longer than it is wide. Around the edge of the pool, at the top, is a border made of the same square tiles. The border is one tile wide.

a. If n represents the number of tiles in the width of the bottom of the pool, write an expression for the total number of tiles in the bottom of the pool.

b. Write an expression for the number of tiles in both the bottom of the pool and the border.

Expand and simplify each expression.

14. $(10 - k)(10 + k)$ **15.** $(3h - 5)(3h + 5)$

16. $(0.4 - 2x)(0.4 + 2x)$ **17.** $\left(\frac{1}{5} + k\right)\left(\frac{1}{5} - k\right)$

Write each expression as the product of binomials.

18. $4x^2 - 1$ **19.** $16 - 25x^2$ **20.** $x^2 - y^2$

Write each product as a difference of squares, and use this form to calculate the product.

21. $35 \cdot 45$ **22.** $27 \cdot 33$

23. $207 \cdot 193$ **24.** $111 \cdot 89$

Look back at your work in Problem 13 of Problem Set E on page 395. Combine the difference-of-squares method with addition to calculate each product.

25. $12 \cdot 9$ **26.** $37 \cdot 25$

impactmath.com/self_check_quiz

Connect & Extend

Expand and simplify each expression.

27. $\left(\frac{x}{2} + \frac{y}{2}\right)^2$

28. $(3 - xy)^2$

29. $(xy - x)^2$

30. $(2xy - 1)^2 + (2xy - y)^2$

31. Write an expression without parentheses for the area of the shaded triangle.

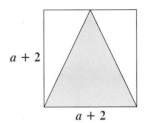

$a + 2$

$a + 2$

32. Challenge For what values of x and a is $(x + a)^2 > x^2 + a^2$? Justify your conclusion. (Hint: Expand the expression on the left side of the inequality.)

Expand and simplify each expression.

33. $(x^2 - y^2)(x^2 + y^2)$

34. $(1 - y^3)(1 + y^3)$

35. $(xy - x)(xy + x)$

Find the values of a and b that make each equation true.

36. $2x^2 + 7x + 3 = (a + bx)(3 + x)$

37. $20 - x - x^2 = (a + x)(b - x)$

38. $21 - 23x + 6x^2 = (3 - ax)(7 - bx)$

39. Physical Science An object is thrown straight upward from a height of 4 feet above the ground. The object's initial velocity is 30 feet per second. The equation relating the object's height in feet to the time t in seconds since it was released is $h = 30t - 16t^2 + 4$.

a. Find a and b to make this equation true:

$$30t - 16t^2 + 4 = (2 - t)(a + bt)$$

b. The statement $xy = 0$ is true whenever one of the factors, x or y, is equal to 0. For example, the solutions to $(k - 1)(k + 2) = 0$ are 1 and $^-2$, because these values make the factors $k - 1$ and $k + 2$ equal to 0. Use this fact and your result from Part a to find two solutions to the equation $30t - 16t^2 + 4 = 0$.

c. One of your solutions tells you at what time the object hit the ground. Which solution is it, and how do you know?

In y o u r

own
words

Write two multi-plication problems that look difficult but that you can solve easily in your head using differences of squares. Explain how using differences of squares can help you find the products.

40. This rectangle is divided into three triangular regions. Find an expression without parentheses for each of the three areas.

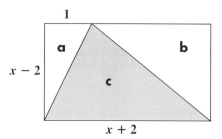

41. Challenge Katie and Gilberto used paper and scissors to convince themselves that $(a + b)(a - b)$ really is equal to $a^2 - b^2$.

a. Gilberto started with this paper rectangle. He wrote labels on the rectangle to represent lengths.

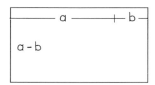

Gilberto then cut the rectangle and rearranged it.

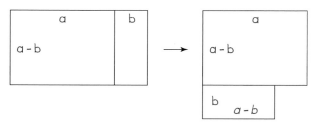

Explain why this shows that $(a + b)(a - b)$ is equal to $a^2 - b^2$.

b. Katie started with this paper square.

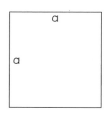

Then she cut the square and rearranged it.

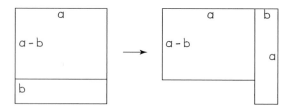

Explain why this shows that $(a + b)(a - b)$ is equal to $a^2 - b^2$. Hint: The areas of the two diagrams above must be the same, so expressions representing those areas must be equal.

42. Challenge Prove that for a quadratic equation, $y = ax^2 + bx + c$, the second differences in a table with consecutive inputs must be constant. Hint: Suppose the inputs for the table are x, $x + 1$, $x + 2$, $x + 3$, and so on. What are the corresponding outputs?

Mixed Review

Simplify each fraction.

43. $\dfrac{21}{99}$ **44.** $\dfrac{15}{75}$ **45.** $\dfrac{63}{210}$

Each table describes either a linear relationship, an exponential relationship, or an inverse variation. Write an equation describing each relationship.

46.

x	y
$^-2$	500
$^-1$	200
0	80
1	32
2	12.8

47.

x	y
$^-6$	$^-4$
$^-3$	$^-8$
$^-2$	$^-12$
2	12
3	8

48.

x	y
$^-4$	10
$^-3$	3
$^-1$	$^-11$
2	$^-32$
3	$^-39$

49. Sketch a graph of $y = \dfrac{1}{x}$.

50. Georgia brought a package containing 16 identical pieces of clay to school. She wanted to share the clay with some friends during recess.

     a. Georgia wants to divide the clay evenly among her friends. If she invites three friends to join her, how many pieces will each friend receive (including Georgia)?

     b. If Georgia invites five friends to join her, how many pieces will each friend receive (including Georgia)?

     c. Write a formula giving the number of pieces n each friend will receive (including Georgia), if Georgia invites f friends.

     d. Use your expression in Part c to find n when f is 3 and when f is 5. If your answers do not agree with those for Parts a and b, find any mistakes you have made and correct them.

6.4

Working with Algebraic Fractions

In earlier lessons, you discovered several tools for working with algebraic expressions more efficiently. Now you will expand your tool kit as you learn how to work with fractions that involve algebraic expressions. You saw fractions like these in Chapter 2, when you studied inverse variation.

Think & Discuss

Just before summer vacation, Adriana borrowed $100 from her favorite aunt to buy a pair of in-line skates for herself. She enjoyed them so much that she borrowed another $100 to buy a pair for her younger sister. She agreed to repay d dollars per month, and her aunt agreed not to charge interest.

Write an algebraic fraction to express how many months it will take Adriana to repay the first $100.

Which of these expressions show how many months it will take Adriana to repay the entire debt?

$$\frac{d}{200} \qquad \frac{d}{100} \qquad \frac{100}{2d} \qquad \frac{200}{2d}$$

$$\frac{100}{d} \qquad \frac{200}{d} \qquad \frac{100}{d} + \frac{100}{d}$$

Investigation  Making Sense of Algebraic Fractions

When you use expressions involving algebraic fractions, the expressions might not make sense for all values of the variables. In this investigation, you will explore some situations in which this is important.

MATERIALS

graphing calculator

Problem Set A

The denominator of the right-hand side of this equation has four factors.

$$y = \frac{24}{(x-1)(x-2)(x-3)(x-4)}$$

1. What is the value of y if you let $x = 5$? If you let $x = 6$?

2. What happens to the value of y if $x = 1$? If $x = 2$? Are there other values of x for which this happens?

3. Choose a number less than 5 for which y does have a value. What is the value of y using your chosen value of x?

4. Look again at the equation above.

a. Use your calculator to make a table for the relationship, starting with $x = 0$ and using an increment of 0.25. Copy the results into a table on your paper. Use the calculator to help fill your table for x values up to $x = 5$.

How does your table show which values of x do not make sense?

b. Now use your calculator to graph the relationship, using x values from $^-1$ to 5 and y values from $^-100$ to 100. Make a rough sketch of the graph.

What happens to the graph at the x values that do not make sense?

Algebraic fractions don't make *mathematical* sense for values of the variables that make the denominator equal to 0—in other words, they are *undefined* for these values.

Problem Set **B**

Carlota and Ling attended a fundraising auction, hoping to bid for some collector's comic books. They both said they would bid no higher than x dollars per comic book.

Carlota bought some comic books she really wanted, each for $5 less than her set maximum price. She spent $120 in all. The comic books Ling wanted were worth more, and she ended up paying $5 more per comic book than she had intended. She spent $100 in all.

1. Explain what each expression means in terms of the auction story.

  a. x **b.** $x + 5$ **c.** $x - 5$

2. Write an expression to represent the number of comics Ling purchased. Then write an expression to represent the number Carlota bought.

3. This algebraic expression represents the total number of comic books purchased by the friends.

$$\frac{100}{x + 5} + \frac{120}{x - 5}$$

Read each comment about this expression, and decide whether the student is correct. If the student is incorrect, explain his or her mistake.

  a. Mikayla: "Even though you can't go to an auction intending to pay $^-$$10 per comic book, the expression is *mathematically* sensible when x has a value of $^-$10. The expression then has a value of $^-$28."

  b. Ben: "The expression does not make sense at all because when $x = 5$, one of the denominators is 0, and you can't divide by 0."

  c. Héctor: "The expression makes *mathematical* sense for all values of the variable except 5 and $^-$5."

  d. Tamika: "In the auction situation, we can only think about paying some positive number of dollars for a comic book. Therefore, for this story, the expression makes sense for positive values of x only, except 5 of course."

  e. Tala: "We can't use just *any* positive value of x in the expression. For example, if $x = 7$, Ling would have paid $x + 5 = $12 per comic book—which means she would have bought eight and a third comic books."

  f. Kai: "The expression makes *mathematical* sense for any values of x except 5 and $^-$5. However, in the auction situation, there are only a small number of sensible answers."

4. Consider Kai's statement. Find all the possible values for x given the auction situation. Assume x is a whole number.

5. Now use your calculator to make a table and a graph for $y = \frac{100}{x+5} + \frac{120}{x-5}$. Use values of x from $^-10$ to 10 and values of y from $^-100$ to 100. For the table, start with $x = {}^-10$ and use an increment of 1.

How do the graph and the table show the values of x for which the expression does not make *mathematical sense*?

Share & Summarize

1. When you are looking at an expression for a situation, what is the difference between *mathematical sense* and *sense in the context of the situation*? Give examples if it helps you make your point.

2. Consider the auction situation in Problem Set B.

   a. What did you have to think about when trying to determine the values that made *mathematical* sense?

   b. What did you have to think about when trying to determine the values that made sense in the *context* of the situation?

Investigation Rearranging Algebraic Fractions

When you work with numeric fractions, you sometimes want to write them in different ways. For example, to calculate $\frac{1}{2} + \frac{1}{3}$, it's helpful to rewrite the problem as $\frac{3}{6} + \frac{2}{6}$.

Think & Discuss

Tamika tried to write equivalent expressions for three algebraic fractions. Which of these are correct, and which are incorrect? How do you know?

$$\frac{3}{12m} = \frac{1}{4m} \qquad \frac{2}{m+2} = \frac{1}{m+1} \qquad \frac{2x}{x^2} = \frac{2}{x}$$

The expressions Tamika wrote correctly are *simplified* versions of the original fractions. In a simplified fraction, the numerator and the denominator have no factors in common.

There are several strategies for simplifying fractions. For example, to simplify $\frac{15}{18}$, you can factor the numerator and the denominator:

$$\frac{15}{18} = \frac{3 \cdot 5}{3 \cdot 6} = \frac{3}{3} \cdot \frac{5}{6} = \frac{5}{6}$$

Another method is to divide the numerator and the denominator by a common factor, in this case 3:

$$\frac{15}{18} = \frac{\frac{15}{3}}{\frac{18}{3}} = \frac{5}{6}$$

You can also use these strategies to simplify algebraic expressions.

EXAMPLE

• To simplify $\frac{5}{5x + 15}$, factor the numerator and the denominator.

$$\frac{5}{5x + 15} = \frac{5}{5(x + 3)} = \frac{5}{5} \cdot \frac{1}{x + 3} = \frac{1}{x + 3}$$

Or, divide the numerator and the denominator by their common factor.

$$\frac{5}{5x + 15} = \frac{\frac{5}{5}}{\frac{5x + 15}{5}} = \frac{1}{x + 3}$$

• To simplify $\frac{5a^2}{10a}$, factor the numerator and the denominator.

In Two Steps

$$\frac{5a^2}{10a} = \frac{5 \cdot a^2}{5 \cdot 2 \cdot a} = \frac{5}{5} \cdot \frac{a^2}{2 \cdot a} = \frac{a^2}{2a}$$

$$\frac{a^2}{2a} = \frac{a \cdot a}{2 \cdot a} = \frac{a}{a} \cdot \frac{a}{2} = \frac{a}{2}$$

In One Step

$$\frac{5a^2}{10a} = \frac{5 \cdot a \cdot a}{5 \cdot 2 \cdot a} = \frac{5}{5} \cdot \frac{a}{a} \cdot \frac{a}{2} = \frac{a}{2}$$

Or, divide both the numerator and the denominator by their common factors.

In Two Steps

$$\frac{5a^2}{10a} = \frac{\frac{5a^2}{5}}{\frac{10a}{5}} = \frac{a^2}{2a} = \frac{\frac{a^2}{a}}{\frac{2a}{a}} = \frac{a}{2}$$

In One Step

$$\frac{5a^2}{10a} = \frac{\frac{5a^2}{5a}}{\frac{10a}{5a}} = \frac{a}{2}$$

Problem Set C

Simplify each fraction.

1. $\dfrac{6x^2y}{18x}$ **2.** $\dfrac{2}{2a+4}$ **3.** $\dfrac{x}{x^2+2x}$

Write two fractions that can be simplified to the given fraction.

4. $\dfrac{1}{3+a}$ **5.** $\dfrac{x}{2}$ **6.** $\dfrac{5y}{z}$

Find each product. Simplify your answers.

7. $\dfrac{1}{2d}\cdot\dfrac{4}{3}$ **8.** $\dfrac{1}{2}\cdot\dfrac{-2}{d-5}$

9. $\dfrac{-4(d-1)}{3}\cdot\dfrac{-1}{2d}$ **10.** $\dfrac{1}{3(a-4)}\div\dfrac{3a}{5}$

11. $\dfrac{\frac{a}{7}}{\frac{3a}{5}}$ **12.** $\dfrac{\frac{1}{a}}{\frac{1}{a+1}}$

Remember

Dividing $\frac{a}{b}$ by $\frac{c}{d}$ is the same as multiplying $\frac{a}{b}$ by the reciprocal of $\frac{c}{d}$:

$$\frac{\frac{a}{b}}{\frac{c}{d}}=\frac{a}{b}\cdot\frac{d}{c}$$

You will now use what you have learned to analyze several number tricks.

Problem Set D

Brian made up four number tricks. For each, do the following:

- Check whether or not the trick *always* works. If it always works, explain why.

- If it doesn't always work, does it work with only a few exceptions? If so, what are the exceptions? Explain why it works for all numbers other than those exceptions.

- If it never works or works for only a few numbers, explain how you know.

1. *Number Trick 1:* Pick a number, any number. Multiply it by 2 and square the result. Add 12. Then divide by 4 and subtract the square of the number you chose at the beginning. Your answer is 3.

2. *Number Trick 2:* Pick a number, any number. Add 2 to it and square the result. Multiply the new number by 6 and then subtract 24. Divide by your chosen number. Divide again by 6, and then subtract 4. Your answer is your chosen number.

3. *Number Trick 3:* Pick a number, any number. Multiply your number by 3 and then subtract 4. Divide by 2 and add 5. The result is 6.

4. Challenge *Number Trick 4:* Pick a number, any number. Add 6 to it and multiply the result by the chosen number. Then add 9. Now divide by 3 more than the chosen number, and then subtract the chosen number. Your answer is 3.

Share & Summarize

Evan simplified each fraction as shown. Check his answers. If he did a problem correctly, say so. If he didn't, explain what is wrong and how to find the right answer.

1. $\dfrac{3}{x+3} = \dfrac{1}{x+1}$

2. $\dfrac{a}{a+4} = \dfrac{1}{4}$

3. $\dfrac{5a}{3} \div \dfrac{3}{a} = \dfrac{5a}{3} \cdot \dfrac{a}{3} = \dfrac{5a^2}{9}$

4. $\dfrac{12t^2}{35} \cdot \dfrac{21}{16t} = \dfrac{9t}{20}$

Practice & Apply

1. Consider this equation.

$$y = \frac{2 - x}{(x - 2)(x + 1)}$$

   a. For what values of x is y undefined?

   b. Explain how you could use the information from Part a to help you sketch a graph of the equation.

2. Every morning a restaurant manager buys $300 worth of fresh fish at the market. One morning she buys fish that is selling for d dollars per pound. The next morning the price has risen $2 per pound.

   a. Write an expression for the quantity of fish, in pounds, the manager purchased on the first morning.

   b. Write an expression for the quantity of fish, in pounds, the manager purchased on the second morning.

   c. Write an equation for the total quantity of fish the manager purchased on these two days.

   d. For what values of d, if any, does your expression from Part c not make *mathematical* sense?

   e. For what additional values of d, if any, does your expression not make sense in the situation?

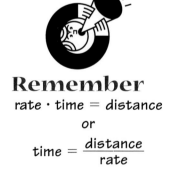

3. Every Friday a delivery person drives her truck 120 miles into the city to make a pickup and then drives back. On one particular Friday, she drove into the city at the posted speed limit, s. However, on the return trip, she was slowed by road construction and had to travel 15 miles per hour below the speed limit.

   a. Write an expression for the time it took her to drive into the city.

   b. Write an expression for the time her return trip took.

   c. Write an equation for her total driving time for the round trip.

   d. For what values of s, if any, does your expression from Part c not make *mathematical* sense?

   e. For what additional values of s, if any, does your expression not make sense in the situation?

Remember

rate · time = distance

or

time = $\dfrac{\text{distance}}{\text{rate}}$

Simplify each fraction.

4. $\dfrac{12m}{2m}$

5. $\dfrac{2x}{4xy}$

6. $\dfrac{20a^2b}{16ab^2}$

7. $\dfrac{3k}{k^2 - 6k}$

Simplify each fraction.

8. $\dfrac{1 + a}{a(1 + a)}$

9. $\dfrac{3(x + 1)}{6}$

10. $\dfrac{nm}{m^2 + 2m}$

11. $\dfrac{3ab}{a^2 b^2 - 3ab}$

Find each product or quotient. Simplify your answers.

12. $\dfrac{1}{3} \cdot \dfrac{1}{a}$

13. $\dfrac{4}{3} \cdot \dfrac{d}{2}$

14. $\dfrac{1}{5a} \cdot \dfrac{3a^2}{2}$

15. $\dfrac{1}{a} \div \dfrac{1}{a}$

16. $\dfrac{m}{4} \div \dfrac{4}{m}$

17. $\dfrac{^-1(x - 2)}{3(2 - x)}$

For the number tricks in Exercises 18 and 19, do the following:

- Check whether the trick *always* works. If it does, explain why.

- If it doesn't always work, does it work with only a few exceptions? If so, what are the exceptions? Explain why it works for all numbers other than those exceptions.

- If it never works or works for only a few numbers, explain how you know.

18. Pick a number. Subtract 1 and square the result. Subtract 1 again. Divide by your number. Add 2. The result is your original number.

19. Pick a number. Add 3 to it, and square the result. Subtract 4. Divide by the number that is 1 more than your chosen number. Subtract 5 from the result. The result is your chosen number.

20. Consider this equation.

$$y = \frac{2k^2 - 3k}{k^2 - k}$$

a. For what values of k, if any, does y not have a value?

b. Explain what will happen to a graph of the equation at the values you found in Part a.

21. Physical Science All objects attract each other with the force called *gravity*. The English mathematician Isaac Newton discovered this formula for calculating the gravitational force between two objects:

$$F = G\left(\frac{Mm}{r^2}\right)$$

In the formula, F is the gravitational force between the two objects, M and m are the masses of the objects, r is the distance between them, and G is a fixed number called the *gravitational constant*.

a. How does the gravitational force between two objects change if the mass of one of the objects doubles? If the mass of one of the objects triples?

b. How does the gravitational force between two objects change if the distance between the objects doubles? If the distance triples?

c. Suppose the masses of two objects is doubled and the distance between the objects is doubled as well. How does this affect the gravitational force between the objects?

Simplify each expression.

22. $\dfrac{4k - 2}{2k^2 + 4k - 2}$

23. $\dfrac{(u - 3)(u + 2)(u - 1)}{{}^{-}1(3 - u)(1 - u)}$

24. Consider this equation.

$$y = \frac{24}{2 - 5x}$$

a. For what values of x does y not have a value? Explain.

b. For what values of x will y be positive? Explain.

c. For what values of x will y be negative? Explain.

d. For what values of x will y equal 0? Explain.

25. The equation $y = \frac{(x + 1)^2}{x + 1}$ can be simplified to $y = x + 1$ for all values of x except $^-1$ (which makes the denominator 0). The graph of $y = \frac{(x + 1)^2}{x + 1}$ looks like the graph of $y = x + 1$, but with an open circle at the point where $x = ^-1$.

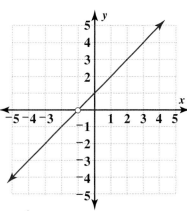

Use this idea to graph each equation.

a. $y = \frac{4x^3}{2x}$ **b.** $y = \frac{4x^2 + 2x}{2x}$

26. For what values of m is it true that $\frac{1}{m} > \frac{1}{m + 1}$? Explain.

Evaluate each expression.

Mixed Review

27. $\frac{2}{3} + \frac{5}{8}$ **28.** $\frac{3}{10} - \frac{1}{4}$ **29.** $\frac{3}{7} - \frac{8}{3}$

30. $\frac{2}{3} \cdot \frac{5}{8}$ **31.** $\frac{3}{10} \div \frac{1}{4}$ **32.** $\frac{3}{7}\left(\frac{8}{3}\right)$

Rewrite each expression using a single base and a single exponent.

33. $27x^3$ **34.** $a^{12} \cdot \left(a^2\right)^{-7}$ **35.** $\frac{32}{c^5}$

36. Statistics Gerry surveyed five fast-food restaurants about the number of calories in the various types of sandwiches they sold. He put his results in a histogram. For example, the first bar in his graph reveals that two of the sandwiches had from 200 to 299 calories.

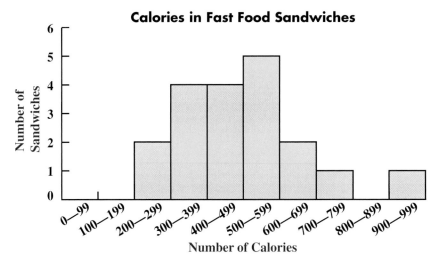

Calories in Fast Food Sandwiches

a. On how many different sandwiches did Gerry gather data?

b. Estimate the median for these data. Explain your answer.

c. What does the histogram tell you about the sandwiches at these restaurants?

6.5

Adding and Subtracting Algebraic Fractions

You have added and subtracted fractions in which the numerator and denominator are both numbers. In this lesson, you will apply what you know to add fractions involving variables.

Think & Discuss

Consider these fractions and mixed numbers.

$$1\frac{1}{2} \qquad \frac{2}{3} \qquad \frac{3}{8} \qquad \frac{3}{10} \qquad 2\frac{5}{12}$$

- Choose any two of the numbers and add them. Describe how you found the common denominator and the sum.

- Choose two of the remaining fractions, and subtract the lesser from the greater. Describe how you found the common denominator and the difference.

- Finally, subtract the remaining fraction from 10. Describe how you found the common denominator and the difference.

Investigation Combining Algebraic Fractions

You will now use what you know about fractions with numbers to add and subtract algebraic fractions—fractions that involve variables.

Problem Set A

1. Copy and complete this addition table.

+			$\frac{3}{4}$
$\frac{1}{2}$		2	
			1
$\frac{3}{5}$	$\frac{11}{5}$		

Find each sum or difference.

2. $\dfrac{100}{w} + \dfrac{100}{w}$ **3.** $\dfrac{100}{w} - \dfrac{100}{w}$

4. $\dfrac{100}{w} + \dfrac{100}{w} + \dfrac{100}{w}$ **5.** $\dfrac{1}{2x} + \dfrac{2}{3x}$

6. $\dfrac{m}{3} + \dfrac{m}{6}$ **7.** $\dfrac{y}{p} + \dfrac{1}{2p}$

8. $\dfrac{5}{8} - \dfrac{5}{6}$ **9.** $\dfrac{3}{2x} - \dfrac{3}{2y}$

10. Dave can type an average of n words per minute. Write an expression for the number of minutes it takes Dave to type

    a. 400 words.

    b. 200 words.

    c. 1,000 words.

    d. Add your expressions from Parts a, b, and c. What does the sum represent in terms of the typing situation?

When you add or subtract algebraic fractions, there are many ways to find a common denominator.

EXAMPLE

Evan, Tala, and Lucita have different methods for adding $\dfrac{14}{8x}$ and $\dfrac{3}{4x}$.

Keep the three methods above in mind as you work on the next problem set.

Problem Set B

1. Copy and complete this addition table.

+			$\dfrac{8}{15x}$
$\dfrac{1}{3x}$		$\dfrac{2}{x}$	
	$\dfrac{5}{4x}$		
	$\dfrac{3}{5x}$	$\dfrac{53}{30x}$	

Find each sum or difference.

2. $\dfrac{6}{3x^2} - \dfrac{2}{2x^2}$

3. $\dfrac{2}{4t} - \dfrac{2t}{3}$

4. $\dfrac{3}{6m} + \dfrac{4m}{8m^2}$

5. Camila and Lakita earn money on the weekends by painting people's houses. It takes Camila $2n$ minutes to paint 1 square meter by herself; it takes Lakita $3n$ minutes.

   a. Write an expression for how much area Camila paints in 1 minute. Do the same for Lakita.

   b. How much area will the friends paint in 1 minute if they work together? Write your expression as a single algebraic fraction.

   c. If the friends are working together to paint a room with 40 m² of wall, how much time will the job take? Show how you found your answer.

Share & Summarize

Consider these five terms.

$$c \qquad 2 \qquad 3 \qquad 2c^2 \qquad 3c^2$$

1. Create four addition or subtraction problems involving fractions whose numerator and denominator are made from these terms (for example, $\dfrac{c}{2} + \dfrac{3}{3c^2}$). Use each term only once in a problem. Then, exchange problems with your partner, and solve your partner's problems.

2. Use the terms to create an addition or a subtraction problem with a sum or difference of 3.

Investigation 2 ▶ Strategies for Adding and Subtracting Algebraic Fractions

Now you will learn more about adding and subtracting algebraic fractions.

Problem Set C

1. Compute each sum without using a calculator.

   a. $\dfrac{1}{1} + \dfrac{1}{2}$ **b.** $\dfrac{1}{2} + \dfrac{1}{3}$ **c.** $\dfrac{1}{3} + \dfrac{1}{4}$ **d.** $\dfrac{1}{4} + \dfrac{1}{5}$

2. Look for a pattern in the sums in Problem 1. Use it to find $\dfrac{1}{5} + \dfrac{1}{6}$ without actually calculating the sum.

3. In each part of Problem 1, how does the denominator of the sum relate to the denominators of the two fractions being added?

4. In each part of Problem 1, how does the numerator of the sum relate to the denominators of the two fractions added?

5. Use the patterns you observed to make a conjecture about this sum.

$$\frac{1}{m} + \frac{1}{m + 1}$$

6. Consider again the sum $\dfrac{1}{5} + \dfrac{1}{6}$.

   a. If this sum is equal to $\dfrac{1}{m} + \dfrac{1}{m + 1}$, what is the value of m?

   b. Use your conjecture for Problem 5 and the value of m from Part a to find the sum of $\dfrac{1}{5} + \dfrac{1}{6}$. Does the result agree with your prediction in Problem 2?

   c. To check your result, calculate the sum by finding a common denominator and adding.

7. **Prove It!** Try to prove that your conjecture is true.

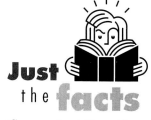

Just the facts

The ancient Egyptians preferred fractions with 1 in the numerator, called *unit fractions*. They expressed other fractions—using hieroglyphics—as sums of unit fractions. For example, $\dfrac{21}{30}$ could be expressed as $\dfrac{1}{3} + \dfrac{1}{5} + \dfrac{1}{6}$.

EXAMPLE

Here's how Lydia thought about the sum of $\frac{1}{m}$ and $\frac{1}{m+1}$.

"When I add $\frac{1}{m}$ and $\frac{1}{m+1}$, I use a common denominator of $m(m+1)$, the product of the two denominators."

$$\frac{1}{m} + \frac{1}{m+1} = \frac{1}{m} \cdot \frac{m+1}{m+1} + \frac{1}{m+1} \cdot \frac{m}{m}$$

$$= \frac{m+1}{m(m+1)} + \frac{m}{m(m+1)}$$

$$= \frac{m+1+m}{m(m+1)}$$

$$= \frac{2m+1}{m(m+1)}$$

Problem Set D

Discuss Lydia's strategy with your partner. Make sure you understand how each line follows from the previous line.

1. Why did Lydia multiply the first fraction by $\frac{m+1}{m+1}$?

2. Why did she multiply the second fraction by $\frac{m}{m}$?

3. Ben says, "I have an easier method for adding $\frac{1}{m}$ and $\frac{1}{m+1}$. Here's what I did."

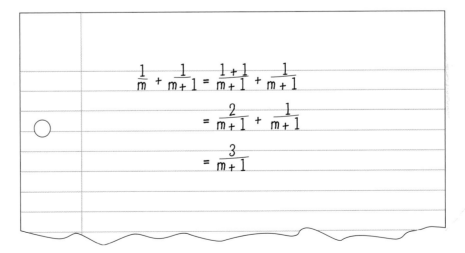

$$\frac{1}{m} + \frac{1}{m+1} = \frac{1+1}{m+1} + \frac{1}{m+1}$$

$$= \frac{2}{m+1} + \frac{1}{m+1}$$

$$= \frac{3}{m+1}$$

Is Ben's method correct? Explain.

When adding or subtracting algebraic fractions, it is often easiest to leave the numerator and the denominator in factored form. Knowing what the factors are allows you to recognize and identify common factors more easily.

Find this sum:

$$\frac{2x}{x(x-1)} + \frac{5}{(x-1)(x+2)}$$

The factored denominators make it easy to simplify the first fraction and then to find a common denominator.

You can simplify the first fraction by dividing the numerator and the denominator by x. Then $(x-1)(x+2)$ can be used as a common denominator of the resulting fractions.

$$\frac{2x}{x(x-1)} + \frac{5}{(x-1)(x+2)} = \frac{2}{x-1} + \frac{5}{(x-1)(x+2)}$$

$$= \frac{2}{x-1} \cdot \frac{x+2}{x+2} + \frac{5}{(x-1)(x+2)}$$

$$= \frac{2(x+2)+5}{(x-1)(x+2)}$$

$$= \frac{2x+9}{(x-1)(x+2)}$$

Problem Set E

Find each sum or difference. Simplify your answers if possible.

1. $\dfrac{1}{m} + \dfrac{2}{m+1}$

2. $\dfrac{4}{m} - \dfrac{1}{m-1}$

3. $\dfrac{4}{b+2} + \dfrac{b}{b+3}$

4. $\dfrac{2(x+1)}{x(x+1)} - \dfrac{1}{x-3}$

5. $\dfrac{10}{x+4} + \dfrac{3x}{9x^2}$

6. $\dfrac{2x}{x-1} - \dfrac{x+1}{x+3}$

7. Consider these subtraction problems.

$$\frac{1}{2} - \frac{1}{3} \qquad \frac{1}{3} - \frac{1}{4} \qquad \frac{1}{4} - \frac{1}{5}$$

a. Find each difference.

b. Use the pattern in your answers to Part a to solve this subtraction problem without actually calculating the difference.

$$\frac{1}{5} - \frac{1}{6}$$

c. Use the pattern to find this difference.

$$\frac{1}{m} - \frac{1}{m + 1}$$

d. Prove It! Use algebra to show that your answer to Part c is correct.

Share & Summarize

In the Example on page 415, Lydia explains how she thinks about adding algebraic fractions. Compute the following difference, and explain how *you* think about figuring it out.

$$\frac{1}{x + 1} - \frac{1}{2x}$$

Investigation ▶3 Solving Equations with Fractions

You have already solved equations involving algebraic fractions. For example, in earlier grades, you have solved proportions such as these:

$$\frac{x}{2} = \frac{9}{6} \qquad \frac{50}{3.6} = \frac{11}{m}$$

Think & Discuss

Describe some ways you could solve each of these equations.

$$\frac{x}{2} = \frac{9}{6} \qquad \frac{50}{3.6} = \frac{11}{m}$$

Problem Set **F**

Solve each equation using any method you like. You may want to use different methods for different equations.

1. $\dfrac{3x - 6}{4} = x - 8$

2. $\dfrac{t}{2} + \dfrac{t}{3} = {}^{-}1$

3. $\dfrac{4a}{5} - \dfrac{2 - a}{4} = 30$

4. $\dfrac{p}{5} - p = {}^{-}0.4$

5. What fraction added to $\dfrac{2x - 1}{4}$ equals $\dfrac{x^2 - 4}{4}$?

6. What fraction subtracted from $\dfrac{k + 3}{5}$ equals $\dfrac{k - 3}{15}$?

7. Evan estimated the solution of the equation $\dfrac{n + 7}{2} + \dfrac{n}{3} = 10$ by finding the intersection of the graphs of these two equations.

$$y = \dfrac{n + 7}{2} + \dfrac{n}{3} \qquad\qquad y = 10$$

   a. Explain why Evan's method works.

   b. Graph both equations in the same window of your calculator, and use the graphs to estimate the solution.

   c. Check your estimate by solving the original equation using whatever method you prefer.

All the equations in Problem Set F contain one or more fractions with variables in the numerator. When you solve equations with variables in the denominators, you need to check that the "solutions" you find do not make any denominators in the original equation equal to 0.

Solve the equation $\frac{5x + 10}{x + 2} = \frac{3}{x + 1}$.

One way to solve this equation is to "clear" the fractions by multiplying both sides by a common denominator of all the fractions in the equation—

$$\frac{(x + 2)(x + 1)}{1} \cdot \frac{5x + 10}{x + 2} = \frac{3}{x + 1} \cdot \frac{(x + 2)(x + 1)}{1}$$

—and then simplify the resulting equation:

$$\frac{x + 2}{x + 2} \cdot (x + 1)(5x + 10) = 3(x + 2) \cdot \frac{x + 1}{x + 1}$$
$$1 \cdot (x + 1)(5x + 10) = 3(x + 2) \cdot 1$$
$$5x^2 + 15x + 10 = 3x + 6$$

Rearranging the equation to set one side equal to 0 will allow you to solve it by graphing.

$$5x^2 + 12x + 4 = 0$$

In this case, the simplified equation is quadratic. You can estimate a solution by graphing $y = 5x^2 + 12x + 4$ and finding the points where $y = 0$.

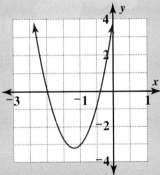

The solutions are approximately $^-2$ and $^-0.4$. Check these values *in the original equation.* Since $^-2$ makes the denominator of $\frac{5x + 10}{x + 2}$ equal to 0, it is *not* a solution of the original equation. The number $^-0.4$ makes the original equation true, so $^-0.4$ is a solution, $^-2$ is not.

As you solve the equations that follow, be sure to check that your solutions don't make any of the denominators in the original equation equal 0.

MATERIALS

graphing calculator
(optional)

Problem Set G

Solve each equation using whatever method you like. You may want to use different methods for different equations. For some, you may need to use a graph to estimate solutions.

1. $\dfrac{10}{7} = \dfrac{k + 1}{k - 3}$

2. $\dfrac{6 - 2x}{x - 3} = 8$

3. $\dfrac{2}{g + 1} - \dfrac{2}{g - 1} = 4$

4. $\dfrac{20 - a}{a^2 - 4} = \dfrac{5}{a - 2} + \dfrac{3}{a + 2}$

5. $0 = \dfrac{2}{s + 3} + \dfrac{s}{s + 2}$

6. $\dfrac{{}^-60 - 12z}{z + 5} = {}^-120$

Share & Summarize

Choose one problem from Problem Set G and explain how you solved it so that a student who was absent could understand.

On Your Own Exercises

Practice & Apply

Find each sum or difference.

1. $\dfrac{9}{8} - \dfrac{8}{9}$

2. $\dfrac{x}{4} + \dfrac{y}{2}$

3. $\dfrac{2xy}{3} - \dfrac{1}{6}$

4. $\dfrac{1}{x} + \dfrac{2}{x^2}$

5. $\dfrac{c}{a} - \dfrac{a}{c}$

6. Marcus and his sister Annette have part-time jobs after school at the grocery store. When she's stacking cans in a display at the end of an aisle, Annette can put 500 cans up in z minutes. Marcus works half as fast as Annette, stacking 500 cans in $2z$ minutes.

   a. Write an expression for how many cans Marcus stacks in 1 minute.

   b. Write an expression for how many cans Annette stacks in 1 minute.

   c. Working together, how many cans can Marcus and Annette stack in 1 minute? Express your answer as a single algebraic fraction.

   d. The store manager has asked the two to create a display using 750 cans. How long will it take them?

7. Esperanza and Jasmine are making the same 300-mile drive in separate cars. Esperanza drives an average of n miles per hour. Jasmine drives 1.5 times as fast.

   a. Write an expression for the time it takes Jan to drive the 300 miles.

   b. Assume Jasmine and Esperanza left at the same time. Write an expression for the difference in time between Jasmine's arrival at the final destination and Esperanza's arrival.

   c. Write an expression for the total time both women spent traveling.

8. Copy and complete this addition table.

+	$\dfrac{5}{2x}$	$\dfrac{4}{x}$	$2x$
$\dfrac{1}{4x}$			
$-\dfrac{2}{3x}$			
$\dfrac{3+x}{2}$			

Find each sum or difference.

9. $\dfrac{1}{m} - \dfrac{2}{m+1}$

10. $\dfrac{4}{m} + \dfrac{1}{m+1}$

11. $\dfrac{3}{d} + \dfrac{4}{d+1}$

12. $\dfrac{3}{c} - \dfrac{4}{c-1}$

13. $\dfrac{a}{a+4} + \dfrac{3a}{5}$

14. $\dfrac{x^2}{x^2-1} - \dfrac{1}{x^2-1}$

15. $\dfrac{5}{k} - \dfrac{5}{k+1}$

16. $\dfrac{2y-1}{4} - \dfrac{y}{2}$

Solve each equation using any method you like.

17. $\dfrac{2x}{3} + \dfrac{1}{4} = x - 1$

18. $\dfrac{v-2}{3} + \dfrac{v}{2} = 10$

19. $\dfrac{n+1}{n-1} = 3$

20. $\dfrac{2-u}{u+1} = 5$

21. $\dfrac{8}{w+5} - \dfrac{2}{w+5} = \dfrac{2}{w} + \dfrac{1}{w+5}$

22. $\dfrac{3}{c-1} + \dfrac{3}{c+1} = \dfrac{21-c}{c^2-1}$

23. What fraction added to $\dfrac{r+1}{r}$ equals 1?

24. What fraction subtracted from $\dfrac{2-x}{7}$ is equal to $\dfrac{x}{14}$?

25. What fraction subtracted from $\dfrac{1}{v}$ is equal to $\dfrac{3v}{2}$?

26. Economics Meg earns $70 for w hours of picking fruit. Her friend Rashid, who is more experienced and works faster, earns $80 for w hours. Together they earn $1,000 in a week.

Write a brief explanation for each expression.

a. $\dfrac{70}{w}$ **b.** $\dfrac{80}{w}$ **c.** $\dfrac{70}{w} + \dfrac{80}{w}$

d. $\dfrac{150}{w}$ **e.** $1,000 \div \dfrac{150}{w}$ **f.** $1,000 \div \dfrac{70}{w}$

g. $1,000 \div \dfrac{80}{w}$

Find each sum or difference.

27. $\dfrac{c}{ab} - \dfrac{a}{bc}$ **28.** $\dfrac{2x}{2y} + \dfrac{y}{x}$

29. $\dfrac{G+1}{G-1} - \dfrac{2}{G+1}$ **30.** $\dfrac{4-2y}{6} + \dfrac{y}{4}$

31. $\dfrac{1}{x^2 y} + \dfrac{1}{xy}$ **32.** $\dfrac{1}{p} + \dfrac{1}{p^2} + \dfrac{1}{y}$

33. $\dfrac{1}{xc} + 1 - \dfrac{1}{c}$ **34.** $\dfrac{a+1}{1} + \dfrac{1}{a-1}$

35. $2 - \dfrac{2}{s+1} - \dfrac{s}{s+1}$

36. $\dfrac{1}{m} + \dfrac{1}{m+1} + \dfrac{1}{m+2}$

37. $\dfrac{1}{m} - \dfrac{1}{m+1} - \dfrac{1}{m+2}$

38. Ms. Diaz drove 135 miles to visit her mother. She knew the speed limit increased by 10 mph after the first 75 miles, but she couldn't remember what the speed limits were.

a. Write an expression representing the amount of time it will take Ms. Diaz to drive the first 75 miles if she travels the speed limit of x mph.

b. Write an expression representing the amount of time it will take her to drive the remaining 60 miles at the new speed limit.

c. Write an expression for Ms. Diaz's total driving time. Combine the parts of your expression into a single algebraic fraction.

Solve each equation using any method you like.

39. $\dfrac{p+2}{2} + \dfrac{p-1}{5} = p + 1$

40. $\dfrac{r-8}{3} + \dfrac{r-5}{2} = r - 5$

41. $\dfrac{T-1}{4} + \dfrac{2-T}{3} + \dfrac{T+1}{2} = 3$

42. $\dfrac{v-2}{4} + \dfrac{2}{v-1} + \dfrac{1}{2} = \dfrac{v^2-9}{4v-4}$

43. $\dfrac{Z-5}{2} - \dfrac{3}{Z+5} = \dfrac{(Z+1)(Z-3)}{2Z+10}$

44. $\dfrac{3}{x-3} + \dfrac{4}{x+3} = \dfrac{21-x}{x^2-9}$

45. Jing earns x dollars per hour bagging groceries at the local market. She also works as a math tutor. Her hourly tutoring rate is $2 more than twice her hourly rate at the market.

Last week Jing earned $51.75 at the market and $40.50 tutoring. She realized that if she had spent all her working hours that week tutoring, she would have earned $162! Write and solve an equation to find Jing's hourly rate for each job.

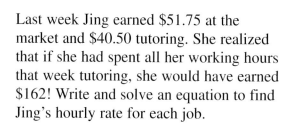

46. For two numbers A and B,

$$\dfrac{5x+1}{x^2-1} = \dfrac{A}{x+1} + \dfrac{B}{x-1}$$

a. What is a common denominator for $\dfrac{A}{x+1}$ and $\dfrac{B}{x-1}$?

b. Find the sum $\dfrac{A}{x+1} + \dfrac{B}{x-1}$ using the common denominator you found in Part a. Write the sum without parentheses.

c. Explain why you can use this system to find the values of A and B:

$$A + B = 5$$
$$^{-}A + B = 1$$

d. Solve the system to find A and B.

Mixed Review

Evaluate without using a calculator.

47. $\sqrt{(^{-}18)^2}$ **48.** $^{-}\sqrt{7^2}$ **49.** $^{-}(\sqrt{64})^2$ **50.** $(^{-}\sqrt{49})^2$

51. Which of the following are equal? Find all matching pairs.

    a. 12 **b.** 12^{-1} **c.** $4\sqrt{\frac{1}{9}}$

    d. $4\left(\frac{1}{3}\right)^{-1}$ **e.** $\sqrt{\frac{4}{9}}$ **f.** the reciprocal of 12

52. Geometry This figure has both reflection symmetry and rotation symmetry.

    a. How many lines of symmetry does it have?

    b. What is the angle of rotation?

53. A figure has an area of a cm^2. An enlargement of the figure using a scale factor of f would have what area?

Rewrite each equation in the form $y = mx + b$.

54. $2(y + x) + 1 = 3x - 2y + 3$

55. $6y + \frac{3}{7}x - 2 = 0$

56. $8 = {}^-(3x + 4) + (4 - y) - (2y + 10)$

57. In 1999 the United States began making a series of quarters whose reverse sides are designed by different states. Chris began collecting the quarters, separating them from the rest of his spare change.

By the time the fourth state's quarters were minted, Chris already had several Delaware, New Jersey, and Pennsylvania quarters. Counting them, he discovered he had 3 times as many Delaware quarters as Pennsylvania quarters. The number of New Jersey quarters he had was 7 more than half the number of Delaware quarters.

    a. Choose a variable, and use it to express the number of quarters of each type that Chris had.

    b. Altogether Chris had saved $12.75 in quarters. Write and solve an equation to find how many of each type of quarter he had.

Just the facts

The U.S. Mint decided to issue 5 state quarters each year in the order in which they ratified the Constitution and joined the Union. The quarter for New York, the 11th state admitted to the Union, was the first issued in the year 2001.

Chapter Summary

This chapter focused on expanding and simplifying algebraic expressions. Two main themes were using the distributive property and working with algebraic fractions.

You solved problems that involved multiplying binomials, requiring you to use the distributive property to expand the product and then to simplify by combining like terms. Turning your attention to algebraic fractions, you discovered you could simplify, add, and subtract them using the same methods you use for numeric fractions.

You concluded the chapter by solving equations that required you to apply all your new skills.

Strategies and Applications

The questions in this section will help you review and apply the important ideas and strategies developed in this chapter.

Using geometric models to expand expressions

1. Draw a rectangle diagram that models the expression $(x + 3)(x + 6)$. Use it to expand the product of these binomials.

2. Draw a rectangle diagram that models the expression $(3t - 1)(t + 1)$. Use it to expand the product of these binomials.

3. Consider this graph.

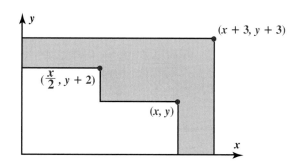

a. Write an expression for the area of the unshaded region.

b. Write an expression, with and without parentheses, for the area of the shaded region.

impactmath.com/chapter_test

Using the distributive property to expand expressions

4. Describe the steps required to simplify $x(1 - x) + (1 - x)(3 - x)$. Give the simplified expression.

5. Simplify the expression $(x^2 - 1)(y^2 - 1) - (1 + xy)(1 + xy)$.

6. *Molding* is a strip of wood placed along the base of a wall to give a room a "finished" look. The diagram shows a floor plan of a room with 1-inch-thick molding along the edges. Measurements are in inches.

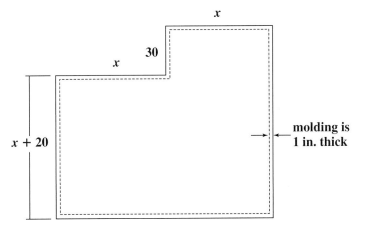

a. Write an expression for the floor area before the molding was installed. Simplify your expression as much as possible.

b. Write an expression for the remaining floor area after the molding was installed. Simplify your expression as much as possible.

Expanding expressions of the forms $(ax + b)^2$, $(ax - b)^2$, and $(ax + b)(ax - b)$

7. Consider expressions of the form $(ax + b)^2$.

a. Describe a shortcut for expanding such expressions, and explain why it works.

b. Use your shortcut to expand $(x + 13)^2$.

8. Consider expressions of the form $(ax - b)^2$.

a. Describe a shortcut for expanding such expressions, and explain why it works.

b. Use your shortcut to expand $\left(\frac{x}{2} - 1.5\right)^2$.

9. Consider expressions of the form $(ax + b)(ax - b)$.

a. Describe a shortcut for expanding such expressions, and explain why it works.

b. Use your shortcut to expand $(xy + 3)(xy - 3)$.

10. Rewrite $24 \cdot 26$ as the product of two binomials, and then use the "difference of squares" pattern to compute the product.

Simplifying expressions involving algebraic fractions

Simplify each expression, and explain each step.

11. $\dfrac{-3x}{9 - 6x}$ **12.** $\dfrac{15xy}{3x^3y^3}$

Solving equations involving algebraic fractions

13. Consider the equation $\dfrac{1}{x + 1} + \dfrac{2}{x - 1} = \dfrac{8}{x^2 - 1}$.

  a. Describe the first step you would take to solve this equation.

  b. Solve the equation.

14. Consider the equation $\dfrac{k}{k - 1} - \dfrac{5}{2} = \dfrac{1}{k - 1}$.

  a. Solve the equation.

  b. Explain why it is especially important to check your solutions when solving equations containing algebraic fractions.

15. Every week a freight train delivers grain to a harbor 156 miles away. Last week the train traveled to the harbor at an average speed of s miles per hour. On the return trip, the train had empty cars and was able to travel an average of 16 miles per hour faster.

  a. Write an expression for the time it took the train to reach the harbor.

  b. Write an expression for the time the return trip took.

  c. The round trip took 10.4 hours. Write an equation to find the value of s. Solve your equation.

Demonstrating Skills

Rewrite each expression without parentheses.

16. $^-6(x + 1) + 2(5 - x) - 9(1 - x)$

17. $^-a(1 - 3a) - (2a^2 - 5a)$

18. $(2b + 1)(4 + b)$

19. $(x + 1)(5 - x)$

20. $(2c - 8)(c - 2)$

21. $(2x + y)(y - xy)$

22. $(L - 8)^2$

23. $(x - xy)(x + xy)$

Simplify each expression.

24. $\dfrac{14x^2y^2}{2xy}$

25. $(5xy - 2)(1 - 2x)$

26. $(d + 2)(1 - d) + (1 - 2d)(3 - d)$

27. $\dfrac{5n}{n - 1} + \dfrac{3n}{2n - 2}$

28. $\dfrac{b}{3} + \dfrac{b - 1}{b + 1} - \dfrac{b}{2b + 2}$

29. $\dfrac{1}{k - 1} - \dfrac{1}{k + 2}$

Solve each equation.

30. $\dfrac{3}{x + 4} = \dfrac{2}{x - 4}$

31. $\dfrac{^-2x + 2}{x - 1} = {}^-3$

CHAPTER 7

Solving Quadratic Equations

Real-Life Math

Super Models Most computer programmers depend on mathematical equations and expressions in the software they design. Many computer and video games, for example, need to be able to model the paths of things flying through the air: balls for sports like baseball, football, soccer, and golf; arrows and other missiles—even water balloons! Such paths are called *trajectories,* and quadratic equations can be used to model them.

Think About It Imagine the path of a football that is kicked for a field goal. Can you describe the shape of its trajectory?

Family Letter

Dear Student and Family Members,

Our next chapter in mathematics deals with solving quadratic equations. Quadratic equations involve the square of the main variable and can be written in the form $ax^2 + bx + c = 0$, where a, b, and c are constants.

Quadratic equations are an important topic in mathematics and science. They are used to describe the movement of objects in space, such as the motion of basketballs, automobiles, satellites, and rockets. They are also used to determine the shapes of radar antennae, satellite dishes, and mirrors used in telescopes.

We will learn and practice three important methods for solving quadratic equations: solving perfect squares, solving equations by factoring, and using the quadratic formula.

- We will learn to recognize perfect square quadratics, which are equivalent to a simple linear expression multiplied by itself:

$$x^2 + 4x + 4 = (x + 2)(x + 2) = (x + 2)^2$$
$$4x^2 - 12x + 9 = (2x - 3)(2x - 3) = (2x - 3)^2$$

- Another important method is solving quadratic equations by factoring. This is useful when a quadratic equation is equivalent to the product of two different linear expressions:

$$x^2 - 8x + 15 = (x - 5)(x - 3)$$
$$2x^2 - 8x - 10 = (2x + 2)(x - 5)$$

- A third method is to use the quadratic formula. The quadratic formula is useful because it can be used to solve any quadratic equation written in the form $ax^2 + bx + c = 0$, not just those that are perfect squares or easily factored.

Vocabulary Along the way, we'll be learning about two new vocabulary terms:

factoring   **trinomial**

What can you do at home?

Knowledge of quadratic equations will allow your student to master one of the most important topics in algebra. The work is abstract, but the power that comes from mastery will help students in all their future algebraic work. Encourage your student to show you the problems we are working on and to explain the methods he or she is learning to use.

Solving by Backtracking

As you know, backtracking is a step-by-step process of undoing operations. To solve a linear equation by backtracking, you must know how to undo addition, subtraction, multiplication, and division. To use backtracking to solve nonlinear equations, however, you have to undo other operations as well.

Think & Discuss

How would you check to make sure that one operation really *does* undo another? Give an example to illustrate your thinking.

What would you do to undo each of these operations?

• finding the square roots of a number

• finding the reciprocal of a number

• changing the sign of a number

• raising a positive number to the *n*th power, such as 2^3

Just the facts

Quadratic equations are used in many contexts. For example, $h = 1 + 2t - 4.9t^2$ might give the height in meters at time t seconds of someone jumping on a trampoline with an initial velocity of 2 m/s.

Investigation Backtracking with New Operations

In this investigation, you will try out the ideas about undoing operations from the Think & Discuss.

Problem Set A

Ben is solving the equation $\sqrt{2x - 11} = 5$.

Kai said Ben's flowchart should really be like this:

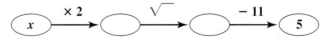

1. What is wrong with Kai's flowchart?

2. What equation would you solve by backtracking with Kai's flowchart?

3. Consider the equation $\sqrt{3x + 7} = 8$.

    a. Draw a flowchart for the equation.

    b. Use backtracking to find the solution. Check your answer by substituting it back into the equation.

Remember
The $\sqrt{}$ sign refers to the nonnegative square root of a number, if one exists.

4. This flowchart is for the expression $\frac{24}{s-2}$.

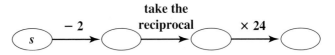

a. Try using the flowchart with a few numbers to see how it works. Record your results.

b. This flowchart is for the equation $\frac{24}{k-2} = 8$.

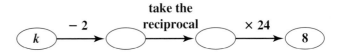

Backtrack to find the solution. Check that your solution is correct by substituting it into the equation.

5. This flowchart is for the equation $3 - p = 1$. The symbol $+/-$ means to take the opposite of the value—that is, to change its sign.

a. Solve the equation using backtracking.

b. Ben made this flowchart for $3 - p = 1$, but he got stuck when he tried to backtrack. Why can't you use his flowchart to solve the equation?

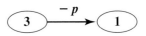

6. This flowchart is for the expression $\frac{2(3-t)}{4}$.

a. Try the flowchart with a few numbers to see how it works.

b. This flowchart is for the equation $\frac{2(3-t)}{4} = 5$.

Backtrack to find the solution, and check that it is correct.

c. What operation undoes the "change sign" operation?

7. You can think of changing the sign of a number as multiplication by $^-1$. For example, $^-x = ^-1 \cdot x$.

a. How do you usually undo multiplication by a number?

b. What does your answer to Part c of Problem 6 suggest about another way to undo multiplication by $^-1$?

Problem Set B

Solve each equation by backtracking.

1. $\dfrac{4}{x} = 0.125$

2. $\dfrac{8 - z}{2} = 9$

3. $\dfrac{7 - m}{2} = 3$

4. $5\left(20 - \dfrac{a}{4}\right) = 85$

5. Consider this equation.

$$\dfrac{12}{3s - 1} = 6$$

    a. Draw a flowchart for the equation.

    b. Solve the equation by backtracking, and check your solution.

6. Katie drew this flowchart.

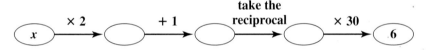

    a. What equation can be solved using Katie's flowchart?

    b. Solve the equation, and check your solution.

Share & Summarize

In this investigation, you learned how to undo taking the square root of a number, taking the reciprocal of a number, and changing the sign of a number.

1. Write an equation that can be solved by backtracking that uses all three of these operations. Find the solution of your equation.

2. Exchange equations with a partner, and try to solve your partner's equation. Check your answer by substitution.

Investigation 2 ▶ Backtracking with Powers

In this investigation, you will extend the types of equations for which you can use backtracking.

Think & Discuss

This flowchart is for the equation $x^2 = 9$. To solve this equation by backtracking, you must undo the "square" operation by taking the square root.

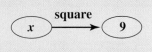

- How many solutions are there for $x^2 = 9$? How do you know?

Now consider the equation $(d - 2)^2 = 25$.

- Write the operations, in order, that you would use to evaluate the expression $(d - 2)^2$ for some value of d.

- Draw a flowchart for the equation $(d - 2)^2 = 25$.

- This equation has two solutions. Which step in your flowchart makes it possible for there to be two answers? Explain.

- As you backtrack beyond the step that makes two answers possible, there are two possible values for each oval. Think of a way you might show two values at each step, and then find both of the equation's solutions.

Problem Set C

1. Consider the equation $(a + 5)^2 = 25$.

   a. Draw a flowchart for the equation.

   b. Solve $(a + 5)^2 = 25$ by backtracking, and check your answer by substitution. Can you find more than one backtracking path (and so more than one solution)?

For each equation, draw a flowchart. Solve the equation using backtracking, and check your answers.

   2. $2(b - 4)^2 + 5 = 55$

   3. $3(c - 5)^2 - 5 = 7$

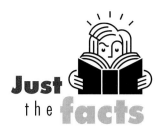

Solve each equation.

4. $(d - 2)^2 - 20 = 44$

5. $\sqrt{(2p - 3)^2 - 5} = 2$

6. $3(6 - T)^3 - 1 = 23$

7. $(e - 3)^4 = 81$

8. When filmmakers make movies outside at night, they often use floodlights to brighten the actors and the scenery. The relationship between the brightness of an object F (measured in *foot-candles*) and its distance d (measured in feet) from the light source follows an *inverse square law.* For a particular 2,000-watt floodlight, the formula might be

$$F = \frac{360,000}{d^2}$$

a. Why do you think this is called an inverse square law?

b. Find the brightness for $d = 10, 20, 30,$ and 50.

c. Explain the effect on an object's brightness of moving closer to or farther from the light source.

d. Draw a flowchart for the brightness formula.

e. Use backtracking to solve the equation

$$120 = \frac{360,000}{d^2}$$

What does the answer tell you?

Problem Set **D**

1. Consider this equation.
$$(x - 3)^2 - 5 = 0$$

a. Show that $3 + \sqrt{5}$ and $3 - \sqrt{5}$ are solutions of this equation.

b. The values $3 + \sqrt{5}$ and $3 - \sqrt{5}$ are *exact* solutions of the equation. Because $\sqrt{5}$ is an irrational number, when you write it in a decimal form, you are giving an *approximation,* no matter how many decimal places you use.

Write each solution from Part a as a decimal accurate to two places.

For each equation, give the exact solutions and the approximate solutions correct to two decimal places.

2. $h^2 - 5 = 45$

3. $(2m - 3)^2 + 7 = 9$

4. $3(J + 5)^2 - 2 = 7$

Share & Summarize

Use backtracking to solve this equation. Explain what you did at each step, and make note of any places you had to be particularly careful.
$$2(3b - 4)^2 + 1 = 19$$

On Your Own Exercises

Practice
& Apply

1. Consider the equation $^-\sqrt{2x-1} = ^-7$.

    a. Draw a flowchart for the equation.

    b. Solve the equation by backtracking.

Solve each equation.

2. $\sqrt{3x+1} = 4$

3. $\dfrac{2}{3p-1} = 5$

4. $\sqrt{a} = 1.5$

5. $\sqrt{2-q} = 2.5$

6. $5\sqrt{\dfrac{z}{5} - 1} = 4$

7. $\dfrac{9}{4-7d} = 18$

8. $2(x-4)^2 + 5 = 7$

9. $b^2 - 5 = 44$

10. $c^2 - 20 = 44$

11. $(L-2)^2 - 5 = 44$

12. $(q-2)^2 + 8 = 44$

13. $y^3 = 27$

14. $3(2w-3)^2 - 5 = 70$

15. $(2t-3)^2 - 20 = 44$

16. $y^3 = ^-27$

17. $(x+2)^3 = 64$

Connect
& Extend

18. A man controlling a robot with a camera eye has just sent it into a burning house to retrieve a safe full of money. He has given it this set of commands:

```
Forward 20 feet.
Right turn.
Forward 15 feet.
Left turn.
Forward 30 feet.
Right turn.
Forward 25 feet.
Pick up safe.
```

The robot is now standing in the burning house holding the safe, but the robot controller has been overcome by smoke and you have been asked to get the robot back. Use what you know about backtracking to write a set of commands that will bring the robot back out of the burning house with the safe.

In your
own
words

Explain how you
can decide
whether an equa-
tion can be direct-
ly solved by back-
tracking. Then give
an example of an
equation that can
be solved directly
by backtracking
and an example of
one that cannot.

19. Mary Ann used her calculator to find $\sqrt{2}$, wrote down the result, and then cleared the calculator. She then used her calculator to square the number she had written down, and got 1.99998.

Mary Ann concluded that squaring doesn't exactly undo taking the square root. Do you agree with her? What would you tell her?

20. Many equations cannot be solved directly by backtracking. Some have the variable stated more than once; others involve variables as exponents.

Here are some equations that can't be solved directly by backtracking.

$$f^2 = f + 1 \qquad x = \sqrt{x} + 1 \qquad k^2 + k = 0$$

$$1.1^B = 2 \qquad \frac{1}{x} = x^2 + 2$$

For each equation below, write *yes* if it can be solved directly by backtracking and *no* if it cannot.

a. $5 = \sqrt{x - 11}$ **b.** $4^d = 9$

c. $3g^2 = 5$ **d.** $\sqrt{x + 1} = x - 4$

21. Use backtracking to solve the equations in Parts a and b.

a. $(3x + 4)^2 = 25$

b. $(3x + 4)^2 = 0$

c. How many solutions did you find for each equation? Can you explain the difference?

22. Use backtracking to solve the equations in Parts a and b.

a. $(\sqrt{x})^2 = 5$

b. $\sqrt{x^2} = 5$

c. How many solutions did you find for each equation? Can you explain the difference?

Mixed Review

Expand each expression.

23. $3(3a - 7)$ **24.** $^-2b(8b - 0.5)$

25. $9c(^-8 + 7c)$ **26.** $(d + 3)(d + 6)$

27. $(2e - 4)(e - 6)$ **28.** $(3f + 10)(9f - 1)$

29. $(g + 7)^2$ **30.** $(3h - 1)^2$

31. $(2j + 2)^3$ **32.** $(3k - 2m)^2$

Geometry Find the area of each rectangle.

33.

```
┌──────────┐
│          │ 2d
└──────────┘
     4d
```

34.

```
┌─┐
│ │ 1.75d
└─┘
0.9d
```

35. Life Science The table lists the approximate amount of dry plant material that each type of habitat produces in 1 year.

Habitat	Plant Material (grams) Produced per Square Meter
Coral reef	2,500
Tropical rainforest	2,200
Temperate rainforest	1,250
Savannah	900
Open sea	125
Semidesert	90

Source: *Ultimate Visual Dictionary of Science.* London: Dorling Kindersley Limited, 1998.

a. One gram is equivalent to 0.035 ounce, and 1 meter is equivalent to 1.09361 yards. Determine how many ounces of plant material per square yard the typical coral reef produces in 1 year. Show how you found your answer.

b. Write two different statements comparing the amount of plant material produced in a tropical rainforest to that produced in a temperate rainforest.

c. The Baltic Sea covers approximately 422,000 square kilometers. Approximately how many kilograms of plant material are produced in the Baltic Sea in a year? Show how you found your answer.

Remember
1 km = 1,000 m
1 kg = 1,000 g

7.2 Solving by Factoring

Some quadratic equations can be solved easily by backtracking. A second solution method, factoring, can be used to solve other quadratic equations fairly easily.

When a quadratic equation consists of a product of two factors on one side of the equal sign and 0 on the other side, such as

$$(x - 2)(x + 5) = 0$$

the solutions can be found exactly. This is because 0 has a special property.

Think & Discuss

If the product of two factors is 0, what must be true about the factors?

Find all the values of k that satisfy the equation $k(k - 3) = 0$. Explain how you know you've found them all.

Now find all the values of x that satisfy $(x - 2)(x + 5) = 0$. Explain how you found them.

Investigation 1 Factoring Quadratic Expressions

You will now use the ideas from the Think & Discuss to solve some equations written as a product equal to 0.

Problem Set A

Find all the solutions of each equation.

1. $(t - 1)(t - 3) = 0$ **2.** $(s + 1)(2s + 3) = 0$

3. $x(3x + 7) = 0$ **4.** $(p + 4)(p + 4) = 0$

5. Using the same idea, find the solutions of this equation.

$$(2x + 1)(x + 8)(x - 1) = 0$$

You have seen how easily you can solve equations that are written as a product of factors equal to 0. Sometimes you can rewrite an equation in this form to make the solution easy to find.

In Chapter 6, you learned how to *expand* a product of two binomials, such as $(x + 3)(x - 2)$. The reverse of this process—rewriting an expression as a product of factors—is called **factoring.** For example, the expression $x^2 + x - 6$ can be factored to $(x + 3)(x - 2)$.

VOCABULARY
factoring

Think & Discuss

You may recall from Chapter 6 that the expression $(3x + 4)(3x - 4)$ can be rewritten as $9x^2 - 16$ when it is expanded. An expression such as $9x^2 - 16$ is called a *difference of two squares*. Can you explain why?

Now think about reversing the expansion. How would you factor $4a^2 - 25$? That is, how would you rewrite it as the product of two factors?

In Chapter 6, you also learned how to square a binomial and rewrite it as a **trinomial,** an expression with three unlike terms. For example:

$$(x - 5)^2 = x^2 - 10x + 25 \qquad (b + 5)^2 = b^2 + 10b + 25$$

Rewrite each trinomial below as the square of a binomial.

$$c^2 + 4c + 4 \qquad 16d^2 - 8d + 1$$

A trinomial such as $16d^2 - 8d + 1$ is called a *perfect square trinomial*. Can you explain why?

Which of these four trinomials are perfect squares?

$$x^2 + 6x + 9 \qquad k^2 - 8k + 25$$

$$4y^2 + 4y + 4 \qquad 49s^2 - 28s + 4$$

Just by looking at the coefficients, how can you tell whether *any* trinomial in the form $ax^2 + bx + c$ is a perfect square?

Remember
Subtracting a term means the coefficient is negative. For example, in the expression $16x^2 - 8x + 1$, the coefficient of x is ⁻8.

Problem Set B

In Problems 1–8, determine whether the quadratic expression to the left of the equal sign is the difference of two squares, or a perfect square trinomial. If it is, rewrite it in factored form and solve the equation. If it isn't in one of these special forms, explain how you know it isn't.

1. $x^2 - 64 = 0$ **2.** $p^2 + 64 = 0$

3. $x^2 - 16x - 64 = 0$ **4.** $k^2 - 16k + 64 = 0$

5. $9y^2 - 1 = 0$ **6.** $9m^2 + 6m + 1 = 0$

7. $9g^2 - 4g - 1 = 0$ **8.** $y^2 + 9 = 0$

Each equation below has two variables. If the quadratic expression is the difference of two squares, or a perfect square trinomial, rewrite the equation in factored form and solve for a. If it is in neither special form, explain how you know it isn't.

9. $a^2 + 9b^2 = 0$

10. $a^2 - 4ab + 4b^2 = 0$

11. $4a^2 - b^2 = 0$

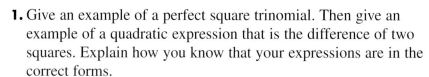

Share & Summarize

1. Give an example of a perfect square trinomial. Then give an example of a quadratic expression that is the difference of two squares. Explain how you know that your expressions are in the correct forms.

2. Explain why the only solutions of $4x^2 - 9 = 0$ are $x = 1.5$ and $x = {}^{-}1.5$.

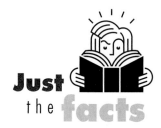

Just the facts

Quadratic equations are often used to describe how the position of a moving object changes over time. For example, the equation $d = 25 - 3t^2$ might describe how many meters an accelerating hyena is from a rabbit after t seconds if it began running toward the rabbit from 25 meters away.

Investigation 2 ▶ Practice with Factoring

If a quadratic expression is equal to 0 and can be factored easily, finding its factors is an efficient way to solve the equation. In this investigation, you will learn some new strategies for determining whether a quadratic expression can easily be factored and for factoring it when you can. For example, consider this expression:

$$x^2 + 8x + 12$$

If such an expression can be factored, it can be rewritten as the product of two linear expressions:

$$(x + m)(x + n)$$

Multiplying terms gives

$$x^2 + (m + n)x + mn$$

You can use this idea to help factor any quadratic expression for which a, the coefficient of the squared variable, is equal to 1.

EXAMPLE

Can $x^2 + 8x + 12$ be factored? If so, solve the equation $x^2 + 8x + 12 = 0$.

First compare the expanded form of $(x + m)(x + n)$ with the given expression:

$$x^2 + 8x + 12$$
$$x^2 + (m + n)x + mn$$

If $x^2 + 8x + 12$ can be factored into the form $(x + m)(x + n)$, the product of m and n must be 12, and their sum must be 8. The only two numbers that fit these conditions are 6 and 2. This means that the expression *can* be factored and that the equation can be rewritten as

$$(x + 2)(x + 6) = 0$$

So, the equation $x^2 + 8x + 12 = 0$ has two solutions, $^-2$ and $^-6$.

In this investigation, you will consider only cases in which m and n are integers.

Problem Set C

For Problems 1–6, do the following:

• Think of the expression as a special case of $(x + m)(x + n)$, and state the values of m and n.

• Use the fact that $(x + m)(x + n) = x^2 + (m + n)x + mn$ to expand the expression.

1. $(x + 7)(x + 1)$ **2.** $(x + 2)(x + 5)$

3. $(x - 4)(x - 5)$ **4.** $(x + 2)(x - 3)$

5. $(x - 2)(x + 3)$ **6.** $(x + 5)(x - 4)$

For Problems 7–10, use the fact that $(x + m)(x + n) = x^2 + (m + n)x + mn$ to do the following:

• Determine what $m + n$ and mn equal. From this, find the values of m and n.

• Rewrite the expression as a product of two binomials. You may want to expand the product to check your result.

7. $x^2 + 7x + 6 = (x + \_\_)(x + \_\_)$

8. $x^2 - 7x + 6 = (x - \_\_)(x - \_\_)$

9. $x^2 - 4x - 12 = (x - \_\_)(x + \_\_)$

10. $x^2 + 4x - 12 = (x - \_\_)(x + \_\_)$

Use the method demonstrated in the Example on page 445 to solve these equations.

11. $x^2 - 10x + 16 = 0$ **12.** $x^2 + 6x - 16 = 0$

Just the facts

The quadratic equation $x = 10t + \frac{1}{2}(2.5)t^2$ describes the distance in meters traveled by a motorcycle that began from a certain point with velocity 10 m/s and increased its speed, or accelerated, at a rate of 2.5 m/s^2.

Think & Discuss

Kai organized the possibilities for factoring trinomials.

Use Kai's approach to factor these expressions.

$$x^2 + 11x + 10 \qquad\qquad x^2 - 7x + 10 \qquad\qquad x^2 - 3x - 10$$

Explain what happens if you use Kai's approach with $x^2 + 6x + 10$.

Problem Set D

Factor each quadratic expression using Kai's method, or state that it can't be factored using his method.

1. $x^2 + 6x + 5$

2. $b^2 + 4b - 5$

3. $w^2 - 2w + 1$

4. $t^2 + 9t - 18$

5. $s^2 - 10s - 24$

6. $c^2 - 4c + 5$

7. Use Kai's approach to solve the equation $w^2 + 4w - 12 = 0$.

If every term of a quadratic expression has a common factor, rewriting it can make factoring easier. For example, $2x^2 + 12x + 10$ can be rewritten as $2(x^2 + 6x + 5)$, which factors to $2(x + 1)(x + 5)$.

Find the common factor for each expression below, and then factor the expression as much as possible.

8. $3a^2 + 18a + 15$

9. $2b^2 + 8b - 10$

10. $4x^2 - 8x + 8$

11. $5t^2 + 25t - 70$

Challenge Sometimes a quadratic expression can be factored even though the coefficient of x^2 is not 1 and the terms do not have a common factor. For example, $2x^2 - 9x + 9$ can be factored as $(2x - 3)(x - 3)$. Use strategies like those you've used before to factor these expressions.

12. $3x^2 - 11x - 4$

13. $8x^2 + 2x - 3$

If you are given a quadratic expression to factor, what steps would you follow to factor it or to determine that it can't be factored using integers?

Investigation Solving Quadratics by Factoring

Sometimes you must rearrange the terms in a quadratic expression to see how—or if—the expression can be factored.

Think & Discuss

Mrs. Torres gave his class a number puzzle.

Why is Tamika's suggestion a good one? What quadratic equation should the class find when they finish rearranging?

Can the quadratic expression be factored? If so, what are the factors?

What numbers could Mrs. Torres have started with? Check your answer in his original number puzzle.

Remember

When solving an equation involving algebraic fractions, always check that the apparent solutions don't make any denominators in the original equation equal to 0.

Problem Set E

Rearrange each equation so you can solve by factoring. Find the solutions.

1. $4a + 3 = 6a + a^2$

2. $b^2 - 12 = 4b$

3. $c(c + 4) + 3c + 12 = 0$

4. $d + \frac{6}{d} = 5$

5. $\frac{(x + 3)(x - 2)^2}{x - 2} = 3x - 3$ (Hint: Simplify the fraction first.)

6. Kenyon challenged his teacher, Ms. Hiroshi, with a number puzzle: "I'm thinking of a number. If you multiply my number by 2 more than the number, the result will be 1 less than four times my number."

    a. Write an equation for Kenyon's puzzle, and then use factoring to solve it. Check that your answer fits the puzzle.

    b. Kenyon expected his teacher to find two solutions to his puzzle. Why didn't she?

7. A rectangular rug has an area of 15 square meters. Its length is 2 meters more than its width.

    a. Write an equation to show the relationship between the rug's area and its width.

    b. Solve your equation. Explain why only one of the solutions is useful for finding the rug's dimensions.

    c. What are the dimensions of the rug?

8. When 20 is added to a number, the result is the square of the number. What could the number be? Show how you found your answer.

9. The sum of the squares of two consecutive integers is 145. Find all possibilities for the integers. Show how you found your answer.

10. Gabriela was trying to solve the equation $(x + 1)(x - 2) = 10$. This is how she reasoned:

> Two factors of 10 are 5 and 2.
> So, $x + 1 = 5$ must be one solution of the equation.
> That means $x = 4$.
> I'll check: $(4 + 1)(4 - 2) = 5 \cdot 2 = 10$.
> It checks!

a. What would have happened if Gabriela had guessed that $x - 2 = 5$?

b. Do you think Gabriela's method is an efficient way to solve quadratic equations? Explain.

c. Solve Gabriela's equation, $(x + 1)(x - 2) = 10$. Start by expanding and then rearranging. Check each solution.

d. Solve $(x + 5)(x - 2) = 30$.

Just
the **facts**

The quadratic equation $K = \frac{1}{2}(64)v^2$ gives the kinetic energy of a skydiver with a mass of 64 kg (about 141 lb) falling through the sky with velocity v in m/s.

Share & Summarize

1. Make up a problem involving area that requires solving a quadratic equation.

2. Try to solve your problem by factoring. If you can, give the solutions of the equation and then answer the question. If not, explain why the expression can't be factored.

On Your Own Exercises

Solve each equation.

1. $(x + 5)(x + 7) = 0$ **2.** $(x - 5)(x + 7) = 0$

3. $(x - 5)(x - 7) = 0$ **4.** $(x + 5)(x - 7) = 0$

In Exercises 5–14, determine whether the expression on the left of the equal sign is a difference of squares or a perfect square trinomial. If it is, indicate which and then factor the expression and solve the equation for x. If the expression is in neither form, say so.

5. $x^2 - 49 = 0$ **6.** $x^2 + 49 = 0$

7. $x^2 + 14x - 49 = 0$ **8.** $x^2 - 14x + 49 = 0$

9. $49 - x^2 = 0$ **10.** $x^2 + 14x + 49 = 0$

11. $a^2x^2 + 4ab + b^2 = 0$ **12.** $a^2x^2 + 4abx + 4b^2 = 0$

13. $m^2x^2 - n^2 = 0$ **14.** $m^2x^2 + n^2 = 0$

Factor each quadratic expression that can be factored using integers. Identify those that cannot, and explain why they can't be factored.

15. $d^2 - 15d + 54$ **16.** $g^2 - g - 6$

17. $z^2 + 2z - 6$ **18.** $h^2 - 3h - 28$

19. $2x^2 - 8x - 10$ **20.** $3c^2 - 9c + 6$

Solve each equation by factoring using integers, if possible. If an equation can't be solved in this way, explain why.

21. $k^2 + 15k + 30 = 0$ **22.** $n^2 - 17n + 42 = 0$

23. Challenge $2b^2 - 21b + 10 = 0$

24. Challenge $8r^2 + 5r - 3 = 0$

25. $4x + x^2 = 21$ **26.** $h^2 + 12 = 3h$

27. $14e = e^2 + 24$ **28.** $g^2 + 64 = 16g$

29. $u^2 + 5u = 36$ **30.** $(x + 3)(x - 4) = 30$

31. $\dfrac{(x + 1)^3}{x + 1} = 5x + 5$ (Hint: Simplify the fraction first.)

32. Carlos multiplied a number by itself and then added 6. The result was five times the original number. Write and solve an equation to find his starting number.

33. Because 7 isn't a perfect square, the expression $4x^2 - 7$ doesn't look like the difference of two squares. But 7 *is* the square of *something*.

    a. What is 7 the square of?

    b. How can you use your answer to Part a to factor $4x^2 - 7$ into a product of two binomials?

34. Geometry Each of these expressions represents one of the shaded areas below.

    i. $D^2 - d^2$ **ii.** $\pi(r + w)^2 - \pi r^2$

    iii. $(d + w)^2 - d^2$ **iv.** $4r^2 - \pi r^2$

 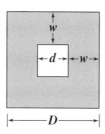 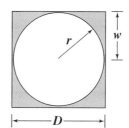 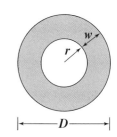

      Figure A   **Figure B**   **Figure C**   **Figure D**

    a. Match each expression with one of the shaded areas so that *every* figure is matched to a different expression.

    b. Write each expression in factored form. Factor out common factors, if possible.

Factor the expression on the left side of each equation as much as possible, and find all the possible solutions. It will help to remember that $x^4 = (x^2)^2$, $x^8 = (x^4)^2$, and $x^3 = x(x^2)$.

35. $x^4 - 1 = 0$ **36.** $x^8 - 1 = 0$

37. $x^3 - 16x = 0$ **38.** $x^3 - 6x^2 + 9x = 0$

39. $x^4 - 2x^2 + 1 = 0$ **40.** $x^4 + 2x^2 + 1 = 0$

Solve each equation. Be sure to check your answers. (Hint: Try factoring the numerator first.)

41. $\dfrac{x^2 + 6x + 9}{x + 3} = 10$ **42.** $\dfrac{16x^2 - 81}{4x + 9} = 31$

In your
own
words

List the steps you
would follow to
solve a quadratic
equation by factor-
ing, or to decide
that it can't be
solved that way.

43. Challenge When you simplify algebraic expressions, sometimes the simplified expression is not equivalent to the original for all values of the variable. For example, consider this expression:

$$\frac{5a + 10}{a^2 - 4}$$

a. Factor the denominator. For what values of a is the expression undefined? That is, for what values is the denominator equal to 0?

b. Now write the expression above using factored forms for both the numerator and denominator. Be sure to look for common factors in the terms.

c. Simplify the fraction.

d. Now try to evaluate the fraction using each value that made the original expression undefined. (You found those values in Part a.)

e. You should have seen in Part d that the simplified fraction is not equivalent to the original fraction for *all* values of a. Explain why this happened.

f. When you simplify an algebraic fraction, you should note any values of the variable that make the simplified fraction unequal to the original. For example, the fraction $\frac{x(x + 1)}{3x}$ can be simplified as $\frac{x + 1}{3}$, where $x \neq 0$.

Simplify the fraction $\frac{2m + 1}{4m^2 - 1}$.

44. The Numkenas built a small, square patio from square bricks with sides of length 1 foot. They bought just enough bricks to build the patio, but after they built it they decided it was too small.

To extend the length and width of the patio by d feet, they had to buy 24 more bricks. The original side length of the patio was 5 feet.

a. Draw a diagram to represent this situation. Be sure to show both the original patio and the new one.

b. Write an equation to represent this situation.

c. Simplify your equation and solve it to find d, the amount by which the patio's length and width were increased. Check your answer.

45. The *triangular numbers* are a sequence of numbers that begins

$$1, 3, 6, 10, \ldots$$

The numbers in this sequence represent the number of dots in a series of triangular shapes.

Triangle 1   Triangle 2   Triangle 3   Triangle 4

T, the number of dots in Triangle *n,* is given by this quadratic equation:

$$T = \tfrac{1}{2}(n^2 + n)$$

Some of the following numbers are triangular numbers. For each possible value of *T,* set up an equation and try to solve it for *n* by factoring. If an equation cannot be factored using integers, *T* cannot be a triangular number. Indicate which numbers are not triangular.

a. 55   **b.** 120   **c.** 150   **d.** 200   **e.** 210

46. Geometry In Chapter 2, you learned that the number of diagonals in an *n*-sided polygon is given by the equation $D = \dfrac{n^2 - 3n}{2}$. Some examples are shown below.

Triangle   Quadrilateral   Pentagon

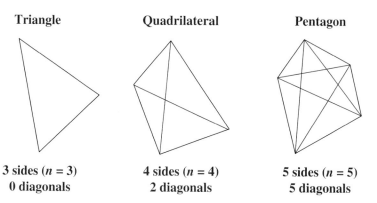

3 sides (*n* = 3)   4 sides (*n* = 4)   5 sides (*n* = 5)
0 diagonals   2 diagonals   5 diagonals

Some of the following numbers are the number of diagonals in a polygon. For each possible value of *D,* set up an equation and try to solve it for *n* by factoring. If an equation can't be factored using integers, *D* cannot be the number of diagonals in a polygon. Indicate which values cannot be the number of diagonals in a polygon.

a. 20   **b.** 30   **c.** 35   **d.** 50   **e.** 54

Mixed Review

Solve each equation by doing the same thing to both sides.

47. $\frac{3k-5}{5} + k = 5 + k$ **48.** $7.5a - 6 = 5a + 4$

Determine whether the points in each set are collinear. Explain how you know.

49. $(^-2, 13), (1.5, ^-4.5), (3, ^-12)$ **50.** $(^-1, ^-4.2), (3, 0.6), (4, 1.6)$

Determine whether the values in each table could represent a linear relationship, a quadratic relationship, or neither. Explain your answers.

51.

x	−3	−2	−1	0	1	2	3
y	−12.6	−9.2	−5.8	−2.4	1	4.4	7.8

52.

x	−3	−2	−1	0	1	2	3
y	−24	−13	−6	0	4	5	6

53.

x	−3	−2	−1	0	1	2	3
y	0	−2	−2	0	4	10	18

54. Astronomy A light-year, the distance light travels in one year, is 5.88×10^{12} miles. Answer these questions without using your calculator.

a. The star Alpha Centauri is about 4 light-years from Earth. Write this distance in miles, using scientific notation.

b. The star Betelgeuse is about 500 light-years from Earth. Write this distance in miles, using scientific notation.

c. Suppose a light beam went from Earth to Betelgeuse, and another light beam went from Earth to Alpha Centauri. How much farther did the beam going to Betelgeuse have to travel?

Tell whether each figure has reflection symmetry, rotation symmetry, or both.

55.

56.

57.

7.3 Completing the Square

Factoring is a very useful tool for solving equations. But factors and squares are not always easy to find. In Lessons 7.3 and 7.4, you will learn some techniques that will enable you to solve *every* quadratic equation.

You have used "doing the same thing to both sides" to solve linear equations. In this strategy, you write a series of equivalent equations that have the same solutions as the original equation but are easier to solve. You can also use this strategy with equations that contain square roots.

EXAMPLE

Solve $\sqrt{3m + 7} = 5$.

$3m + 7 = 25$	after squaring both sides
$3m = 18$	after subtracting 7 from both sides
$m = 6$	after dividing both sides by 3

Think & Discuss

Why are both sides squared in the first solution step of the Example?

In general, what kinds of "same things" do you know you can do to both sides to solve an equation?

Would you get an equivalent equation if you added 1 to the numerator of the fractions on both sides of an equation? Try it with $\frac{x}{2} = \frac{2x}{4}$.

What happens to an equation—such as $x = 2$—when you multiply both sides by x? Does the new equation have the same solutions?

What happens to the set of solutions when you multiply both sides of an equation—such as $x = 2$—by 0?

What effect would squaring both sides have on an equation? For example, begin with $x = 5$.

When you take the square root of both sides of an equation, what should you do to keep the same set of solutions? For example, begin with $w^2 = 36$.

Investigation Finding Perfect Squares

If you can rearrange a quadratic equation into a form with a quadratic expression that is a perfect square on one side and a constant on the other side, you can solve the equation by taking the square root of both sides.

EXAMPLE

Solve the equation $x^2 + 2x + 1 = 7$.

First, notice that $x^2 + 2x + 1$ is equal to $(x + 1)^2$, so the equation $x^2 + 2x + 1 = 7$ can be solved by taking the square root of both sides:

$$(x + 1)^2 = 7$$
$$\sqrt{(x + 1)^2} = \sqrt{7} \text{ or } ^-\sqrt{7}$$

To write "$\sqrt{7}$ or $^-\sqrt{7}$" more easily, use the $\pm$ symbol: $\pm\sqrt{7}$ refers to both numbers, $\sqrt{7}$ and $^-\sqrt{7}$.

$$\sqrt{(x + 1)^2} = \pm\sqrt{7}$$
$$x + 1 = \pm\sqrt{7}$$
$$x = ^-1 \pm \sqrt{7}$$

So the solutions are $^-1 + \sqrt{7}$ and $^-1 - \sqrt{7}$.

Remember

An *exact solution does* not involve approximations. For example, $x = \sqrt{2}$ is an exact solution of $x^2 = 2$, while $x = 1.414$ is an approximate solution to the nearest thousandth.

Problem Set A

Find exact solutions of each equation, if possible, using any method you like.

1. $(x - 3)^2 = 36$

2. $(k - 1)^2 - 25 = 0$

3. $2(r - 7)^2 = 32$

4. $(a - 4)^2 + 2 = 0$

5. $2(b - 3)^2 + 5 = 55$

6. $3(2c + 5)^2 - 63 = 300$

7. $(x - 4)^2 = 3$

8. $2(r - 3)^2 = ^-10$

9. $4(x + 2)^2 - 3 = 0$

10. Find approximate solutions of the equations in Problems 7 and 9 to the nearest hundredth.

To use the solution method demonstrated in the Example on page 457, you need to be able to recognize quadratic expressions that can be rewritten as perfect squares. You worked with such *perfect square trinomials* in the last lesson.

Problem Set B

1. Which of these are perfect squares?

a. $x^2 + 6x + 9$ **b.** $b^2 + 9$ **c.** $x^2 + 6x + 4$

d. $m^2 + 12m - 36$ **e.** $m^2 - 12m + 36$ **f.** $y^2 + y + \frac{1}{4}$

g. $r^2 - 16$ **h.** $1 + 2r + r^2$ **i.** $y^2 - 2y - 1$

2. Which of these are perfect squares?

a. $4p^2 + 4p + 1$ **b.** $4q^2 + 4q + 4$ **c.** $4s^2 - 4s - 1$

d. $4t^2 - 4t + 1$ **e.** $4v^2 + 9$ **f.** $4w^2 + 12w + 9$

3. Describe how you can tell whether an expression is a perfect square, without factoring it.

Suppose you know the x^2 and x terms in a quadratic expression, and you want to make it into a perfect square trinomial by adding a constant. How can you find the missing term?

EXAMPLE

If $x^2 + 20x +$ ___ is a perfect square, then

$$x^2 + 20x + \_\_\_ = (x + ?)^2$$

Since the middle term of the expansion is twice the product of the coefficient of x and the constant term in the binomial being squared,

$$20 = (2)(1)(?)$$
$$10 = ?$$

So the perfect square must be $(x + 10)^2$, or $x^2 + 20x + 100$.

Just the **facts**

The quadratic equation $W = \frac{1}{2}kx^2$ describes the amount of work (in a unit called *Joules*) needed to stretch a spring x cm beyond its normal length. The value of k depends on the spring's strength.

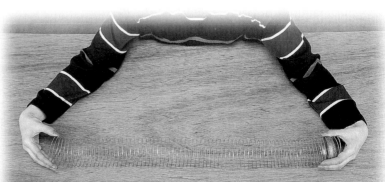

Complete each quadratic expression to make it a perfect square. Then write the completed expression in factored form.

1. $x^2 - 18x +$ ___   **2.** $x^2 + 22x +$ ___

3. $k^2 - 3k +$ ___   **4.** $25m^2 + 10m +$ ___

5. $16r^2 - 8r \;\square$ ___   **6.** $4z^2 - 12z \;\square$ ___

Share & Summarize

1. Why is it useful to look for perfect squares in quadratic expressions?

2. How can you recognize a perfect square trinomial?

Investigation 2 ▶ Solving Quadratics by Completing the Square

In Problem Set A, you learned that it is easy to solve a quadratic equation with a perfect square on one side and a constant on the other. You also solved equations that had a perfect square with a constant added or subtracted. A technique called *completing the square* can be used to rearrange quadratic equations into this form.

In Problem Set C, you found the constant that should be added to transform a quadratic expression into a perfect square. Using the same idea, some expressions that are not perfect squares can be rewritten as perfect squares with a constant added or subtracted.

EXAMPLE

$x^2 + 6x + 10$ is not a perfect square. For $x^2 + 6x +$ ___ to be a perfect square, the added constant must be 9, because $x^2 + 6x + 9$ is a perfect square. (Can you see why?)

This means $x^2 + 6x + 10$ is 1 more than a perfect square. We can use this to rewrite the expression as a square plus 1:

$$x^2 + 6x + 10 = (x^2 + 6x + 9) + 1$$
$$= (x + 3)^2 + 1$$

Problem Set D

Rewrite each expression as a square with a constant added or subtracted.

1. $x^2 + 6x + 15 = x^2 + 6x + 9 +$ ___ $= (x + 3)^2 +$ ___

2. $k^2 - 6k + 30 = k^2 - 6k + 9 +$ ___ $= (k - 3)^2 +$ ___

3. $s^2 + 6s - 1 = s^2 + 6s + 9 -$ ___ $= (s +$ ___$)^2 -$ ___

4. $r^2 - 6r - 21 = r^2 - 6r + 9 \;\square$ ___ $= (r \;\square$ ___$)^2 \;\square$ ___

5. $m^2 + 12m + 30$

6. $h^2 - 5h$

7. $9r^2 + 18r - 20$

8. $9n^2 - 6n + 11$

Marcus and Lydia want to solve the equation $x^2 - 6x - 40 = 0$.

This method of solving equations is called *completing the square.*

Problem Set E

Find exact solutions of each equation by completing the square.

1. $x^2 - 8x - 9 = 0$

2. $w^2 - 8w + 6 = 0$

3. $9m^2 + 6m - 8 = 0$

What can you do if the coefficient of the squared variable is not a square? One approach is to first do the same thing to each side to produce an equivalent equation with 1—or some other square number—as the coefficient of the squared variable.

Problem Set F

1. To solve $2x^2 - 8x - 1 = 0$, you could divide both sides by 2, which gives the equivalent equation $x^2 - 4x - \frac{1}{2} = 0$. Complete the solution by solving this equivalent equation.

2. Use the method from Problem 1 to solve $2m^2 - 12m + 7 = 0$.

3. Consider the equation $18x^2 - 12x - 3 = 0$.

   a. Try dividing the equation by 18 to make the coefficient of x^2 equal to 1.

   b. Now think about the coefficient 18. Find another number you could divide 18 by to get a perfect square. Divide the equation by that number.

   c. Use your answer to Part a or Part b to solve the equation.

4. Explain why $x^2 + 64 = 16x$ has only one solution.

5. Explain why $g^2 - 4g + 11 = 0$ has no solutions.

Share & Summarize

1. Give an example of a quadratic equation that is not a perfect square but that is easy to solve by completing the square. Solve your equation.

2. Suppose you have an equation in the form $y = ax^2 + bx + c$ for which the coefficient of x^2 is not a perfect square. How can you solve the equation? Illustrate your answer with an example.

Practice & Apply

Solve each equation.

1. $(x + 3)^2 = 25$

2. $(r - 8)^2 + 3 = 52$

3. $(2m + 1)^2 - 4 = 117$

4. $3(x - 3)^2 = 30$

5. $^-2(y - 7)^2 + 4 = 0$

6. $4(2z + 3)^2 - 2 = ^-1$

Complete each quadratic expression so that it is a perfect square. Then write the completed expression in factored form.

7. $x^2 - 8x \ \square$ ___

8. $b^2 + 9b \ \square$ ___

9. $81d^2 - 90d \ \square$ ___

Rewrite each expression as a square with a constant added or subtracted.

10. $r^2 - 6r + 1 = r^2 - 6r + 9 +$ ___ $= (r \ \square$ ___$)^2 -$ ___

11. $r^2 + 6r + 6 = (r \ \square$ ___$)^2 \ \square$ ___

12. $p^2 - 16p + 60$

13. $g^2 - 3g - 1$

14. $a^2 + 10a + 101$

15. $4x^2 + 4x + 2$

Solve each equation by completing the square.

16. $m^2 + 2m - 11 = 0$

17. $b^2 - 3b = 3b + 7$

18. $x^2 - 6x = ^-5$

19. $a^2 + 10a + 26 = 0$

20. $2x^2 + 4x - 1 = 0$

21. $2u^2 + 3u - 2 = 0$

Just the facts

The acceleration, in m/s^2, of a bicyclist coasting down a hill might be given by the quadratic equation $a = 0.12 - 0.0006v^2$, where v is the bike's velocity in m/s.

impactmath.com/self_check_quiz

22. Stephen, Consuela, and Kwame each made up a number puzzle for their teacher, Mr. Karnowski.

- Stephen said, "I'm thinking of a number. If you subtract 1 from my number, square the result, and add 5, you will get 4."

- Consuela said, "I'm thinking of a number. If you subtract 1 from my number, square the result, and add 1, you will get 1."

- Kwame said, "I'm thinking of a number. If you double the number, subtract 5, square the result, and add 1, you will get 10."

After thinking about the puzzles, Mr. Karnowski said, "One of your puzzles has one solution, one of them has two solutions, and one doesn't have a solution."

Whose puzzle is which? Write an equation for each puzzle, and explain your answer.

23. Sports Brianna and Lucita are playing tennis. On one volley, the height of the ball h, in feet, could have been described with the following equation, where t is the time in seconds since Brianna hit the ball:

$$h = {}^{-}16(t - 1)^2 + 20$$

Assuming Lucita will let the ball bounce once, when will it hit the ground? Write and solve an equation to help you answer this question. Give your answer to the nearest hundreth of a second.

24. When you start the process of completing the square for an equation, you may be able to tell whether the equation has solutions without solving it.

a. Express each of these using a perfect square plus a constant. Without solving, decide whether the equation has a solution, and explain your answer.

   i. $x^2 + 6x + 15 = 0$

   ii. $x^2 + 6x + 5 = 0$

b. State a rule for determining whether an equation of the form $(x + a)^2 + c = 0$ has solutions. Explain your rule.

25. Geometry A rectangular painting has an area of 25 square feet. One side is 2 feet longer than the other.

a. Quickly estimate approximate values for the lengths of the painting's sides. Do your estimates give an area that is too large or too small?

b. Write an equation relating the sides and the area of the painting and solve it exactly by completing the square.

c. Compare your answer in Part b to your approximation in Part a.

26. History When the famous German mathematician Gauss was a young boy, he amazed his teacher by rapidly computing the sum of the integers from 1 to 100. He realized that he could compute the sum without adding all the numbers, by grouping the 100 numbers into pairs.

To see a shortcut for finding this sum, look at two lists of 1 to 100, one in reverse order.

1	2	3	4	5	6	7	...	50	...	94	95	96	97	98	99	100
100	99	98	97	96	95	94	...	51	...	7	6	5	4	3	2	1

a. What is the sum of each pair?

b. How many pairs are there?

c. What is the sum of all these pairs?

d. How many times is each of the integers from 1 to 100 counted in this sum?

e. Consider your answers to Parts c and d. What is the sum of the integers from 1 to 100?

f. Explain how you can use this same reasoning to find the sum of the integers from 1 to n for any value of n. Write a formula for s, the sum of the first n positive integers.

g. Chloe added several consecutive numbers, starting at 1, and found a sum of 91. Write an equation you could use to find the numbers she added. Solve your equation by completing the square. Check your answer with the formula.

In y o u r
own
words

Why are perfect squares useful in solving quadratic equations?

Mixed Review

Identify the values of *a*, *b*, and *c* in each equation by rearranging it into the form of the general quadratic equation, $ax^2 + bx + c = 0$.

27. $2x^2 - 7x = {}^-5$

28. $8a + 2 = 9a^2$

29. $4.5k^2 + 3k = {}^-3 + k + k^2$

30. ${}^-m - 2 = m^2 - m - 3$

31. $4 = {}^-p^2$

32. $7 - w^2 = w + 2.5w^2$

Graph each inequality on a separate grid like the one shown.

33. $y \geq x - 3$

34. $y < 3 - x$

35. $y \leq 1.5x + 3$

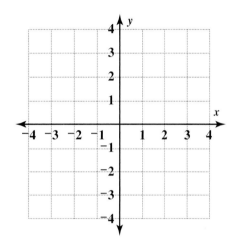

36. Geometry Match each solid to its name. (Hint: For the ones you are unsure of, think about what the term might mean.)

a. square pyramid

b. cone

c. cylinder

d. triangular prism

e. oblique prism

f. hexagonal prism

g. octahedron

h. tetrahedron

i. hemisphere

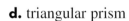
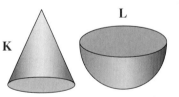

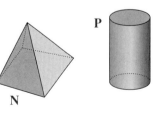
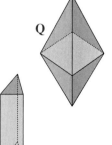

7.4 The Quadratic Formula

You have seen that some quadratic equations are easier to solve than others. Some can be solved quickly by factoring or by taking the square root of both sides. Any quadratic equation can be solved by completing the square, but it is not always obvious what has to be done.

Look again at the process of solving an equation by completing the square. The general quadratic equation

$$ax^2 + bx + c = 0$$

is solved below by completing the square. To help you see each step more easily, a specific quadratic equation is solved alongside the general equation.

General Equation	**Specific Equation**
$ax^2 + bx + c = 0$	$2x^2 + 8x + \frac{1}{2} = 0$

Step 1: Divide by a.

$$x^2 + \frac{b}{a}x + \frac{c}{a} = 0 \qquad\qquad x^2 + 4x + \frac{1}{4} = 0$$

Step 2: Complete the square.

$$x^2 + \frac{b}{a}x + \frac{b^2}{4a^2} + \frac{c}{a} - \frac{b^2}{4a^2} = 0 \qquad\qquad x^2 + 4x + 4 + \frac{1}{4} - 4 = 0$$

Step 3: Rearrange.

$$x^2 + \frac{b}{a}x + \frac{b^2}{4a^2} = \frac{b^2}{4a^2} - \frac{c}{a} \qquad\qquad x^2 + 4x + 4 = \frac{15}{4}$$

$$\left(x + \frac{b}{2a}\right)^2 = \frac{b^2 - 4ac}{4a^2} \qquad\qquad (x + 2)^2 = \frac{15}{4}$$

Step 4: Take the square root of both sides.

$$x + \frac{b}{2a} = \frac{\pm\sqrt{b^2 - 4ac}}{2a} \qquad\qquad x + 2 = \pm\frac{\sqrt{15}}{2}$$

Step 5: Subtract the constant added to x.

$$x = \frac{^-b \pm \sqrt{b^2 - 4ac}}{2a} \qquad\qquad x = {}^-2 \pm \frac{\sqrt{15}}{2}$$

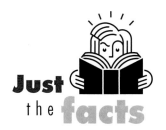

This process gives us a formula that can be used to find the solutions of any quadratic equation.

The Quadratic Formula

The solutions of $ax^2 + bx + c = 0$ are

$$x = \frac{-b \pm \sqrt{b^2 - 4ac}}{2a}$$

That is, the solutions are

$$x = \frac{-b + \sqrt{b^2 - 4ac}}{2a} \quad \text{and} \quad x = \frac{-b - \sqrt{b^2 - 4ac}}{2a}$$

Think & Discuss

Use the quadratic formula to solve this equation.

$$x^2 + 3x - 5 = 0$$

Now solve the equation by completing the square. Do you get the same answer? Which method seems easier?

Investigation ▶ 1 ▶ Using the Quadratic Formula

This investigation will help you learn to use the quadratic formula.

Problem Set A

For Problems 1–8, do the following:

- Identify the values of *a, b,* and *c* referred to in the quadratic formula. You may need to rewrite the equation in the form $ax^2 + bx + c = 0$.

- Solve the equation using the quadratic formula.

1. $2x^2 + 3x = 0$ **2.** $7x^2 + x - 3 = 0$

3. $3 - x^2 + 2x = 0$ **4.** $6x + 2 = x^2$

5. $2x^2 = x - 5$ **6.** $x^2 - 12 = 0$

7. $x^2 = 5x$ **8.** $x(x - 6) = 3$

9. Consider the equation $x^2 + 3x + 2 = 0$.

   a. Solve the equation by factoring.

   b. Now use the quadratic formula to solve the equation.

   c. Which method seems easier?

   d. Which of Problems 1–8 could you have solved by factoring?

Problem Set B

For each problem, do the following:

• Solve the equation using any method you like. Check your answers.

• If you did not solve the equation by factoring, decide whether you could have used factoring to solve it.

1. $x^2 - 5x + 6 = 0$
        2. $w^2 - 6w + 9 = 0$

3. $t^2 + 4t + 1 = 0$
        4. $x^2 - x + 2 = 0$

5. $k^2 + 4k + 2 = 0$
        6. $3g^2 - 2g - 2 = 0$

7. $z^2 - 12z + 36 = 0$
        8. $2e^2 + 7e + 6 = 0$

9. $x^2 + x = 15 - x$
        10. $3n^2 + 14 = 8n^2 + 3n$

Share & Summarize

1. What is the connection between the quadratic formula and the process of completing the square?

2. You have learned several methods for solving quadratic equations: backtracking, factoring, completing the square, and the quadratic formula.

   a. Which of these can be used with only some quadratic equations?

   b. Which can be used with all quadratic equations?

3. When you are given a quadratic equation to solve, how do you choose a solution method?

Investigation 2 ▶ Applying the Quadratic Formula

In Chapter 4 you examined quadratic equations in specific situations and estimated solutions using a graphing calculator. Now you will apply the quadratic formula to solve them exactly.

Problem Set C

In some of these problems, the quadratic formula will give two solutions. Make sure your answers make sense in the problem's context.

1. Josefina, a tapestry maker, has a client who wants a tapestry with an area of 4 square meters. Josefina decides it will have a rectangular shape and a length 1 meter longer than its width.

   a. Write an equation representing the client's requirements and Josefina's decision.

   b. Use the quadratic formula to find the width and the length of the tapestry. Express your answer in two ways, exactly and to the nearest centimeter.

2. Another client wants a rectangular tapestry with an area of 6 m^2, but she insists the length be exactly 2 m longer than the width.

   Write and solve an equation that represents this situation. Give exact dimensions of the tapestry, and then estimate the dimensions to the nearest centimeter.

3. Jesse threw a superball to the ground, and it bounced straight up with an initial speed of 30 feet per second. The height h in feet t seconds after the ball left the ground is given by the formula $h = 30t - 16t^2$. Write and solve an equation to find the value of t when the ball returns to the ground.

Remember

Motion equations such as $h = 30t - 16t^2$ and $h = {}^-16t^2 + 20$ only give estimates of an object's position because they ignore air resistance.

4. When an object is thrown straight upward, its height h in feet after t seconds can be estimated using the formula $h = s + vt - 16t^2$, where s is the initial height (at $t = 0$) and v is the initial velocity. In Problem 3, s was 0 feet and v was 30 feet per second, so the ball's height was estimated by $h = 30t - 16t^2$.

Suppose Jesse threw the ball upward instead of bouncing it, so that the ball's height when $t = 0$ was 5 ft above the ground but the initial velocity was still 30 ft/s.

   a. Write an equation describing the height h of Jesse's ball after t seconds.

   b. Write and solve an equation to find how long it takes the ball to reach the ground.

5. If an object is dropped with an initial velocity of 0, its height can be approximated by adding its starting height to $^-16t^2$, which represents the effect of gravity. For example, the height of a rock dropped from 20 feet above the ground can be approximated by the formula $h = {}^-16t^2 + 20$.

Write and solve an equation to determine how many seconds pass until a rock dropped from 100 ft hits the ground.

Problem Set D

The town of Seaside, which now has only a few small hotels, is considering allowing a large tourist resort to be built along the oceanfront. Some residents are in favor of the plan because it will bring income to the community. Others are against it, saying it will disrupt their lifestyle. The state tourism board has a formula for computing the overall tourism rating T of an area based on two factors: U, the uniqueness rating, and A, the amenities rating.

1. The amenities rating scale, A, is used to assess the attractiveness of a tourist destination, including how easy it is to find a place to stay. Seaside currently has a rating of 5. It is estimated that for every 100 beds the resort opens to tourists, A will increase by 2 points.

   If the resort has p hundreds of beds, what is the estimate for the new amenities rating?

2. The uniqueness rating scale, U, is used to assess the special features that will attract tourists. Seaside currently has a high uniqueness rating, 20, because dolphins are often sighted close to the local beaches. A committee has gathered evidence that an increase in tourists will keep dolphins from coming near shore. They estimate that for every 100 beds in the resort, U will drop by 2 points.

   If the resort has p hundreds of beds, what is the estimate for the new uniqueness rating?

3. The overall tourism rating, T, is computed by multiplying A and U.

    a. What is the town's current tourism rating?

    b. Write an expression in terms of p for the estimated tourism rating if p hundreds of beds are added.

    c. Use your expression from Part b to decide for what values of p there would be no change in the tourism rating. (Hint: Write and solve an equation.)

    d. For what numbers of beds would the resort create a decrease in the tourism rating?

    e. Seaside's town council believes that the disruption to the town's lifestyle could not be justified unless the development resulted in an increase in the tourism rating to at least 140 points.

        i. Use your expression from Part b to decide for what values of p you would expect to achieve a tourism rating of 140 points.

        ii. What values of p would give a tourism rating *over* 140 points?

    f. What would you advise the council to do?

Share & Summarize

Explain the steps involved in using the quadratic formula to solve an equation.

Investigation What Does $b^2 - 4ac$ Tell You?

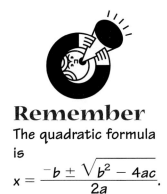

Remember

The quadratic formula is

$$x = \frac{-b \pm \sqrt{b^2 - 4ac}}{2a}.$$

In some situations, you might be more interested in knowing *how many* solutions a quadratic equation has than exactly what the solutions *are*. For example, suppose you are thinking about the height of a thrown ball at various times. You could solve a quadratic equation to find at what time the ball reaches a certain height. But if you want to know only *whether* it reaches that height and don't care about *when* it does, the question you want to answer is

Does this equation have any solutions?

You need only part of the quadratic formula—the expression $b^2 - 4ac$—to answer this question.

Think & Discuss

You have seen examples of quadratic equations that have no solutions. Sometimes this is easy to tell without using the quadratic formula. Give an example of such an equation, and explain how you know that it doesn't have solutions.

Some quadratic equations have exactly one solution. Give some examples.

Of course, many quadratic equations have two solutions. Give an example.

Problem Set E

You will now investigate the relationship between the value of $b^2 - 4ac$ and the number of solutions of $ax^2 + bx + c = 0$.

1. The equation $x^2 + 1 = 0$ has no solutions.

a. Explain why this is true.

b. What is the value of $b^2 - 4ac$ for this equation? Is it positive, negative, or 0?

c. Give another example of a quadratic equation that you know has no solutions. Find the value of $b^2 - 4ac$ for your example: is it positive, negative, or 0?

2. The equation $(x - 3)(x + 5) = 0$ has two solutions.

a. Express this equation in the form $ax^2 + bx + c = 0$ and find the value of $b^2 - 4ac$. Is it positive, negative, or 0?

b. Give another example of a quadratic equation with two solutions. Find the value of $b^2 - 4ac$ for your equation.

3. The expression $x^2 + 2x + 1$ is a perfect square trinomial since it is equal to $(x + 1)^2$. The equation $x^2 + 2x + 1 = 0$ has one solution.

a. Is the value of $b^2 - 4ac$ for this equation positive, negative, or 0?

b. Give another example of a quadratic equation with one solution. Is the value of $b^2 - 4ac$ for your equation positive, negative, or 0?

4. As you know, a quadratic equation can have zero, one, or two solutions. This problem will help you explain the connection between the value of $b^2 - 4ac$ and the number of solutions an equation has.

a. Where does the expression $b^2 - 4ac$ occur in the quadratic formula?

b. What value or values must $b^2 - 4ac$ have for the quadratic formula to give no solutions? Explain.

c. What value or values must $b^2 - 4ac$ have for the quadratic formula to give one solution? Explain.

d. What value or values must $b^2 - 4ac$ have for the quadratic formula to give two solutions? Explain.

Just the facts

The acceleration of the passengers on this carnival ride might be given by the quadratic equation $a = 0.2v^2$, where a is the acceleration toward the center of the ride in m/s², and v is the velocity in m/s.

Problem Set F

Find the number of solutions each equation has.

1. $2x^2 - 9x + 5 = 0$

2. $3x^2 - 7x + 9 = 0$

In Problem Set C, Jesse bounced a superball with a velocity of 30 feet per second as it left the ground. The ball's height is given by the formula $h = 30t - 16t^2$, where t is time in seconds since the ball left the ground.

3. You can use your knowledge of the quadratic formula to find how high the ball will travel. First look at whether the ball will reach 100 feet.

    a. What equation would you solve to find if and when the height of the ball reaches 100 feet?

    b. Write your equation in the form $at^2 + bt + c = 0$.

    c. What is the value of $b^2 - 4ac$ for your equation?

    d. Will the ball reach a height of 100 feet? Explain.

4. Challenge This graph of $h = 30t - 16t^2$ can help you determine just how high the ball will go.

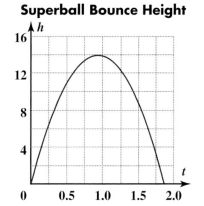

Superball Bounce Height

    a. Suppose M is the maximum height reached by the ball. Write an equation to represent when the ball is at this height.

    b. Write your equation in the form $at^2 + bt + c = 0$.

    c. How many solutions will this equation have? Hint: Look at the graph.

    d. What do you know about the value of $b^2 - 4ac$ for a quadratic equation with the number of solutions that this equation has?

    e. Use your answer to Part d to help you find the value of M. Show how you found your answer.

    f. How high does the ball travel?

    g. Write and solve an equation to find how long it takes the ball to reach this height.

1. Without actually solving it, how can you tell whether a quadratic equation has zero, one, or two solutions?

2. For a quadratic relationship in the form $y = ax^2 + bx + c$, how can you tell whether y ever has a certain value d?

Lab Investigation ▶ The Golden Ratio

MATERIALS
- ruler
- graph paper (optional)

In this investigation, you will work with a ratio that has been important since the time of the early Greeks. The ratio arises in many surprising places, including mathematics, art, music, architecture, and genetics.

What Do You Like?

1. Here are several rectangles. Which do you think is the most "appealing to the eye"? (You don't need reasons for your answers; just say which one you like.)

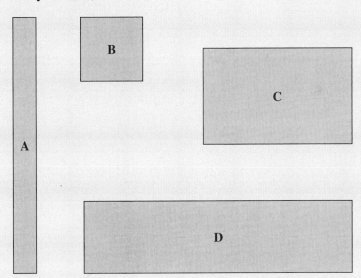

2. Draw some other rectangles that have a shape that is pleasing to you. Explain why you think one shape is more pleasing than another.

Many people think Rectangle C—and rectangles geometrically similar to it—is the most pleasing to the eye. It is called a *golden rectangle,* and the ratio of its sides (the ratio of the long side to the short side) is the *golden ratio.*

One special property of a golden rectangle is that, when you add a square to its longer side to form a new rectangle, the new shape is similar to the original. So, the new rectangle is a golden rectangle and its sides are in the golden ratio.

Remember

In similar figures, corresponding sides have lengths that share a common ratio and corresponding angles are congruent.

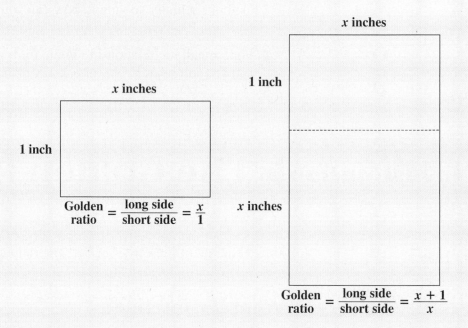

$$\text{Golden ratio} = \frac{\text{long side}}{\text{short side}} = \frac{x}{1}$$

$$\text{Golden ratio} = \frac{\text{long side}}{\text{short side}} = \frac{x+1}{x}$$

Try It Out

3. Measure the dimensions of the two rectangles above and determine whether they have the same $\frac{\text{long side}}{\text{short side}}$ ratio. What is the ratio?

4. Now find the $\frac{\text{long side}}{\text{short side}}$ ratio of each rectangle you drew in Problem 2.

Solve It

5. The ratios of the two rectangles above must be equal. Write an equation setting the ratios equal to each other. That is, complete this equation:

$$\frac{x}{1} = \underline{}$$

6. Now write your equation in the form $ax^2 + bx + c = 0$. (Hint: You will need to think about how to get x out of the denominator.)

7. Find the exact solutions of your equation, and then express the solutions to the nearest thousandth.

8. Do both of your solutions make sense? Explain.

9. What is the value of the golden ratio?

10. Compare the value of the golden ratio with the ratios you measured in Question 3. What do you notice?

11. Now compare the value of the golden ratio to the measurements you made of your own rectangles in Question 4. What do you notice?

Going Further

12. Using graph paper or ordinary paper and a ruler, you can draw a rectangle that's almost a golden rectangle.

a. *Step 1:* Start with a square with side lengths of 1 unit in the top left corner of your page. What is the ratio of the sides?

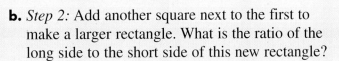

b. *Step 2:* Add another square next to the first to make a larger rectangle. What is the ratio of the long side to the short side of this new rectangle?

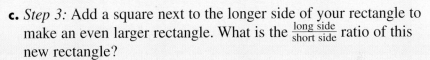

c. *Step 3:* Add a square next to the longer side of your rectangle to make an even larger rectangle. What is the $\frac{\text{long side}}{\text{short side}}$ ratio of this new rectangle?

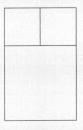

d. Repeat Step 3 as many times as you can on your paper. What is the ratio for the final rectangle you make? Compare this value to the golden ratio you calculated in Problem 9.

Step 4   *Step 5*

13. Look at the dimensions of the rectangles you made in Problem 11.

• The first is 1×1.

• The next four are 1×2, 2×3, 3×5, and 5×8.

• Listing only the smaller dimension in each rectangle gives the *Fibonacci sequence:*

$$1, 1, 2, 3, 5, 8, \ldots .$$

Look for a pattern in the Fibonacci sequence. What are the next two numbers? How did you find them?

14. Compute the sequence of ratios of Fibonacci numbers up to the ratio $\frac{\text{tenth Fibonacci number}}{\text{ninth Fibonacci number}}$, or $\frac{F_{10}}{F_9}$. The first two are computed below.

$$\frac{F_2}{F_1} = \frac{1}{1} = 1 \qquad\qquad \frac{F_3}{F_2} = \frac{2}{1} = 2$$

15. Compare the ratios to the golden ratio. What do you notice?

Finding Out More

16. The golden ratio and the Fibonacci numbers appear in many contexts both inside and outside mathematics. For example, pineapples have scales in sets of 8, 13, and 21 rows.

Look for answers to some of these questions at the library or on the Internet.

• How do the golden ratio and the Fibonacci sequence appear in the natural world?

• How has the golden ratio been used by Leonardo da Vinci and other artists?

• How is the golden ratio applied in architecture?

• How is the golden ratio used in music?

Just the facts

The Fibonacci sequence is named for its discoverer, Leonardo Fibonacci (also known as Leonardo Pisano), who was born about 1170 A.D. in the city of Pisa (Italy). He was one of the first people to introduce the Hindu-Arabic number system—which uses the digits 0 to 9 and a decimal point—into Europe.

On Your Own Exercises

Practice & Apply

Solve each equation using the quadratic formula, if possible.

1. $2x^2 + 5x = 0$

2. $5x^2 + 7x + 4 = 0$

3. $c^2 - 10 = 0$

4. $b^2 + 10 = 0$

5. Solve the equation $9x^2 - 16 = 0$ by factoring and by using the quadratic formula.

6. Geometry The area of a photograph is 320 square centimeters. Its length is 2 cm more than twice its width. Write and solve an equation to find its dimensions.

7. Physical Science Suppose that, at some point into its flight, a particular rocket's height h, in meters, above sea level t seconds after launching depends on t according to the formula $h = 2t(60 - t)$.

    a. How many seconds after launching will the rocket return to sea level?

    b. Write and solve an equation to find when the rocket will be 1,200 m above sea level.

Find the number of solutions to each quadratic equation without actually solving the equation. Explain how you know your answers are correct.

8. $x^2 + 2x + 3 = 0$ **9.** $x^2 - 2x - 3 = 0$ **10.** $9x^2 + 12x + 4 = 0$

11. A ball is thrown upward with a starting velocity of 40 feet per second from 5 feet above the ground. The equation describing the height h of the ball after t seconds is $h = 40t - 16t^2 + 5$.

    a. Will the ball travel as high as 100 feet? Explain.

    b. Will it travel as high as 15 feet? Explain.

    c. Challenge Find the ball's maximum height.

12. When Lourdes solved the equation $2x^2 - 13x = 24$, she was surprised to find that the solutions were exactly 8 and $^-1.5$. Ben said he thought this meant the equation could have been solved by factoring.

  a. Write a quadratic equation in factored form that has the solutions 8 and $^-1.5$.

  b. Expand the factors to write an equation without parentheses. Was Ben correct? (Hint: If your equation contains a fraction, try multiplying by its denominator to get only integers for coefficients.)

  c. Write one advantage and one disadvantage of using the quadratic formula to solve the equation $2x^2 - 13x = 24$.

13. Consider the equation $3x + \frac{1}{x} = 4$.

  a. Do you see any obvious solutions to this equation?

  b. Now solve the equation using the quadratic formula. (Hint: First write an equivalent quadratic equation.) Check your solutions in the original equation.

Challenge Although these equations are not quadratic, the quadratic formula can help you solve them. Try to solve them, and explain your reasoning.

14. $(x^2 - 2x - 2)^2 = 0$ **15.** $x^3 - 2x^2 - 2x = 0$

16. History Here is a problem posed by the 12th-century Indian mathematician Bhaskara:*

The eighth part of a troop of monkeys, squared, was skipping in a grove and delighted with their sport. Twelve remaining monkeys were seen on the hill, amused with chattering to each other. How many were there in all?

That is, take $\frac{1}{8}$ of the entire troop and square the result. That number of monkeys, along with the 12 on the hill, form the entire troop. How many monkeys are there in the troop? Show your work.

*Source: Victor Katz. *A History of Mathematics: An Introduction.* Reading, Mass.: Addison-Wesley, 1998.

In your
own
words

Describe the relationship between the graph of $y = ax^2 + bx + c$ and solutions of the equation $ax^2 + bx + c = d$. If the equation $ax^2 + bx + c = d$ has no solutions, what does that mean about the graph?

17. In Chapter 4, you solved inequalities involving linear relationships. For this problem, use the same ideas to solve inequalities involving quadratic relationships.

   a. First use the quadratic formula to solve $x^2 - 3x - 7 = 0$.

   b. Use the information from Part a to help graph $y = x^2 - 3x - 7$. You may want to plot some additional points.

Use your solutions and graph to solve each inequality.

   c. $x^2 - 3x - 7 < 0$

   d. $x^2 - 3x \geq 7$

   e. $x^2 - 3x \leq 7$

18. Consider the quadratic relationship $y = x(x - 1)$.

   a. For what values of x is $y = 0$?

   b. Is y positive or negative for x values between those you listed in Part a?

   c. Can y ever be equal to $^-1$? Explain.

   d. Can y ever be equal to 1? Explain.

   e. Sketch a graph of this relationship.

   f. Challenge Use your knowledge of the quadratic formula and your graph to find the *minimum* value of y.

19. Challenge You may have solved this problem in Chapter 4. Now you can use your knowledge of the quadratic formula to solve it in another way.

Jermaine wants to construct a large picture frame using a 20-foot strip of wood.

   a. Express the height and area of the frame in terms of its width.

   b. Sketch a graph of the relationship between area and width. Is there a maximum area or a minimum area?

   c. Use the quadratic formula to find the maximum or minimum area for Jermaine's frame. Explain how you found your answer.

   d. What dimensions give this area?

Write an equation to represent the value of B in terms of r.

20.

r	B
0	3
1	0.6
2	0.12
3	0.024
4	0.0048

21.

r	B
0	12
1	4.8
2	1.92
3	0.768
4	0.3072

In Exercises 22 and 23, write an equation to represent the situation.

22. Economics the balance b in a savings account at the end of any year t if \$5,000 is deposited initially and the account earns 8% interest per year

23. Life Science the number of bacteria b left in a sample after 24 hours if one-sixteenth of the remaining colony of c bacteria dies every hour

24. Describe how the y values of each graph change as the x values increase.

   a. Graph a

   b. Graph b

   c. Graph c

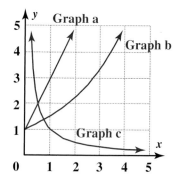

25. Probability Kendra fills a jar with 100 chips, some blue and some orange. She asks Ignacio to guess how many of each color are in the jar.

   a. Ignacio reaches in without looking and removes 10 chips, 4 blue and 6 orange. What reasonable guess might he make for the number of chips of each color in the jar?

   b. Ignacio takes out 5 more chips, and all are orange. What reasonable guess might he now make for the number of chips of each color in the jar?

   c. Kendra tells Ignacio there are actually three times as many orange chips as blue chips in the jar. How many of each are there?

Chapter Summary

Quadratic equations can be solved with several methods. In this chapter, you began with backtracking to solve quadratics of a particular form as well as equations requiring finding reciprocals, taking square roots, and changing signs. You also learned how to solve some quadratic equations by *factoring* and using the fact that when a product is equal to 0, at least one of the factors must be equal to 0.

As these methods don't work well for all quadratic equations, you also learned how to *complete the square* and to use the *quadratic formula:*

$$x = \frac{-b \pm \sqrt{b^2 - 4ac}}{2a}$$

Strategies and Applications

The questions in this section will help you review and apply the important ideas and strategies developed in this chapter.

Using backtracking to undo square roots, squares, reciprocals, and changes of sign

1. Identify the operation that undoes each given operation. Note the cautions, if any, you must take when undoing the given operation.

   a. taking the square root

   b. taking the reciprocal

   c. changing sign

   d. squaring

Indicate whether you can solve each equation directly by backtracking. If so, draw a flowchart and find the solution. If not, explain why not.

2. $\sqrt{2x + 3} - 4 = 7$

3. $\frac{24}{y - 7} = 4$

4. $3a - \sqrt{2a + 3} - 4 = 7$

5. $3(v - 1)^2 + v = 8$

6. $3 - (11w - 3) = 72$

7. $(4n + 5)^2 - 3 = 6$

Solving quadratic equations by factoring

Tell whether you can solve each equation by factoring using integers. If you can, do so, and show your work. If not, explain why not.

8. $g^2 + 3g = {}^-6$

9. $81x^2 + 1 = {}^-18x$

10. $3k^2 - 5k - 12 = 12 + 2k^2$

11. $4w^2 - 9 = 0$

12. $(x + 5)(x - 1) = {}^-8$

13. $2s^2 - 4s + 2 = 0$

Solving quadratic equations by completing the square

14. Explain what it means to solve by "completing the square." Use the equation $4x^2 + 20x - 8 = 0$ to illustrate your explanation.

15. Give an example of a quadratic equation that is possible, but not easy, to solve by completing the square.

Understanding and applying the quadratic formula

16. How was the quadratic formula derived? That is, what technique or method was used and on what equation?

17. Suppose a, b, c, and d are all numbers not equal to 0. Explain why the solutions of $ax^2 + bx + c = d$ are not $x = \dfrac{{}^-b \pm \sqrt{b^2 - 4ac}}{2a}$.

18. How can you determine the number of solutions of a quadratic equation in the form $ax^2 + bx + c = 0$ using the value of $b^2 - 4ac$?

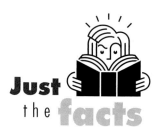

Just the facts

The air resistance of a particular race car (in a unit called Newtons) might be given by the quadratic equation $F = 0.4v^2$, where v is the car's velocity in m/s.

Demonstrating Skills

Factor each expression.

19. $a^2 + 3a$ **20.** $2b^2 - 2$ **21.** $c^2 + 14c + 49$

22. $8d^2 - 8d + 2$ **23.** $e^2 + 8e - 9$ **24.** $f^2 + 7f + 10$

Write an expression equivalent to the given expression by completing the square.

25. $4g^2 + 12g - 3$ **26.** $h^2 - 10h + 7$ **27.** $2j^2 + 24j$

Tell how many solutions each equation has. (Do not solve them.)

28. $k^2 + 10 = 20k - 90$

29. $2m^2 + 3m + 3 = {}^-5$

Solve each equation, if possible.

30. $\sqrt{3n + 1} = 13$

31. $\dfrac{60}{{}^-(2p - 3)} = 12$

32. $(7q + 3)(q - 8) = 0$

33. $(10r + 4)(5r + 4) = {}^-2$

34. $4s^2 + 3s - 40 = 3s - 41$

35. $t^2 - 100 = 0$

36. $2u^2 - 4u = 14$

37. $9v^2 - 3 = 4v^2 + 32$

38. $5w^2 = 8w$

39. $3 - 9x - x^2 = 17$

CHAPTER 8

Functions and Their Graphs

Real-Life Math

Flattening the Globe Creating an accurate map of the world is difficult to do because you must show a three-dimensional surface using only two dimensions. Mathematical functions called *projections* help cartographers create maps. A projection assigns every point on a three-dimensional globe to a point on a two-dimensional surface, in effect *flattening* the globe.

There are many different types of projections, some of which create very interesting maps. The Mercator projection you see in the background of these two pages exaggerates the areas of landmasses farthest from the equator, such as Greenland and Antarctica. On this type of map, Greenland looks like it is almost the size of Africa, when in fact it has only about 7% of Africa's area. Goode's interrupted projection reduces this distortion, but breaks the oceans and Antarctica into pieces.

Think About It In geometry, you studied nets. How are nets of geometric solids similar to projections?

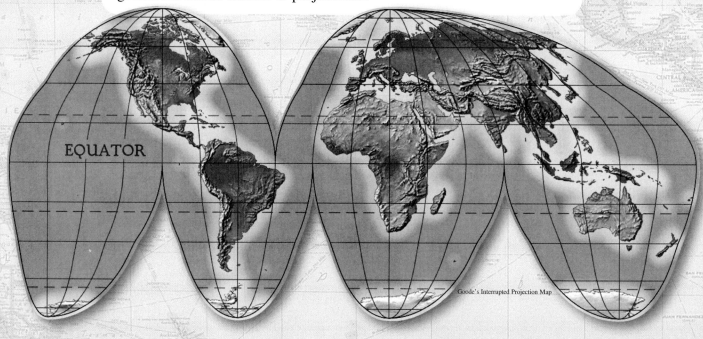

EQUATOR

Goode's Interrupted Projection Map

Family Letter

Dear Student and Family Members,

Our next chapter is about functions and their graphs. The concept of functions is central to algebra and has been a major thread throughout this course, although the term *function* has not yet been used.

One useful way to think about a function is as a machine that takes some input (a number or something else) and produces an output. The output must be *unique*, meaning you get only one output for a particular input. Also, the output must be *consistent*: you get that output every time you use the same input. For example, in this machine, if 3 is the input, it goes into the machine and is multiplied by 5, so the output is 15. Every time you input 3 you will get the same answer, 15.

Functions are often expressed as mathematical sentences. For example, each of these rules describes the function represented by the machine above: multiply by 5.

$$y = 5x \qquad f(x) = 5x \qquad g(t) = 5t$$

Once we have looked at functions using input-output machines, we will go on to use graphs for finding the maximum and minimum values of functions. We will also use functions to solve problems. For instance, if you have 6 meters of fencing for a rabbit pen and you want the greatest area possible, you can use the function $A(L) = L(3 - L)$ to determine that each side of the pen should have 1.5 meters of fencing.

Vocabulary
Along the way, we be learning about these new vocabulary terms:

domain   **range**
function   **x-intercepts**

What can you do at home?

During the next few weeks, your student may show interest in functions and their graphs. You might help him or her think of some situations that can be represented as functions, like the following.

Input: the total restaurant bill   Output: the tip at 15%
Input: the side length of a square   Output: the area of that square
Input: number of adult moviegoers   Output: the total cost for tickets at $8.00 each

8.1 Functions

In your study of algebra, you have analyzed many relationships between variables—relationships like these:

A car traveling along a highway at 55 miles per hour for t hours will cover a distance of $55t$ miles. This can be represented by the equation $d = 55t$.

When a quarterback throws a football, the height of the ball in yards when it has traveled d yards might be described by the equation $h = 2 + 0.8d - 0.02d^2$.

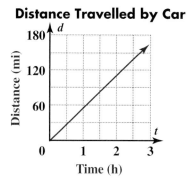

Distance Travelled by Car

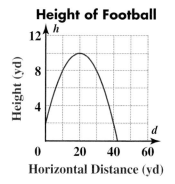

Height of Football

Many of the relationships you have studied, including those above, have a special name: they are called *functions*. In mathematics, a **function** is a relationship between an input variable and an output variable in which there is only one output for each input.

- In the car example, the input variable is the time spent on the highway. The output variable is the distance traveled. Since there can be only one distance traveled for any given time, the relationship is a function. In this case, the distance traveled is a *function of* the time.

- In the football example, the input variable is the horizontal distance the ball has traveled, and the output variable is the ball's height. Since there can be only one height for any given horizontal distance, the relationship is a function. In this case, the height is a *function of* the horizontal distance.

One way to think about a function is to imagine a machine that takes some input— a number, a word, or something else (depending on what the function is)— and produces an output.

For example, suppose you put 10 into a function machine for the football example. Since the machine is a function, the output must be *unique*. If you put 10 into the machine, it can give an output of 8, but it can't give both 8 and some other number.

For a function machine, the output must be consistent. That is, the machine will always give the same output for the same input. If you get an output of 8 for an input of 10, then every time you put 10 into the machine, the output will be 8.

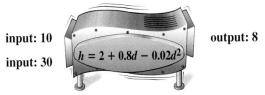

input: 10

input: 30

$h = 2 + 0.8d - 0.02d^2$

output: 8

It *is* possible that two (or more) inputs will produce the same output. For example, the football-height function machine will produce 8 when you put 10 or 30 into it. (Try it!)

If more than one output is possible for a given input, the relationship is *not* a function. For example, a machine that outputs the square roots of a positive number can't be a function, because every positive number has *two* square roots.

Think & Discuss

Here are some examples of functions. For each function, explain why there is only one possible output for each input.

- Input: a number
  Output: twice that number

×2

- Input: the name of a state
  Output: the state's capital

Capital?

- Input: an integer
  Output: classification as even or odd

- Input: a person's social security number
  Output: that person's birth date

- Input: the side length of a square
  Output: the area of that square

- Input: a word
  Output: the first letter of that word

Which of the functions above give the same outputs for different inputs? Explain.

The following relationships are *not* functions. For each, explain why there might be more than one output for some inputs.

- Input: a number
  Output: a number less than that number

- Input: a whole number
  Output: a factor of that number

- Input: a person
  Output: the name of that person's grandparent

- Input: a city name
  Output: the name of the state in which that city can be found

- Input: the side length of a rectangle
  Output: the area of that rectangle

- Input: a word
  Output: that word with the letters rearranged

Investigation ▶ Function Machines

You can describe a function in various ways—such as using words, symbols, graphs, or machines. In this investigation, you will think about functions as machines.

Problem Set A

Two machines that each perform one operation have been hooked together to form a more complicated function called Function A. Function A takes an input, doubles it, and then produces 7 more than that result as an output.

Function A

1. If the input is 5, what is the output?

2. If the input is ⁻4, what is the output?

3. If the output is ⁻10, what could the input have been?

4. Is there more than one answer to Problem 3? Explain why or why not.

5. If the input is some number *x,* what is the output?

6. Function A is called a *linear function.* Explain why that makes sense.

7. Function B is represented by this machine hookup. Is it the same as Function A? Explain.

Function B

8. If possible, describe a hookup that would "undo" Function A. That is, create a hookup so that if you put a number into Function A and then put the output into your hookup, you *always* get back your original number. If it isn't possible, explain why not.

Problem Set B

The "Prime?" machine takes positive whole numbers as inputs and outputs *yes* if a number is prime and *no* if a number is not prime.

1. If the input is 3, what is the output?

2. If the input is 2, what is the output?

3. If the input is 100, what is the output?

4. If the input is 1, what is the output?

5. If the output is *yes*, what could the input have been?

6. Is there more than one answer to Problem 5? Explain why or why not.

7. If possible, describe a machine that would undo the "Prime?" machine. That is, create a machine that takes the output from the "Prime?" machine and always produces the original number. If it isn't possible, explain why not.

Problem Set C

The "3" machine takes numbers as inputs and always outputs the number 3.

1. If the input is 17, what is the output?

2. If the input is ⁻2, what is the output?

3. If the output is 3, what could the input have been?

4. Is there more than one answer to Problem 3? Explain why or why not.

5. Explain why "3" is a function.

6. The function "3" is a *constant function*. Explain why that name makes sense.

7. If possible, describe a machine that would undo the "3" machine. That is, create a machine that takes the output from the "3" machine and always produces your original number. If it's not possible, explain why not.

Share & Summarize

Lucita and Ben are trying to decide whether $y = x^4$ is a function.

Who is correct, Ben or Lucita? Is $y = x^4$ a function? Explain how you know.

Investigation 2 ▶ Describing Functions with Rules and Graphs

Functions, like the one described by this hookup, are a type of rule that assigns one output value to each input value. You can often write such rules as algebraic equations, which is easier than drawing machines.

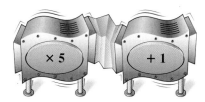

For example, each of these equations describes the same function as the one shown by the hookup: multiply the input by 5, and then add 1.

$$y = 5x + 1 \qquad f(x) = 5x + 1 \qquad g(t) = 5t + 1$$

Letters like f and g are often used to name functions. In the second rule above, the variable x represents the input, f is the name of the function, and $f(x)$ represents the output. The symbol $f(x)$ is read "f of x." It does *not* mean "f times x." Instead, it means "apply rule f to the value x." For example, $f(2) = 5(2) + 1 = 11$. This is illustrated below.

Problem Set D

Kenneth is thinking about a rule to change one number into another number. He is wondering whether his rule is a function.

Double the number, add 1, and square the result.

1. Make an input/output table for Kenneth's rule, showing outputs for at least four inputs.

2. Is Kenneth's rule a function? How can you tell?

3. For Parts a–c, decide which functions describe Kenneth's rule.

a. $y = (2x + 1)^2$
$y = 2x^2 + 1$
$y = 2(x + 1)^2$
$y = (2x)^2 + 1$

b. $m = (2n + 1)^2$
$a = (2b + 1)^2$
$p = (2t + 1)^2$

c. $f(z) = 2(z + 1)^2$
$g(x) = (2x + 1)^2$
$p(t) = (2t + 1)^2$
$j(k) = 1 + (2k)^2$

MATERIALS
graphing calculator

Problem Set E

You can graph a function with the input variable on the horizontal axis (the *x*-axis) and the output variable on the vertical axis (the *y*-axis).

1. Graph Kenneth's rule from Problem Set D on your calculator.

a. What did you enter into the calculator for the rule?

b. Sketch the graph. Remember to label the minimum and maximum values on each axis.

2. Decide which graphs below represent functions. Explain how you decided.

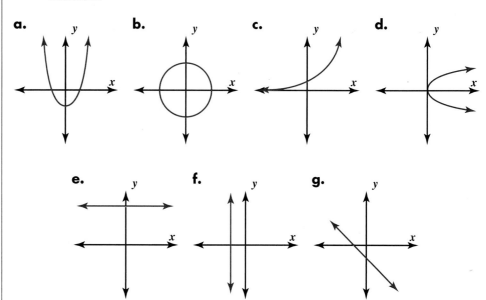

When you have a function such as $f(x) = x^2$, you may want to find the value of the function for different values of x.

EXAMPLE

Consider the function $f(x) = x^2$.

If $x = 3$, then $f(3) = 3^2 = 9$. Finding $f(3)$ is like putting 3 into this machine:

If $x = {}^-10$, then $f({}^-10) = ({}^-10)^2 = 100$.

Remember, $f(2)$ does not mean "f times 2." It means "use 2 as the input to machine f" or "evaluate the function f with the input 2."

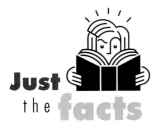

Just the facts

From previous chapters, you might recall that the equation $d = 16t^2$ can also be used to describe the distance traveled by a falling object. The equations $d = 16t^2$ and $d = 4.9t^2$ describe the same relationship, only distance is in feet in the first equation and meters in the second.

Problem Set F

The distance fallen by skydivers, before they open their parachutes, is a function of the time since they fell from the aircraft. The function is approximated by $f(t) = 4.9t^2$, where t is the time in seconds and $f(t)$ is the distance in meters.

1. What does $f(2)$ represent in the skydiving situation? What is the numerical value of $f(2)$?

2. How far has a skydiver fallen after 10 seconds?

3. In the context of this situation, would it make sense to find the value of $f(^-3)$? Explain your answer.

Some functions can have only certain inputs. In the skydiver problem, only positive numbers make sense as inputs, because the function measures how far a skydiver has fallen *after* jumping.

As another example, here is a function you considered earlier:

• Input: an integer
  Output: classification as even or odd

The input is described as "an integer" because non-integers, such as $\frac{3}{4}$ and $^-12.92$, don't make sense as inputs. It isn't reasonable to ask whether such numbers are even or odd.

V O C A B U L A R Y
domain

The set of allowable inputs to a function is called the **domain** of that function. If some numbers are not allowed as inputs, we say they *are not in the domain* of the function.

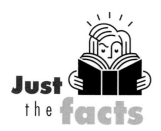

Ants are found all over the world except in the polar regions. It is estimated that there are 10,000 different species of ants and 10 million billion individual ants.

Think & Discuss

Consider this function: $r(x) = \frac{1}{x}$.

What numbers are not in the domain of this function? Why?

Problem Set G

In Problems 1–5, describe the domain of the function.

1. $f(x) = x^2$

2. $g(t) = \sqrt{t}$

3. $R(x) = \frac{1}{1 - x}$

4. $e(n)$ is the number of factors of n.

5. $q(p)$ is *yes* if p is evenly divisible by 3 and *no* if p is not evenly divisible by 3.

6. The number of legs in an ant farm is a function of the number of ants in the farm. Specifically, the number of legs is 6 times the number of ants.

   a. If there are 2,523 ants in the farm, how many legs are there?

   b. What numbers cannot be inputs to this "number of legs" function? Explain your answer.

   c. You can describe this "number of legs" function using algebraic symbols. Let a be the number of ants, and write a function g so that $g(a)$ is the number of legs.

Share & Summarize

A particular function can be described in several ways, including using words, equations, tables, graphs, and machines.

1. Describe, write, or draw three representations of this function.

$$g(x) = 7 - 3x$$

2. Are any numbers not in the domain of $g(x) = 7 - 3x$? If so, which numbers?

Investigation ▶3 Finding Maximum Values of Functions

Graphs are very useful for finding approximate maximum and minimum values of functions. For example, in Chapter 4, you considered the maximum height a thrown or bounced ball might reach. A manufacturer might use a function to predict the price that will give the maximum profit for a product.

MATERIALS

graphing calculator

Problem Set **H**

Tala threw a stone vertically up from the edge of a pier. The height of the stone above the water level is a function of *t*, where *t* is the number of seconds after the stone is thrown. The function, which measures height in meters, is

$$h(t) = 15t - 4.9t^2 + 6$$

At right is a graph of this relationship.

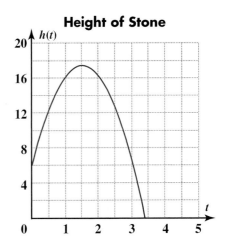

Height of Stone

1. When the stone first leaves Tala's hand, about how high is it above the water? Explain how you can find the answer from the equation or the graph.

2. About how high is the pier? Explain why your answer is reasonable.

3. When is the stone at a height of 15 meters?

4. Between what times is the stone more than 15 meters above the water?

5. To the nearest meter, what is the maximum height the stone reaches?

6. About how long after it is thrown does the stone reach its maximum height?

7. Use your calculator's Trace and Zoom features to better approximate the stone's maximum height. Find the maximum height to the nearest hundredth of a meter.

Just the facts

There are two types of anesthetics. *General anesthetics act on the central nervous system and affect the entire body, making the patient unconscious. Local anesthetics affect only the particular body part where they are injected or applied.*

Problem Set I

A company that manufactures medicine has researched the concentration of a local anesthetic in a patient's bloodstream. They found that the concentration can be approximately calculated with the function

$$C(t) = \frac{21t}{t^2 + 1.3t + 2.9}$$

where t is the number of minutes after the anesthetic is administered and $C(t)$ is the concentration of the anesthetic, measured in grams per liter. A higher concentration means the patient is less likely to feel pain.

1. Find $C(1)$, $C(6)$, and $C(10)$. What does each of these values represent in terms of this situation?

2. The graph shows the relationship between $C(t)$ and t. Use it to estimate the maximum concentration reached by the anesthetic.

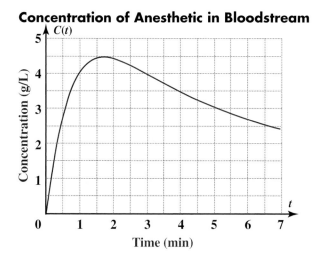

Concentration of Anesthetic in Bloodstream

3. About how long does it take the anesthetic to reach the maximum concentration?

4. Using the equation, draw your own graph on your calculator. Use Zoom and Trace to find the answers to Problems 2 and 3 to the nearest hundredth of the given units (g/L and min).

5. Tests have shown that when the concentration reaches 2 g/L, patients report feeling numbness. About how long after the injection does this happen?

6. A doctor wants to stitch a cut in Jemma's hand. She expects the stitching to take about 3 minutes. How long after she injects Jemma with anesthetic should she wait before she starts? Explain.

Share & Summarize

Decide whether each of these functions has a maximum value. If so, approximate the maximum value and the input that produces it.

1.
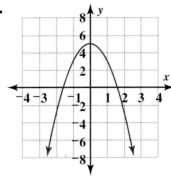

2.
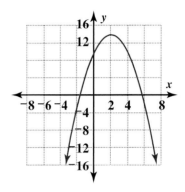

3.
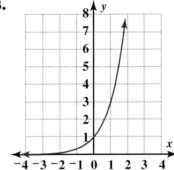

4.
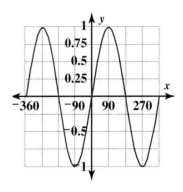

Investigation 4 ▶ Maximum Areas, Minimum Lengths

These shapes have the same perimeter but different areas.

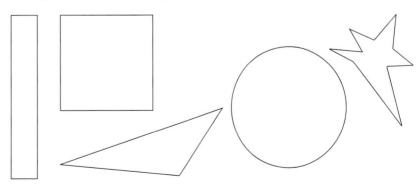

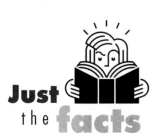
Farmers, builders, and geometers often want to maximize the area of a shape for a given perimeter. In this investigation, you will consider the maximum area for rectangular shapes with a given perimeter. You will also consider the minimum perimeter for a given area.

Problem Set J

Keisha and her twin sister Monifa have bought some guinea pigs. They are building a fenced pen for the animals. They have 6 meters of fencing, and they want to give their pets as much space as possible.

Keisha drew some possible rectangular shapes for the pen.

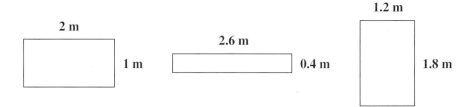

2 m   2.6 m   1.2 m

1 m   0.4 m   1.8 m

1. The twins need to consider two dimensions for a rectangular pen: length and width. Copy and complete the table, which relates possible lengths and widths, both measured in meters. The total perimeter must be 6 meters in each case.

Length	0.5	1	1.5	2	2.5	3
Width	2.5	2				
Perimeter		6				

2. If the length of the rectangle increases by a certain amount, what happens to its width?

3. Write an equation that gives the width *W* for any length *L*.

Your equation shows the relationship between one dimension of the rectangular pen and the other. However, because Keisha and Monifa want to find the greatest rectangular area they can enclose using 6 meters of fencing, the mathematical relationship they need is between one of the dimensions, such as length, and the area.

4. Complete this table, showing dimensions and area of some possible rectangles. All measurements are in meters.

Length	0.5	1	1.5	2	2.5	3
Width	2.5					
Area	1.25					

5. Write an equation for the function *A* giving the area for length *L*.

6. Use your calculator to graph the length of the pen versus its area, using your function from Problem 5. Sketch your graph. Remember to label the minimum and maximum values on each axis.

7. What length and width should the pen be to produce the greatest area from 6 meters of fencing? Use the graph you drew to approximate your answer.

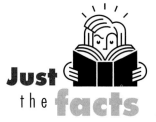

Just the facts

Guinea pigs are native to South America and live an average of eight years.

Problem Set K

A family wants to build a rectangular pen for their chickens, using an existing stone wall as one side and fencing for the other three sides. They decide on an area of 40 m². They want to know what shape rectangle will give this area using a minimum length of fencing. Here are some shapes they are considering.

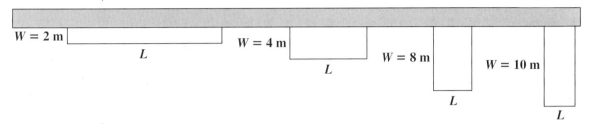

1. Copy and complete the table. Try additional width values, if necessary, to determine the least amount of fencing needed. All measurements are in meters.

Width, W	2	4	8	10
Length, L				
Amount of Fencing				

2. Express the length of the pen in terms of W.

3. Use your expression from Problem 2 to write the amount of fencing as a function of W. Name your function F.

$$F(W) = \underline{\hspace{2cm}}$$

4. Use one of the following methods to find the width that requires the least amount of fencing:

• Use your calculator to graph the amount of fencing versus width, using your function from Problem 3. Approximate the least value for the length.

• Use a calculator to guess-check-and-improve.

Share & Summarize

Héctor was experimenting with his calculator, adding positive numbers and their reciprocals. Here are some examples.

$$5 + \frac{1}{5} = 5.2 \qquad 0.1 + \frac{1}{0.1} = 10.1 \qquad 1.25 + \frac{1}{1.25} = 2.05$$

1. Do you think there is a minimum total he can produce doing this? If so, what is it? If not, explain why not. (Hint: Let x stand for the number, and write an equation to express what Héctor is doing.)

2. Do you think there is a maximum total he can produce? If so, what is it? If not, explain why not.

Lab Investigation ▶ The Biggest Box

MATERIALS

- 5-inch-by-8-inch cards
- ruler
- scissors
- graphing calculator
- tape

Your teacher will give you 5-inch-by-8-inch cards. You can cut squares out of the corners of a card and then fold the sides to make an open box (a box without a top).

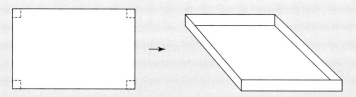

Your challenge is to create the box with the greatest possible volume.

Try It Out

The volume of your box will depend on the side length of the squares you cut from it.

1. Using the method above, try to create the box with the greatest possible volume. Be careful to cut squares of the *same size* from each corner. Record the side lengths of the squares you cut, so you can refer to them later.

2. Compare the greatest volume you found with the greatest volume found by others in the class. Record the side length of the cutout squares for the boxes with the greatest volume.

Remember

For a rectangular prism such as a box, volume is area of base times height.

Analyze the Situation

3. Each dimension of your box depends on the side length of the squares you cut out. Copy and complete the table for squares of different side lengths. All measurements are in inches.

Side Length of Square	0	0.5	1	1.5	2	2.5
Height of Box	0					
Length of Box	8					
Width of Box	5					

4. Add a row to your table, calculating the volume of the box for each side length of the square. Of those boxes listed in the table, which has the greatest volume?

Of course, there are more possible square sizes than the six listed in the table above. You can use functions and graphs to help you check *all* the possibilities.

5. If the side length of the square you cut out is *x*, find each of the following in terms of *x*.

a. the height of the box   **b.** the length of the box

c. the width of the box   **d.** the volume of the box

6. Based on your answer to Part d of Question 5, write an equation for the function relating the box's volume to the side length of the square you cut out. Name your function *v*.

7. Use your calculator to graph the volume function, and sketch the graph. Then use Zoom and Trace to estimate the value of *x* that gives the maximum volume.

What Have You Learned?

You estimated the maximum volume of the open box you can make from a 5-inch-by-8-inch card. Suppose you start with a standard sheet of paper, 8.5 inches by 11 inches, instead.

8. Use what you learned in this lab investigation to answer these questions. Show your work, including sketches of any graphs you make.

a. What cutout size will maximize the volume for an open box made from a standard sheet of paper?

b. What is the greatest possible volume?

9. Use an ordinary sheet of paper and your answers to Question 8 to create the box you think has the greatest volume. Tape the corners to make it strong.

On Your Own Exercises

1. Consider this function machine.

   a. If the input is 10, what is the output?

   b. If the input is $-\frac{2}{3}$, what is the output?

   c. If the input is 1.5, what is the output?

   d. If the input is some number x, what is the output?

   e. If the output is $^-9$, what was the input?

   f. Suppose you want a function machine that will undo this machine. That is, if you put a number first through the "÷ 2" machine and then through your new machine, it *always* returns your original number. What function machine would accomplish this?

2. Consider this function machine, which squares the input.

   a. If the input is $\frac{4}{3}$, what is the output?

   b. If the input is $-\frac{4}{3}$, what is the output?

   c. If the output is 9, what was the input?

   d. Suppose you want a function machine that will undo this machine. That is, if you put a number first through the "Square" machine and then through your new machine, it always returns your original number. What function machine would accomplish this?

3. Consider this hookup, Function F.

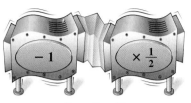

Function F

a. If the input is 1.5, what is the output?

b. If the input is ⁻3, what is the output?

c. If the input is 11, what is the output?

d. If the input is some number *x*, what is the output?

e. If the output is ⁻8, what was the input?

f. Suppose you want a function machine that will undo this machine. That is, if you put a number through the Function F machine and then through your new hookup, it will always return your original number. What function machine would accomplish this?

Tell whether each example below is a function, and explain how you decided.

4. Input: a circle
Output: the ratio of the circumference to the diameter

5. Input: a rugby team
Output: a member of the team

6. Input: a CD
Output: a song from the CD

Just the facts

Rugby is played in more than 100 countries by several million people.

Determine if the relationship represented by each input/output table could be a function.

7.

Input	Output
$^-3$	4
$^-2$	3
$^-1$	2
0	1
1	0
2	$^-1$
3	$^-2$

8.

Input	Output
$^-3$	0
$^-2$	$^-2.828$ and 2.828
$^-1$	$^-2.236$ and 2.236
0	$^-3$ and 3
1	$^-2.236$ and 2.236
2	$^-2.828$ and 2.828
3	0

9.

Input	Output
$^-3$	$\frac{1}{3}$
$^-2$	$\frac{1}{2}$
$^-1$	1
0	undefined
1	1
2	$\frac{1}{2}$
3	$\frac{1}{3}$

10. Consider this rule: *Square a number, subtract 2, and then divide by 2.*

a. Copy and complete the table using this rule.

b. Sketch a graph of the relationship shown in your table.

c. Is this rule a function? How do you know?

Input, I	Output, O
$^-3$	
$^-2$	
$^-1$	
0	
1	
2	
3	

11. When Kai entered math class, the table and functions below were on the board. Kai thought the values in the first column of the table were function inputs and the values in the second column were outputs.

1	3
2	7
3	13
4	21
5	31
6	43
7	57
8	
9	91
10	
11	

$g(t) = 1 + t + t^2$

$f(x) = x^2 + 2x$

$h(z) = z^2 + z + 1$

$b = a^2 + a + 1$

$K(d) = d^2 + d - 1$

$Y = 2x + 1$

$B(x) = 4x - 1$

$F(X) = (x + 1)^2 - x$

a. Which of the functions, if any, might be shown in the table? Explain.

b. Complete the table by finding the missing values of the function.

12. Physical Science A rock falls over the edge of a cliff 600 meters high. The distance in meters the rock falls is a function of time in seconds and can be approximated by the function $s(t) = 4.9t^2$.

a. Find the value of $s(8)$. In this situation, what does $s(8)$ represent?

b. How far has the rock fallen after 9 seconds? After 10 seconds?

c. When does the rock hit the ground?

d. What is the domain of the function $s(t) = 4.9t^2$ in this context?

Describe the domain of each function.

13. $f(x) = 2^x$

14. $g(x) = \frac{1}{x + 1}$

15. $h(x) = \frac{1}{x + 1} + \frac{1}{x - 1}$

16. Which of the following are not graphs of functions? Explain how you know.

a.

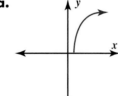

b.

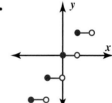

c.

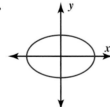

d.

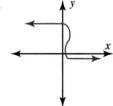

17. Suppose a person throws a stone straight upward so that its height h in meters is given by the function $h(t) = 6 + 20t - 4.9t^2$, where t represents the time in seconds since the stone was released.

a. Find $h(4)$. What does it represent in this situation?

b. Find the height of the stone after 3 seconds.

c. Sketch a graph of the stone's height over time.

d. Use your graph to approximate the stone's maximum height. How long does it take the stone to reach this height?

18. Economics ABC Deli sells several kinds of sandwiches, all at the same price. The weekly profit of this small business is a function of the price of its sandwiches. This relationship between profit, P, in hundreds of dollars and the price per sandwich, s, in dollars is given by the equation

$$P(s) = {}^-s(s - 7)$$

a. Complete the table for this function.

b. Explain the meaning of $(7, 0)$ in terms of the deli's profit.

c. Extend your table to search for the sandwich price that will yield the maximum profit.

d. What is the maximum profit this business can expect in a week?

s	$P(s)$
0	0
1	
2	
3	
4	
5	
6	
7	

Find the maximum value of each function, and then determine the input value that yields that maximum value.

19. $f(t) = 200t - 5t^2$

20. $k(t) = 4 + 4t - 4t^2$

21. Marcus gave his little brother an 8-meter strip of cardboard for making a rectangular fort for his toy soldiers.

a. Copy and complete the table, which relates possible lengths and widths for the fort.

Length (m)	0.5	1	1.5	2	2.5	3	3.5
Width (m)							
Perimeter (m)							

b. Write an equation for the function that gives the width for any length L. Name the function W.

c. Now add a row to your table showing the area of some possible rectangles.

d. Write an equation for the function A giving the area for length L.

e. Use your function from Part d to sketch a graph of the fort's area in terms of its length.

f. What dimensions give the greatest area for the fort?

22. Geometry Roof gutters are designed to channel rainwater away from the roof of a house, protecting the house from excess moisture.

If you cut through a gutter and look at its side view, you see a *cross section.* Here are some cross sections of gutters.

Nicky's Metalworks wants to produce some gutters from a roll of metal that is 39 cm wide. They want the gutters to have vertical sides. Nicky has drawn some possible cross sections.

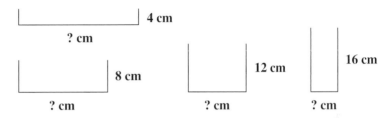

a. To keep the gutters from overflowing during heavy rainfall, the company wants them to have the greatest cross-sectional area possible. Copy and complete the table to show the widths and areas for gutters of various heights.

Height (cm), h	4	8	12	16
Width (cm), w				
Area (cm^2), A				

b. Find a formula for width w in terms of height h.

c. Write an equation for the cross-sectional area A as a function of h. What sort of function is it?

d. Sketch a graph of the area function.

e. Estimate the gutter height that gives the greatest area.

23. Create a function machine that produces 3 more than twice every input as an output.

24. Create a function machine that produces 1 less than one-third every input as an output.

25. Create a function machine that returns an odd number for any whole-number input.

26. Physical Science Think about the relationship between the temperature of a hot cup of coffee and the time (in minutes) since the coffee was poured.

a. Sketch a graph of how you think the relationship between temperature and time might look. (Hint: Think about the rate at which the coffee cools. Does it cool more quickly at first?)

b. Is this relationship a function? If so, explain why.

27. Geometry The sum of the interior angles of a polygon is a function of the number of sides the polygon has. For example, the sum of the interior angles of a triangle is 180°, of a square is 360°, of a pentagon is 540°, and of a hexagon is 720°.

a. What is the sum of the interior angles of a polygon with 12 sides (a dodecagon)? Use the pattern in the angle sums for the polygons mentioned above.

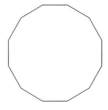

b. Write an equation for the function relating the number of sides to the angle sum. Name the function g, and use s to represent the number of sides.

c. What is the domain of this function? Explain your answer.

In your
own
words

Give one example
of a function that
can be described
by an algebraic
equation and one
example of a func-
tion that can't.
Explain how you
know that both of
your examples are
functions.

Challenge In Exercises 28–30, write an equation for a function f that does *not* have the given numbers in its domain.

28. 3 and $^-3$

29. negative numbers

30. positive numbers

Use the given graph and a table of values, if necessary, to find the minimum value of each function and the input that produces it.

31. $f(x) = x + x^2$

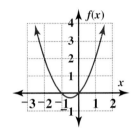

32. $f(x) = 1 - x + x^2$

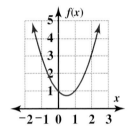

33. You can think of a *sequence* as a function for which the input variables are the counting numbers (1, 2, 3, 4, . . .). For example, the sequence of even whole numbers greater than zero—2, 4, 6, 8, . . .—can be given by the function $f(n) = 2n$, where 1, 2, 3, 4, . . . are the inputs.

   a. List the first seven terms of the sequence described by the function $g(n) = \frac{1}{2^n}$, for n starting at 1.

   b. Add the first five terms of this sequence.

   c. Add the first six terms of this sequence.

   d. Add the first seven terms of this sequence.

   e. Suppose you were to add *all* the terms of this sequence for some large value of n—such as 100 terms. Do you think the sum of this sequence approaches a particular value, or do you think it increases indefinitely?

34. Two numbers add to 1. What is the maximum value of their *product*? Explain.

35. Economics A company that manufactures charcoal pencils for artists has decided to redesign the shipping boxes for the pencils. The pencils are in the shape of rectangular prisms, with a 0.25-inch-by-0.25-inch base and a length of 8 inches. The manufacturer plans to package a dozen pencils in each box.

  a. Calculate the volume of a single pencil. Then find the volume each box must contain—that is, find the volume of 12 pencils.

  b. One dimension of the box must be the length of the pencils, 8 in. Using x, y, and 8 for the dimensions of the box, write a formula for the volume a box can hold.

  c. Use the total volume of the 12 pencils, along with your formula from Part b, to write an equation for y in terms of x.

The company wants to use as little cardboard as possible in making the boxes.

  d. Write a formula for the surface area S of the box, using only x as the input variable. Ignore the area of the flaps that hold the box together. (Hint: You may want to write it using x and y first, and then replace y with an expression in terms of x.)

  e. Make a table of values giving the surface area of the box for different values of x. Since the pencils are 0.25 in. wide, the dimensions of the box must be multiples of 0.25 in.—for example, 0.25 in., 0.5 in., and 0.75 in.

  f. What dimensions should the box be so that it uses the least amount of cardboard?

Mixed Review

Set up and solve a proportion to answer each question.

36. 32.2 is 92% of what number?

37. What percent of 125 is 90?

38. What is 81% of 36?

Write each expression in the form 7^b.

Remember

A *translation* is a transformation that moves a figure a specific amount in a specific direction. It does not change the figure's size or orientation.

39. $\dfrac{7^{23}}{7^{15}}$ **40.** $(7^3)^{10}$ **41.** $\left(\dfrac{1}{7}\right)^{11}$

42. A rule for translating a figure on a coordinate grid changes the original coordinates (x, y) to the image coordinates $(x - 2, y + 3)$. On a coordinate grid, show the translation vector for this rule.

Match each equation with one of the graphs.

43. $y = x^2 - 3$

44. $y = {}^-x^2$

45. $y = x^2$

46. $y = x^2 - 5x + 4$

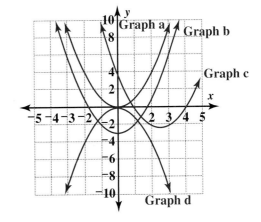

Rewrite each expression as a square with a constant added or subtracted.

47. $x^2 + 12x + 17 = x^2 + 12x + 36 -$ ___ $= (x + 6)^2 -$ ___

48. $k^2 - 14k + 70$

49. $b^2 + 5b - \frac{3}{4}$

50. Geometry This is a map of Golden Gate Park in San Francisco.

Remember
There are 5,280 feet in 1 mile and 12 inches in 1 foot.

a. Find the park's area on this map, in square inches. (Hint: The park is very close to being a rectangle. What are the lengths of its sides?)

b. The area of Golden Gate Park is about 1,017 acres, or about 1.59 square miles. Find the scale factor from this map to the actual park. (Hint: A scale factor is a comparison of the same linear, not square, units.)

c. About how many miles long is the northern (top) border of Golden Gate Park, along Fulton Street?

Graphs of Functions

Ms. Torres drew the graphs and table below to show his class that the graphs of $f(x) = x^2$ and $g(x) = (x - 3)^2$ are related.

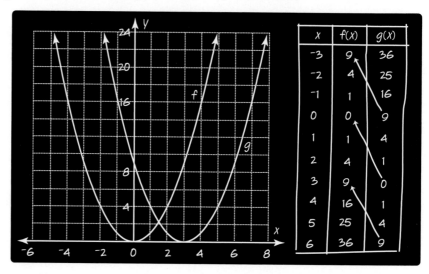

x	f(x)	g(x)
-3	9	36
-2	4	25
-1	1	16
0	0	9
1	1	4
2	4	1
3	9	0
4	16	1
5	25	4
6	36	9

Ms. Torres then asked the class why the graphs look as they do.

MATERIALS

- tracing paper
- graphing calculator

Remember

A *translation* is a transformation that moves a figure a specific amount in a specific direction. It does not change the figure's size or orientation.

Think & Discuss

Which of the four students' comments do you find the most helpful in understanding why the graphs look as they do? Explain.

Describe in your own words why it seems reasonable for the graph of $g(x) = (x - 3)^2$ to be 3 units to the right of the graph of $f(x) = x^2$.

Trace the graph of $f(x) = x^2$. Place your tracing on top of the graph of $g(x) = (x - 3)^2$, lining up the parabolas. Are the parabolas congruent?

The graph of $g(x) = (x - 3)^2$ is related to that of $f(x) = x^2$ by a translation. What are the direction and distance of the translation?

Predict what the graph of $h(x) = (x + 4)^2$ looks like. Test your prediction by graphing with your calculator.

Is the graph of h related to the graph of f by a translation? If so, specify the direction and the distance of the translation.

Investigation 1 ▶ Comparing Graphs of Functions

In this investigation, you will explore sets of related functions.

MATERIALS

graphing calculator

Remember

When you make sketches of graphs from your calculator, label the graphs with their names (such as *j, f, g, h*), and label the minimum and maximum values on each axis.

Problem Set A

For Problems 1 and 2, do Parts a–c. Work in pairs or groups of four. Your group will need two graphing calculators, one for each problem.

    a. Graph the four equations in the same window, and make a quick sketch of the graphs. Don't erase your graphs for Problem 1 when you go on to Problem 2; you will need both sets for Problem 3.

    b. Describe how the four graphs in the set are alike and different. Use the concept of translation in your comparisons.

    c. Write equations for two more functions that belong in the set.

1. $j(x) = (x + 1)^2$
    $f(x) = x^2$
    $g(x) = (x - 1)^2$
    $h(x) = (x - 2)^2$

2. $j(x) = \frac{1}{x + 1}$
    $f(x) = \frac{1}{x}$
    $g(x) = \frac{1}{x - 1}$
    $h(x) = \frac{1}{x - 2}$

3. Describe how the two sets of graphs are alike and different.

4. On which graph would you find the point (4, 9)? Explain.

LESSON 8.2   Graphs of Functions   **515**

Problem Set B

Work in pairs or groups of four on this problem set.

1. Consider the function $f(x) = 2^x$.

   a. Write equations for three functions, g, h, and j, so that their graphs have the same shape as the graph of f, but

     i. g is translated 2 units to the right of f.

     ii. h is translated 3 units to the right of f.

     iii. j is translated 3 units to the left of f.

   b. Graph the four functions in the same window, and make a quick sketch of the graphs.

2. Consider the function $f(x) = 2x^2$.

   a. Write equations for three functions, g, h, and j, so that their graphs have the same shape as the graph of f, but

     i. g is translated 1 unit to the right of f.

     ii. h is translated 2 units to the left of f.

     iii. j is translated 3 units to the right of f.

   b. Graph the four functions in the same window, and make a quick sketch of the graphs.

3. On which graph from Problems 1 and 2 would you find the point $(^-3, 1)$? Explain how you found your answer.

Problem Set C

Work in pairs or groups of four again. Your group will need two graphing calculators. For Problems 1 and 2, do Parts a–c.

   a. Graph the four equations in the same window, and make a quick sketch of the graphs, remembering to label them. Don't erase your graphs for Problem 1 when you go on to Problem 2.

   b. Describe how the four graphs in the set are alike and different. Use the concept of translation in your comparisons.

   c. Write two more functions that belong in the set.

1. $j(x) = 2^x - 1$

   $f(x) = 2^x$

   $g(x) = 2^x + 1$

   $h(x) = 2^x + 2$

2. $j(x) = \frac{1}{x} - 1$

$f(x) = \frac{1}{x}$

$g(x) = \frac{1}{x} + 1$

$h(x) = \frac{1}{x} + 2$

3. Describe how the two sets of graphs are alike and how they are different.

4. On which graph would you find the point $\left(3, \frac{4}{3}\right)$? Explain.

Share & Summarize

1. Suppose you have the graph of a function f. You create a new function g by using the rule for f but replacing the variable x by the expression $x + h$, for some constant h. If $f(x) = 2x$, for example, you might replace x with $x + 3$ to get $g(x) = 2(x + 3)$. If $f(x) = 3x^2 - 2$, you might replace x with $x - 5$ to get $g(x) = 3(x - 5)^2 - 2$.

Write a sentence or two describing the differences and similarities between the graphs of f and g. You may want to draw sketches to help you explain.

2. Suppose you create a function g by adding a constant h to f—for example, $f(x) = 2x$ and $g(x) = 2x + 3$, or $f(x) = 3x^2 - 2$ and $g(x) = 3x^2 - 7$. Describe the differences and similarities between the graphs of f and g. You may want to include sketches.

3. Suppose you want to know whether the point (a, b) is on the graph of a function. How could you find out?

Investigation 2 Working with Graphs

Earlier you saw that the maximum or minimum value of a quadratic function can be found by looking at the vertex of its graph, which is a parabola. You also learned that parabolas are *symmetric*—they can be folded on a line of symmetry so that the two sides match.

You will now examine connections between the graph and the equation of a quadratic function. You'll also learn what the range of a function is and how it relates to the maximum or minimum point.

Think & Discuss

Look at this graph of $f(x) = (x - 2)^2 + 1$.
Find the values of x for which

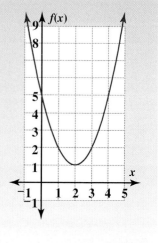

$f(x) = 1$ $f(x) = 2$ $f(x) = 5$

Now find values of x for which

$f(x) = 0$ $f(x) = {}^-1$ $f(x) = {}^-5$

Describe all possible values for $f(x)$.

Describe all values $f(x)$ can never be.

V O C A B U L A R Y
range

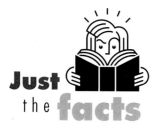

Just the facts

A biologist might need to know the range of a function modeling the temperature of a body of water over time to study how water temperature affects an organism living there.

All the possible *output* values of a function f are the **range** of the function. For the function graphed above, the range is $f(x) \geq 1$. No matter what you substitute for x, the value of $f(x)$ will always be greater than or equal to 1—and every value greater than or equal to 1 has an input value. Numbers less than 1 are not in the range of $f(x) = (x - 2)^2 + 1$.

MATERIALS

graphing calculator
(optional)

Problem Set D

For each function, specify the domain (possible inputs) and range (possible outputs). For some functions, it may help to make a graph with a calculator.

1. $g(x) = 4^{x+2}$ (Hint: Can $g(x)$ be negative? Zero?)

2. $h(s) = \frac{1}{s+3}$

3. $c(x) = {}^{-}3x + 4$

4. Input: a state
Output: the capital of that state

5. Input: a number
Output: the integer part of that number (For example, if the input is 4.5, the output is 4; if the input is $^{-}3.2$, the output is $^{-}3$.)

It's not always easy to determine the range of a function. You can test a few logical input values and draw some conclusions from the outputs, or you can graph the function. In Problem Set E, you will see how the range of a quadratic function is related to its vertex. To begin, consider how to find the vertex.

Remember

A parabola has its maximum or minimum value at the vertex.

EXAMPLE

This graph is of the function $f(x) = (x-2)^2 + 1$.

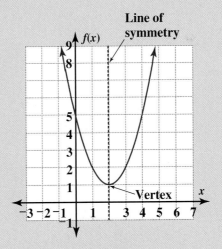

For this parabola, the line of symmetry is the line $x = 2$.

The turning point, or *vertex*, is the point on the graph where $x = 2$. When $x = 2$, $f(x) = (2-2)^2 + 1 = 1$, so the vertex has coordinates $(2, 1)$.

Problem Set E

1. Below are the graphs of these functions.

$$f(x) = 3x^2 \qquad\qquad g(x) = 3(x - 2)^2 + 4$$

i.

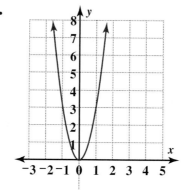

ii.

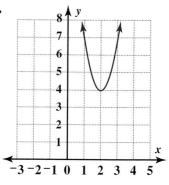

Remember

The *range* is the set of all possible outputs for a function.

a. Without using your calculator, decide which graph represents which function.

b. Sketch the graphs, and draw the line of symmetry for each.

c. What is the vertex of each graph?

d. Specify the range of each function.

e. How is the range of a function related to the vertex?

2. Consider this graph.

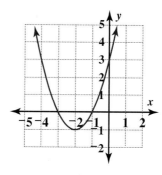

a. What is the graph's line of symmetry? What is its vertex?

b. Which of these functions does the graph represent? Explain how you know.

$$f(x) = (x + 2)^2 + 1 \qquad\qquad g(x) = (x + 2)^2 - 1$$
$$h(x) = {}^-(x - 2)^2 + 1 \qquad\qquad i(x) = (x - 2)^2 - 1$$

c. For each function in Part b, specify the range and the vertex.

d. How is the range related to the maximum or minimum point of a function?

3. Answer these questions about the function $f(x) = (x - 3)^2 - 1$ without drawing the graph.

   a. What is the line of symmetry?

   b. What is the vertex?

4. A parabola has a vertex at the point (3, 4).

   a. Write an equation for a quadratic function whose graph has this vertex. Check your answer by graphing.

   b. Are there other parabolas with this vertex? If so, state two more. How many are there?

5. Suppose you have graphs of these quadratic functions.

$$f(x) = (x - h)^2 + k \qquad g(x) = x^2$$

   a. How is the graph of f related to the graph of g?

   b. What is the vertex of g? What is the vertex of f?

When a quadratic function is written in a form like

$$f(x) = 2(x - 3)^2 + 1$$

you can predict the line of symmetry and the vertex of the parabola without drawing a graph.

It is much harder to visualize the graph when a quadratic function is written in a form like

$$f(x) = 2x^2 - 12x + 19$$

In this case, it is helpful to rewrite the function by completing the square, as you did in Chapter 7.

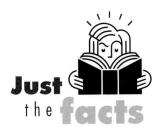

Just the facts

The sparks from a welder's torch travel in paths shaped like parabolas (ignoring the effect of wind and other factors).

<table>
<tr><td>

MATERIALS

graphing calculator

</td></tr>
</table>

Problem Set F

For each function, do Parts a–d.

    a. Complete the square to rewrite $f(x) = ax^2 + bx + c$ in the form $f(x) = a(x - h)^2 + k$.

    b. Find the line of symmetry of the graph of f.

    c. Find the coordinates of the vertex of the parabola.

    d. Use the rewritten form of the function to sketch its graph. Check with a graphing calculator.

  1. $f(x) = x^2 + 8x + 7$

  2. $f(x) = {}^-x^2 + 4x + 1$ (Hint: Factor out a $^-1$ first.)

  3. $f(x) = x^2 - 6x - 3$

Share & Summarize

Describe the relationship between a range of a quadratic function and the vertex of the related parabola.

Investigation 3 ▶ Using x-intercepts

VOCABULARY
x-intercepts

Recall that the y value at which a graph crosses the y-axis is called the y-intercept. In the same way, the x values at which a graph crosses the x-axis are called the **x-intercepts.**

Think & Discuss

How are the x-intercepts of the graph of a function f related to the solutions of $f(x) = 0$? For example, how is the x-intercept of $f(x) = 3x + 7$ related to the solution of $3x + 7 = 0$?

Without making a graph, find the x-intercepts of these functions:

$$h(x) = (3x + 1)(x - 4) \qquad j(x) = x^2 - 7x - 18$$

You will now explore how the x-intercepts of a quadratic function are related to each other and to a parabola's line of symmetry.

31. **Physical Science** A launcher positioned 6 feet above ground level fires a rubber ball vertically with an initial velocity of 60 feet per second. The equation relating the height of the ball over time t is

$$h(t) = 6 + 60t - 16t^2$$

where h is in feet and t is in seconds.

a. Sketch a graph of h.

Another rubber ball is launched 2 seconds later, with the same direction and initial velocity.

b. Suppose you graph the height of the second ball, with time since the *first* ball was launched on the horizontal axis. How will the second graph be related to the first?

c. Write an equation for the height of the second ball over time.

d. Will the second ball collide with the first ball when the first ball is on its way up or on its way down? Explain how you could tell from the graphs of the two functions.

For each function f, write a new function g translated 2 units down and 4 units to the left of f.

32. $f(x) = 2^{x+1} - 1$

33. $f(x) = 2(x - 3)^2 + 1$

34. $f(x) = (x - 1)^3 - x + 1$

35. $f(x) = 1 + \dfrac{1}{x^2 + 1}$

36. **Geometry** A piece of wire 20 cm long is used to make a rectangle.

a. Call the length of the rectangle L. Write a formula for the width W of the rectangle in terms of its length.

b. Write a function for the area A of the rectangle in terms of the length L.

c. Complete the square of the quadratic expression you wrote for Part b. Use your rewritten expression to find the coordinates of the vertex of the graph of Function A.

d. What are the dimensions of the sides of the rectangle with the maximum area? What is the area of that rectangle?

37. The expression $\frac{2x^2}{x}$ is equivalent to $2x$ for all x values except 0 ($\frac{2x^2}{x}$ is undefined for $x = 0$). The graph of $g(x) = \frac{2x^2}{x}$ looks like the graph of $h(x) = 2x$ with a hole at $x = 0$.

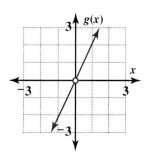

The domain of g is all real numbers except 0. The range of g is also all real numbers except 0.

Now consider the function $f(x) = \frac{(x - 2)(x + 1)}{x - 2}$. Sketch a graph of f, and give its domain and range.

In Exercises 38–40, do Parts a–c.

    a. Write the equation in the form $f(x) = ax^2 + bx + c$. Then complete the square to rewrite it in the form $f(x) = a(x - h)^2 + k$.

    b. Find the line of symmetry of the graph of f.

    c. Find the coordinates of the parabola's vertex.

38. $f(x) = 2x^2 - 8x + 2x^2 - 1$

39. $f(x) = 1 + 4x - 2x^2$

40. $f(x) = {}^-x^2 - x - 1 - x - (2 + x)$

41. Sketch a graph and use it to explain why the equation $3^x + 2 = 0$ has no solutions.

42. The cubic function $c(x) = x^3 + 2x^2 - x - 2$ can be rewritten as $c(x) = (x + 2)(x + 1)(x - 1)$.

    a. Find the x-intercepts of c.

    b. Find the y-intercept of c.

    c. Use the intercepts to draw a rough sketch of c.

43. Consider the equation $(x + 2)^2 - 2 = {}^-x^2 + 4$.

    a. Use the method of graphing two functions to estimate solutions of the equation.

    b. Use the quadratic formula to find the exact solutions of this equation. How close were your estimates?

44. Sketch a graph of $y = \frac{1}{x}$. Use your sketch to think about these questions.

    a. How many solutions of $\frac{1}{x} = 5$ are there?

    b. How many solutions of $\frac{1}{x-5} = 5$ are there?

    c. How many solutions of $\frac{1}{x} = x$ are there? Of $\frac{1}{x} = {}^{-}x$?

    d. How many solutions of $\frac{1}{x} = x^2$ are there?

    e. Use the method of graphing two functions to show the solutions of $\frac{1}{x} = (x-3)^2$. Use your graph to estimate those solutions.

45. Use the method of graphing two functions and locating the points of intersection to find at least four values of x for which each inequality is satisfied.

    a. $x^2 - 2x - 7 < 2x - 3$

    b. $x^2 - 2x - 7 > 2x - 3$

46. Geometry The radius of the cylindrical container is 1 unit less than the side length of the square base of the rectangular container.

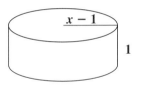 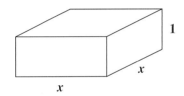

    a. For each container, write a function for the volume.

    b. Use the volume functions to make a graph comparing the volumes of the two containers. Keep in mind that the value of x must be greater than 1, or the cylindrical container would not exist.

    c. For what value of x do the containers hold the same amount?

47. Geometry Recall that a two-dimensional figure that can be folded into a closed three-dimensional figure is called a *net*. For example, this is a net for a cube.

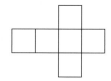

Draw another net for a cube.

In Exercises 48–51, tell which of the following descriptions fits the relationship:

• a direct variation

• linear but not a direct variation

• nonlinear

48. $r = 25v + 32$

49. $a = -\frac{5}{6}j$

50. $k = \frac{3}{n}$

51.

x	2	4	6	8	10	12
y	25	36	47	58	69	80

Tell whether the points in each set are collinear.

52. $(3, 1)$; $(8, 12)$; $(^-1, ^-10)$

53. $(^-2, 9)$; $(2, 2)$; $(4, ^-1.5)$

54. $(15, 22)$; $(0, 1)$; $(5, ^-6)$

Solve each equation.

55. $3 - \sqrt{7s + 2} = ^-7$

56. $3 - \frac{1}{z + 7} = 2$

57. $x^2 + 3x = 6$

58. $16k^2 + 1 = ^-8k$

59. Prove that this number trick always gives 3: *Choose any number except 0. Multiply the number by 9 and add 6. Then divide by 3 and subtract 2. Divide by the number you started with.*

60. Statistics Following are the areas of the 50 U.S. states, in square miles.*

1,545	2,489	5,543	8,721	9,350	9,614	10,555	10,931	12,407	24,230
32,020	35,385	36,418	40,409	42,143	42,774	44,825	46,055	48,430	51,840
52,419	53,179	53,819	54,556	56,272	57,914	59,425	65,498	65,755	69,704
69,898	70,700	71,300	77,116	77,354	82,277	83,570	84,899	86,939	96,716
97,814	98,381	104,094	110,561	113,998	121,589	147,042	163,696	268,581	663,267

The Washington Monument in Washington, D.C.

a. Create a histogram to display these data. Use intervals of 20,000 for the bars. Because two states are much larger than all the others, you may want to exclude them or make a special bar, such as "over 200,000."

b. Complete the stem-and-leaf plot, shown below, of these data. The "stems" in this plot represent ten-thousands.

Stem	Leaf
0	1545 2489 5543 8721
1	0555 0931 2407
2	4230 Key: 1|0555 = 10,555

c. Now create a box-and-whisker plot for these data. Recall that to create the plot, you need the maximum and minimum values along with the median and the first and third *quartiles*. The quartiles can be thought of as medians of the lower and upper halves of the data. For example, consider this data set:

$$2 \quad 3 \quad 4 \quad 5 \quad 6 \quad 7$$

The median is 4.5, the first quartile is 3, and the third quartile is 6. The box-and-whisker plot of this small data set is shown below.

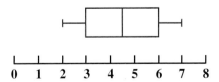

*Source: *World Almanac and Book of Facts 2003.

Chapter Summary

VOCABULARY
domain
function
range
x-intercepts

This chapter focused on a particular type of mathematical relationship called a *function*. A mathematical function produces a single output for each input and can be described with a graph or an equation.

You worked with graphs and equations to find the maximum and minimum values of functions. You used these extreme values to identify the range of a function and to solve problems involving maximum height or maximum area.

You studied in depth the graphs of quadratic functions. You found the line of symmetry and the coordinates of the vertex by inspecting graphs and by completing the square of quadratic expressions. You also solved equations of the form $f(x) = 0$ to find the x-intercepts of quadratic functions.

Finally, you solved equations of the form $f(x) = g(x)$ by locating the points where the graphs of f and g intersect.

Strategies and Applications

The questions in this section will help you review and apply the important ideas and strategies developed in this chapter.

Understanding functions and describing the domain and range of a function

1. Explain how you can determine whether a relationship could be a function by examining a table of inputs and outputs.

2. Give an example of a relationship that is *not* a function.

3. Explain how you can tell whether a relationship is a function by looking at its graph. Give an example of a graph that is not a function.

4. Give an example of a function in which negative numbers do not make sense as part of the domain.

5. Describe the range of the function $k(n) = 3n^2 - 4$.

Finding the maximum and minimum values of quadratic functions

6. Explain two ways to find the maximum or minimum value of a quadratic function.

7. Consider all the possible rectangles with a perimeter of 22 centimeters.

    a. If the length of one such rectangle is x cm, write an equation for a function A for the area of the rectangle.

    b. Use your answer to Part a to find the maximum possible area of the rectangle.

    c. What dimensions give the maximum area?

8. Assume the function $H(t) = 100t - 4.9t^2$ gives the height in meters of a rocket launched vertically from ground level, where t is time in seconds. Estimate the maximum height of the rocket, and tell how many seconds after its launch it attains this maximum height.

Understanding and using graphs of quadratic functions

9. Explain how the graphs of g and h are related to the graph of $f(t) = 2t^2$.

$$g(t) = 10 + 2(t + 2)^2 \qquad h(t) = 2(t - 2)^2 - 3$$

10. Which of these quadratic functions has its vertex at $(3, {}^{-}3)$?

$$g(t) = 3(t + 3)^2 - 3 \qquad h(t) = 4(t - 3)^2 - 3 \qquad k(t) = 3 - 3(t - 3)^2$$

11. Write the equation for a quadratic function with vertex $({}^{-}6, 1)$.

12. Explain how the range of a quadratic function is related to the vertex of its parabola.

13. Explain two methods for finding the vertex and the line of symmetry for the graph of $g(x) = (x + 2)(x + 4)$. Give the vertex and the line of symmetry.

14. Explain how you can use the x-intercepts of a quadratic function f to find its vertex.

Solving equations involving two functions

15. Explain how to use the method of graphing two functions to solve the equation $x^3 = 2x^2 - 1$.

16. Determine how many solutions this equation has, and explain how you found your answer.

$$x^2 + 2x - 3 = x - 2$$

Demonstrating Skills

Copy and complete each table for the given function.

17. $g(x) = x^2 + 3x - 1$

Input	Output
$^-2$	
$^-1$	
0	
1	
2	

18. $h(x) = \frac{1}{4 - x}$

Input	Output
$^-4$	
$^-2$	
0	
2	
4	

19. This is a graph of $f(x) = 2^x$.

  a. Sketch a graph of $g(x) = 2^{x-2}$.

  b. Sketch a graph of $h(x) = 2^{x+3}$.

  c. Sketch a graph of $j(x) = 2^x - 2$.

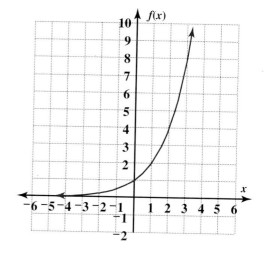

Tell whether each function has a minimum or maximum value, and give the coordinates of this point.

20. $f(x) = ^-x^2 + 2x - 2$

21. $j(x) = ^-5 + x + x^2$

22. $k(x) = 3 - 4(1 - x)^2$

Write the equation for a function g that is the same shape as f but translated 2 units to the left and 1 unit down.

23. $f(x) = {}^-1 + \dfrac{1}{x^3 + 1}$

24. $f(x) = 3^{x+1} - 2$

25. $f(x) = x(x - 2)$

26. Determine the vertex and the line of symmetry of $f(x) = (x + 5)^2 + 9$ without graphing.

For each quadratic function, complete the square and find the vertex and the line of symmetry of its parabola without graphing.

27. $Q(x) = 2x^2 + 2x - 6$

28. $m(x) = {}^-x^2 + \dfrac{7}{2}x - 3$

29. $r(x) = x(x + 3)$

30. Consider the function $f(x) = {}^-x^2 + 8x - 7$.

a. Find the x-intercepts of the graph of f.

b. What is the line of symmetry and the vertex of the graph of f?

c. Use the x-intercepts and the vertex to sketch a graph of f.

31. Solve this equation by graphing.

$$x^2 + 1 = 0.5x + 2.5$$

32. Explain how to solve the equation $x^2 + x = {}^-x - 1$ without graphing. Solve for x.

CHAPTER 9

Probability

Old West Action Crosswords, coded messages, and hidden words are all types of word puzzles. Another type of word puzzle is an *anagram*. Anagrams are particularly fun because you don't need someone to create a puzzle for you. Just choose a word or phrase, and then try to rearrange the letters to form another word or phrase. For example, *Old West action* is an anagram of *Clint Eastwood* (a star of many western films).

While a short word such as *star* has only a few possible combinations— 24, actually, including *rats, tars,* and *tsar*—a word doesn't have to be much longer to have thousands or millions of possible rearrangements. For example, the letters in *Clint Eastwood* can be arranged in 1,556,755,200 different ways.

Think About It List as many anagrams of the word *stop* as you can.

Family Letter

Dear Student and Family Members,

In the next few weeks, we will be looking at many new situations that involve probability. In finding the probability that something will occur, you must first find all the *possible* outcomes. For example, if you are drawing blocks from a bag containing 3 blue, 2 green, and 5 white blocks, there are 10 possible outcomes: the 10 blocks. The probability of drawing a green block is the ratio of the number of green blocks, 2, to the number of possible outcomes: $\frac{2}{10}$.

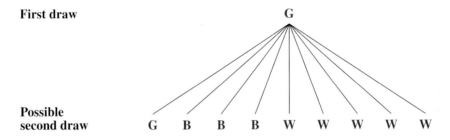

Our class will consider more complicated situations soon. For example, imagine that you draw a block and keep it, and then draw another block. We will learn how to calculate such probabilities as the probability of drawing a green block first and a blue block second, using methods from the mathematical field called *combinatorics*—which includes finding the possible combinations of items. One such method is to use a tree diagram, like the one below, to record all 10 possible first draws, and then for each first draw, the 9 possible second draws. This diagram shows the branches for a tree diagram when green is the result of the first draw:

At the end of this chapter, we will apply what we have learned to analyze the fairness and probabilities of complicated games, such as sports playoffs and state lotteries. We will answer questions like these:

- What are the chances of winning any of several lotteries?
- Is one team favored by a particular playoff structure?
- Which playoff structure is the fairest in a given situation?

Vocabulary There is only one new vocabulary term in this chapter—*sample space*. We'll find the sample space to determine the probability of various events.

What can you do at home?

During the next few weeks, your student may show interest in the topic of probability. You might help him or her think about common occurrences of this topic such as lotteries, or play a game with your student that involves the use of dice or spinners and probability.

9.1

Counting Strategies

A *probability* is a number between 0 and 1 that indicates how likely something is to happen. Often the key to determining the probability that something will occur is to first find all the possible *outcomes*.

When you toss a coin, the two possible outcomes—heads and tails—are equally likely. So, the probability of getting heads is 1 out of 2, or $\frac{1}{2}$, or 0.5, or 50%.

Think & Discuss

Suppose you spin the two spinners below. Each spinner has an equal chance of landing on white, blue, or orange.

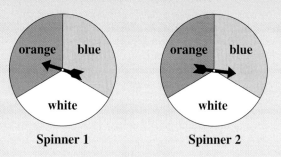

Spinner 1   Spinner 2

List all the possible outcomes. For example, one outcome is Spinner 1 landing on white and Spinner 2 landing on orange. You can use the notation white/orange, or WO, to represent this outcome.

How many possible outcomes are there?

In how many outcomes does Spinner 1 land on blue and Spinner 2 land on orange? What is the probability of this happening?

In how many outcomes does one spinner land on blue and the other land on orange? What is the probability of this happening?

In how many outcomes does Spinner 2 land on blue? What is the probability of this happening?

In some situations, counting outcomes is not as easy as it sounds. In this lesson, you will investigate some counting strategies.

Lab Investigation ▶ Pizza Toppings

Paula's Pizza Place offers four toppings for its vegetarian cheese pizzas.

Customers can order a pizza with any combination of toppings, from no topping at all (just cheese) to all four toppings. However, a topping can be used only once—a customer can't order a pizza with two helpings of green peppers, for example.

MATERIALS

- cubes in 5 colors
- circles to represent pizzas

Make a Prediction

1. How many different pizzas do you think can be made using these four toppings?

Try It Out

You can explore this problem by making a model. Use colored cubes to represent the toppings, and circles to represent pizzas. Create a few pizzas by placing the cubes on the circles. Remember, use only one helping of a topping on any one pizza.

See whether you can make all the possible pizzas. Try to find a systematic way to organize the pizzas so you can be sure you found them all.

2. List all the different pizzas that are possible. How many possibilities are there?

Try It Again

You will now solve the problem again, but with fewer toppings available to choose from. As you work, look for a pattern in your results that may confirm your answer to the four-topping pizza problem.

3. How many different pizzas can you make if only one topping is available?

4. How many different pizzas can you make if only two toppings are available?

5. How many different pizzas can you make if only three toppings are available?

6. Organize your results from Questions 3–5 in a table.

Toppings	1	2	3
Different Pizzas			

7. Do you see a pattern in your results? If so, does the number of different pizzas you made from four toppings fit the pattern? If not, check your work for Questions 2–5.

8. Suppose Paula adds pineapple to the list of toppings. How many combinations do you predict are now possible? Use a fifth color of cube to represent pineapple, and make enough pizzas to see a pattern and check your prediction.

What Did You Learn?

Review your results for all the pizza problems. You should see a pattern in them. If you don't, check your work.

9. Use the pattern in your results to extend your table up to at least 12 toppings. You may use a calculator, but try to complete the table without using paper-and-cube pizza models.

10. Make a list of all the pizza toppings you have ever heard of. In a short report, explain how to determine the number of combinations that can be made from your list of toppings. In your report, also answer this question:

If you order a different type of pizza every day, how many days, weeks, or months will pass before you will have ordered all the possibilities?

Investigation One-on-One Basketball

Ally, Brevin, Carol, and Doug are playing one-on-one basketball. To decide the two players for each game, they put their names into a hat and pull out two at random.

To find the probability that Brevin and Carol will play the next one-on-one game, you might start by first listing all the possible pairs of the four friends. Each pair is an *outcome* in this situation. The set of all possible outcomes—in this case, the set of all possible pairs—is called the **sample space.**

VOCABULARY
sample space

There are many ways to find the sample space for a particular situation, but you need to be careful. If there are lots of possible outcomes, it can be difficult to determine whether you have listed them all or have listed an outcome more than once.

Problem Set A

You will now use a systematic method to find the sample space for drawing pairs of names for the one-on-one basketball situation.

1. List all the possible pairs of names that include Ally.

2. List all the possible pairs that include Brevin but *not* Ally (since you already listed that pair in Problem 1).

3. Now list all the pairs that include Carol but *not* Ally or Brevin.

4. Review your answers to Problems 1–3.

    a. Are there any pairs you have listed more than once or that you have overlooked? If so, correct your errors.

    b. How many pairs are there in all? List them.

5. Brevin wants to play Carol in the next game.

    a. How many pairs match Brevin with Carol?

    b. What is the probability that Brevin will play Carol in the next game?

In Problem Set A, you calculated that the probability that Brevin and Carol will play in the next game is $\frac{1}{6}$. This does not necessarily mean that in the next six draws, the pair Brevin/Carol will be chosen exactly once. It's possible—though not likely—that the pair will be drawn all six times or that Carol's name won't be drawn even once.

Probabilities do not tell you what will definitely happen. They tell you what you can expect to happen *over the long run*.

M A T E R I A L S
• 4 identical slips of paper
• container

Problem Set B

Write the four names—*Ally, Brevin, Carol,* and *Doug*—on identical slips of paper and put them into a container.

1. Suppose you randomly—that is, without looking or trying to choose one name over another—draw 12 pairs of names from the container, putting the pair back after each draw. Based on the probability you found in Problem Set A, how many times would you expect to draw the pair Brevin/Carol?

Draw two names at random and record the results. Return the names to the container. Repeat this process until you have drawn 12 pairs.

2. How many times was the pair Brevin/Carol drawn? How do these experimental results compare with your answer to Problem 1?

3. Each group in your class drew 12 pairs of names. How many total draws occurred in your class? In this number of draws, how many times would you expect the pair Brevin/Carol to be drawn?

4. Each group in your class should now record how many times they drew the pair Brevin/Carol. How many times in all was the pair Brevin/Carol drawn? How does this compare with your answer to Problem 3?

Next you will investigate what happens when a fifth player is added to the one-on-one basketball situation.

Problem Set C

1. Suppose Evan joins Ally, Brevin, Carol, and Doug.

   a. How many new pairs can now be made that could not be made before Evan joined?

   b. What is the size of the new sample space?

   c. To verify the size of the new sample space, systematically write down all the possible pairs. (Rather than writing out the entire names, just use each player's first initial.)

2. Look at the new sample space you listed in Problem 1.

   a. How many of the pairs involve Evan? Which are they?

   b. What is the probability that Evan will be involved in the next game?

In Problem Set C, you found the probability that the following event would occur: *Evan is in the pair.* This particular event has four outcomes in the sample space: Evan/Ally, Evan/Brevin, Evan/Carol, and Evan/Doug.

If you know that the outcomes in a sample space are *equally likely*—that is, that each outcome has the same chance of occurring—it is easy to calculate the probability of a particular event.

EXAMPLE

The names *Ally, Brevin, Carol, Doug,* and *Evan* are put into a hat, and one pair of names is pulled out at random. What is the probability that the pair will include either Ally or Brevin (or both)?

The sample space consists of 10 outcomes:

Ally/Brevin	Ally/Carol	Ally/Doug	Ally/Evan
Brevin/Carol	Brevin/Doug	Brevin/Evan	Carol/Doug
Carol/Evan	Doug/Evan		

Of these 10 outcomes, 7 include Ally or Brevin:

Ally/Brevin	Ally/Carol	Ally/Doug	Ally/Evan
Brevin/Carol	Brevin/Doug	Brevin/Evan	

Since each pair has the same chance of being drawn, the probability that the pair will include either Ally or Brevin is $\frac{7}{10}$.

Problem Set D

The names *Ally, Brevin, Carol, Doug,* and *Evan* are put into a hat, and one pair of names is pulled out at random.

1. What is the probability that the pair includes Doug?

2. What is the probability that the pair includes Carol or Evan (or both)?

3. What is the probability that the pair does not include Brevin?

4. What is the probability that the pair includes Ally or Brevin or Doug?

5. What is the probability that the pair includes Carol but not Doug?

6. What is the probability that the pair includes Ally and Evan?

7. Make up a question like those in Problems 1–6 that involves an event with a 2-in-10 chance of occurring.

8. Make up another question like those in Problems 1–6, and give the answer to your question.

Share & Summarize

If all the outcomes in a sample space are equally likely, how can you find the probability that a particular event will occur?

Investigation 2 ▶ More Counting Strategies

You have seen that to find the probability of an event, you must find the size of the sample space. In Investigation 1, you used a systematic strategy to list all the possible pairs for a one-on-one matchup. In this investigation, you will discover some other useful counting strategies.

Think & Discuss

Jesse and Marcus have the five CDs by their favorite band, X Squared:

- *Algebraic Angst*

- *Binary Breakdown*

- *Chalkboard Blues*

- *Dog Ate My Homework*

- *Everyday Problems*

The friends want to listen to all five CDs. Predict the number of different orders in which they can play the five CDs.

Just the facts

The table and tree diagram show the ways these CDs can be arranged or listed. A listing in which order is important is called a *permutation*.

To find how many ways a group of CDs can be ordered, you can list all the possibilities. With only one CD, there is obviously only one order. With two CDs—call them A and B—there are two possible orders: AB and BA. With three CDs, there are six orders:

ABC   ACB   BAC   BCA   CAB   CBA

To be certain you haven't missed any possibilities, you need a systematic method of counting and recording. You could list the possibilities for three CDs in a table.

A First	B First	C First
ABC	BAC	CAB
ACB	BCA	CBA

Or you could organize them using a *tree diagram*.

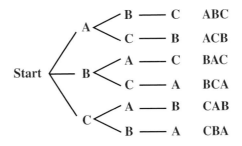

The tree diagram works in much the same way as the table. It shows that there are three ways to start: A, B, or C. Then there are two choices for the second CD and one choice for the third.

Problem Set E

Consider the case of four CDs: A, B, C, and D.

1. Predict the number of ways these four CDs can be ordered.

2. Make an organized list of all the possibilities in which A is played first. To do this, you may want to make a tree diagram like the one started at right.

A First
ABCD
ABDC
ACBD
⋮

3. The number of orders of four CDs in which A is first is the same as the total number of ways three CDs can be ordered. Explain why.

4. Why is the number of orders of four CDs in which B is first the same as the number of orders of four CDs in which A is first?

5. List all the orders in which B is first. Then list all the orders in which C is first. Finally, list all the orders in which D is first. Use your answer to Problem 4 to verify that you have listed all the possibilities.

6. How many possible orders are there altogether? If you know how many entries are in each list, how can you determine the total number of entries without counting them all?

7. Is your estimate from Problem 1 greater than or less than the actual number of orders?

8. Now return to the problem presented in the Think & Discuss on page 551: In how many different orders can Jesse and Marcus play their five CDs? Try to find your answer *without* listing all the possibilities. Show how you found it.

If you consider the orders of the CDs as the outcomes in a sample space, you can calculate probabilities of specific events.

EXAMPLE

Consider all the orders of four CDs: A, B, C, and D. If one of these orders is selected at random, what is the probability that B will be before D?

In Problem Set E, you found that there are 24 outcomes in the sample space for this situation. B appears before D in 12 of these 24 outcomes.

ABCD	BACD	CABD
ABDC	BADC	CBAD
ACBD	BCAD	CBDA
	BCDA	
	BDAC	
	BDCA	

So, the probability that B will be played before D is $\frac{12}{24}$, or 50%.

In Problem Set F, you will find the probability of other events that involve the order in which four CDs are played.

Problem Set F

In Problems 1–7, determine the probability that the given event will occur. Before you begin, be sure you have a complete list of the 24 outcomes in the sample space for this situation.

1. C immediately follows B.　　**2.** A is played last.

3. C is not played first.　　**4.** B is played before A.

5. D is played first *and* A is played last.

6. The CDs are played in the order CBAD.

7. A is played first *and* C is not played last.

8. The probability of A being played last is the same as the probability of B being played last. Explain why.

9. Why are the chances of A being played last the same as the chances of A being played first?

10. How can you use the chances that C is played first to check your answer to Problem 3?

11. If the probability of an event is $\frac{1}{4}$, what is the probability that the event will *not* happen? Explain how you found your answer.

12. If the probability of an event is *p,* where *p* is between 0 and 1, what is the probability that the event will not happen?

Describe an event related to the CD problem with the given probability.

13. $\frac{6}{24}$ **14.** $\frac{22}{24}$ **15.** $\frac{24}{24}$ **16.** $\frac{3}{24}$

Share & Summarize

1. Kai found a shortcut for determining the number of ways to put CDs in order. He said, "For five CDs, multiply 5 by the number of ways to put four CDs in order. In fact, for *n* CDs, just multiply *n* by the number of ways to put *n* − 1 CDs in order."

a. There is only one way to order one CD. Use Kai's method to find the number of ways to order two CDs. Did it work?

b. Use Kai's method and your answer to Part a to find the number of ways to order three and four CDs. Did it work?

c. Explain why Kai's method makes sense.

d. How many ways are there to order seven CDs?

2. Lucita thought she could use the strategy from Problem Set E to find the number of one-on-one pairs of Ally, Brevin, Carol, and Doug. She started listing the possibilities.

Ally First
Ally/Brevin
Ally/Carol
Ally/Doug

Lucita said, "There are three outcomes on this list, and there will be four lists—one for each friend—so there are 12 pairs in all." Is she correct? Explain.

Investigation Counting Strategies Using Patterns

In Investigation 1, you found the size of a sample space by listing all the possibilities (all the one-on-one pairs). In Investigation 2, you saw that you can sometimes discover a pattern that allows you to find the total number of outcomes without listing them all. In this investigation, you will explore other counting strategies.

Remember

Whole numbers are nonnegative integers, or the numbers 0, 1, 2, 3, ...

Two whole numbers add to 12. What might the numbers be?

List every pair of whole numbers with a sum of 12. Make sure you've listed all the possibilities. How many pairs are there? (In this situation, order does not matter. The pair 3-9 is the same as the pair 9-3.)

Predict how many whole-number pairs have a sum of 100.

One strategy for finding the number of whole-number pairs with a sum of 100 is to first consider some simpler problems and look for a pattern.

Problem Set G

1. The sum of two whole numbers is 10. Three possible pairs are 0-10, 1-9, and 2-8. What are the other pairs? How many pairs are there in all?

2. Now write down all the whole-number pairs with a sum of 11. How many pairs are there?

3. Look back at the Think & Discuss above. How many whole-number pairs have a sum of 12?

4. Copy and complete the table to show the number of whole-number pairs with each sum.

Sum	10	11	12	13	14	15	16
Number of Pairs							

5. Copy this table, and use any patterns you have observed to find the number of whole-number pairs with each given sum.

Sum	20	27	40	80	100	275
Number of Pairs						

6. Look at the even and odd sums and the number of pairs that produce them.

    a. Write two expressions—one for even sums and one for odd sums—that describe the relationship between the sum S and the number of whole-number pairs that produce that sum.

    b. Explain why the expressions for even sums and odd sums are different.

Once you know the size of the sample space, you can calculate the probability of an event occurring.

Problem Set H

Use the patterns and answers you found in Problem Set G to determine these probabilities.

1. All the whole-number pairs with a sum of 20 are put into a hat, and one is drawn at random.

   a. What is the size of the sample space in this situation?

   b. What is the probability that one of the numbers in the pair selected is greater than 14? Explain how you found your answer.

2. All the whole-number pairs with a sum of 100 are put into a hat, and one is drawn at random.

   a. What is the size of the sample space in this situation?

   b. In how many pairs are both numbers less than 60? List them.

   c. What is the probability that both of the numbers in the selected pair are less than 60?

3. All the whole-number pairs with a sum of 55 are put into a hat, and one is drawn at random.

   a. What is the size of the sample space in this situation?

   b. How many pairs include a number greater than 48? List them.

   c. What is the probability of choosing a pair in which neither number is greater than 48?

So far, you have used three strategies to find the size of a sample space:

- Systematically list all the possibilities.

- Begin a systematic list of possibilities, and look for a pattern that will help you find the total number without completing the list.

- Start with simpler cases and look for a logical pattern that you can extend to the more complicated cases.

In Problem Set I, you will find the size of a sample space by breaking it into manageable parts. The problem you will work on is similar to those in Problem Set G, but a bit more complicated.

Problem Set I

Consider this problem: *Three whole numbers have a mean of 3. How many such whole-number triples exist? How can you be sure you've found them all?* Here's one way to think through this problem.

1. If the mean of three whole numbers is 3, what is their sum? Why?

Now you can think about the problem as finding all the combinations of three numbers with a sum of 9.

2. When you list combinations of three whole numbers with a sum of 9, does the *order* of the numbers matter? For example, is the triple 1-2-6 considered the same as or different from the triple 6-2-1?

3. One way to break the problem into manageable parts is to start by thinking about all the whole-number triples in which at least one number is 0. List all such triples with a sum of 9. How many are there?

4. To continue listing the combinations, you might next decide to find all those that contain at least one 1 but no 0s.

a. Why would you exclude 0s from your triples at this stage?

b. List the triples that contain at least one 1 but no 0s. How many are there?

5. Continue this process. List all the triples that contain at least one 2 but not 0 or 1, and then all the triples that include at least one 3 but not 0, 1, or 2. Complete the table.

Smallest Number in Triple	0	1	2	3
Number of Triples	5			

6. Why are no more columns needed in this table? In other words, explain why you don't need to consider triples in which the smallest number is 4, 5, or any greater number.

7. How many different triples are there in all?

8. If all the triples are put into a hat, what is the probability of drawing a combination whose smallest number is 0? Whose smallest number is 4?

9. Use the strategy of breaking the problem into smaller parts to find the number of whole-number triples with a mean of 4.

Share & Summarize

A set of four numbers has a mean of m. Explain how you would find the number of whole-number sets that this could describe.

Practice  **Apply**

1. Three friends—Avery, Batai, and Chelsea—get together to play chess. To determine who will play whom, they put their names into a hat and draw out two at random.

   a. List all the possible pairs of names.

   b. What is the probability that Batai and Chelsea will play the next game?

   c. Three more friends—Donae, Eric, and Fran—join the group. Now how many possible pairs are there? List them.

   d. What is the probability that the next match will involve Avery or Eric (or both)?

   e. What is the probability that the next match will *not* involve Batai?

2. **Sports** You have been asked to organize the matches for the singles competition at your local tennis club. There are seven players in the competition, and each must play every other player once.

   a. The matches Player A must play are listed below. Notice that this list already includes the match of Player A against Player B. Copy the table, and in the row for Player B, list all the other matches Player B must play. (In other words, list all the matches that include Player B but *do not* include Player A.)

Player	Matches to Play	Number of Matches
A	AB, AC, AD, AE, AF, AG	6
B		
C		
D		
E		
F		
G		

   b. Predict the number of matches Player C must play that *do not* include Players A or B. Write your prediction in the "Number of Matches" column.

   c. Predict the number of matches for the remaining players. In each case, consider only those matches that *do not* include the players listed above that player.

impactmath.com/self_check_quiz

d. Check your predictions by listing the matches for each player.

e. Describe the pattern in the "Number of Matches" column. Why do you think this pattern occurs?

f. Find the total number of matches that will be played in the singles competition.

g. Use what you have discovered in this problem to predict the total number of matches in a singles competition involving eight players. Explain how you found your answer.

3. Petra wants to make a withdrawal from an automated teller machine, but she can't remember her personal identification number. She knows that it includes the digits 2, 3, 5, and 7, but she can't recall their order. She decides to try all the possible orders until she finds the right one.

a. How many orders are possible?

b. Petra remembers that the first digit is an odd number. Now how many orders are possible?

c. Petra then remembers that the first digit is 5. How many orders are possible now?

4. The "Shuffle" button on Tamika's CD player plays the songs in a random order. Tamika puts a four-song CD into the player and presses "Shuffle."

a. How many ways can the four songs be ordered?

b. What is the probability that Song 1 will be played first?

c. What is the probability that Song 1 will *not* be played first?

d. Songs 2 and 3 are Tamika's favorites. What is the probability that *one* of these two songs will be played first?

e. What is the probability that Songs 2 and 3 will be the first two songs played (in either order)?

5. All the whole-number pairs with a sum of 26 are put into a hat, and one is drawn at random.

   a. List all the possible whole-number pairs with a sum of 26.

   b. What is the size of the sample space in this situation?

   c. What is the probability that at least one of the numbers in the pair selected is greater than or equal to 15? Explain how you found your answer.

   d. What is the probability that both numbers in the pair selected are less than 15? Explain.

6. **Challenge** All the whole-number pairs with a sum of 480 are put into a hat, and one is drawn at random.

   a. What is the size of the sample space in this situation?

   b. What is the probability that one of the numbers in the pair selected is greater than 300? Explain how you know.

7. Three whole numbers have a mean of 5.

   a. List all the whole-number triples with a mean of 5, and explain how you know you have found them all.

   b. How many such whole-number triples exist?

   c. Suppose all the whole-number triples with a mean of 5 are put into a hat, and one is drawn at random. What is the probability that at least two of the numbers in the triple are the same?

Connect &
Extend

8. The Alvarez family—Amelia, Bernie, Carlos, Dina, Eduardo, and Flora—want to form two teams of three to play charades. They put their names into a hat and choose three names to form one team. The remaining three players will form the other team.

   a. How many teams are possible? List them all.

   b. What is the probability that the three names drawn will include Amelia or Eduardo (or both)?

   c. What is the probability that the three names drawn will include both Amelia and Eduardo?

   d. What is the probability that the three names drawn will include neither Amelia or Eduardo?

   e. What is the probability that Amelia and Eduardo will be on the same team? Explain how you found your answer.

9. Five seventh grade friends—Anya, Ben, Calvin, Dan, and Ezra—challenged five eighth grade friends—Vic, Wendi, Xavier, Yvonne, and Zac—to a backgammon tournament. They put the names of the seventh graders into one hat and the names of the eighth graders into another. To determine the two players for each match, they pulled one name from each hat.

   a. What is the size of the sample space in this situation? That is, how many different pairs of names are possible? Explain.

   b. What is the probability that the next match will involve Anya and either Xavier or Yvonne?

   c. What is the probability that the next pair drawn will not involve Calvin?

   d. Suppose all 10 names are put into one hat, and two are drawn at random. What is the probability that the pair will include one seventh grader and one eighth grader? Explain.

10. Kai is helping to plan a school picnic. Each picnic lunch will include a sandwich, a side item, and a dessert. The possible choices for each are given below.

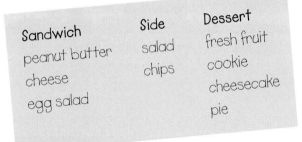

Sandwich
peanut butter
cheese
egg salad

Side
salad
chips

Dessert
fresh fruit
cookie
cheesecake
pie

   a. How many different lunch combinations are possible?

   Kai and his co-workers make an equal number of each combination, but they forgot to mark the bags. Assume that when a person takes a bag, each combination is just as likely to be in the bag as any other combination.

   b. Bharati doesn't care what dessert she gets, but she really wants an egg sandwich and a salad. What is the probability that her lunch will include these two items?

   c. Evan doesn't like eggs. What is the probability that he will choose a bag that *does not* include an egg salad sandwich?

11. In this exercise, you will think about the different ways a number of people can be seated along a bench and around a circular table.

  a. How many ways can three people—call them A, B, and C—be seated along a bench? List all the possibilities.

  b. If the three people are arranged around a circular table, there will be no starting or ending point. So, for example, these two arrangements are considered the same.

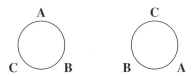

How many *different* ways can three people be arranged around a circular table? Sketch all the possibilities.

  c. Copy and complete the table to show how many ways the given number of people can be arranged along a bench and around a circular table.

People	Row Arrangements	Circle Arrangements
1	1	1
2		
3		
4		

  d. Describe at least one pattern you see in your table.

  e. Five people can be arranged along a bench in 120 ways. Use the patterns in your table to predict the number of ways five people can be seated around a circular table.

12. A programmer wrote some software that composes pieces of music by randomly combining musical segments. For each piece, the program randomly chooses 4 different segments from a group of 20 possible segments and combines them in a random order.

How many different musical pieces can be created in this way? (Hint: How many choices are there for the first segment? For each of those, how many choices are there for the second segment?)

In y o u r
own
words

Explain two counting strategies you can use to find the size of a sample space.

13. Ms. McDonald raises only chickens and pigs on her farm. If you know how many legs are in Ms. McDonald's barn, you can find all the possible combinations of pigs and chickens. For example, if there are 6 legs, there could be 3 chickens, or 1 chicken and 1 pig.

a. Copy and complete the table to show the possible combinations for different numbers of legs. The notation 3C-0P means 3 chickens and no pigs.

Legs	Combinations	Number of Combinations
2	1C-0P	1
4		
6	3C-0P, 1C-1P	2
8		
10		
12		
14		

b. Predict the number of combinations for 16 legs and for 18 legs. Check your predictions by listing all the possibilities.

c. Challenge Write two expressions that describe the number of chicken-pig combinations for L legs. One of your expressions should be for L values that are multiples of 4; the other should be for L values that are not multiples of 4.

d. There are 42 legs in the barn. Assuming each possible combination of pigs and chickens is equally likely, what is the probability that there are 8 pigs and 5 chickens in the barn?

Mixed Review

Write an equation of a line that is parallel to the given line.

14. $2(y - 3) = {}^-7x + 1$

15. $x = {}^-2 - y$

Rewrite each expression using a single base and a single exponent.

16. $2^5 \cdot 2^{-8} \cdot 2^{2p}$

17. $({}^-3^{3m})^6$

18. $k^7 \cdot 2^7$

Solve each inequality, and graph the solution on the number line.

19. $5(9 - x) \le 4(x + 18)$

20. $3x - 9 < {}^-4.5x + 6$

Factor each expression.

21. ${}^-3x^2 + 3x + 18$

22. $0.5a^2 - 2a - 16$

Copy each expression, adding a constant to complete the square.

23. $a^2 + 0.4a + \underline{\hspace{1cm}}$

24. $b^2 - 12b + \underline{\hspace{1cm}}$

25. Suppose a certain type of cell divides into two cells every half hour.

 a. Make a table showing how many cells there will be at the end of every hour, starting with one cell, for a 4-hour period.

Time (h)	0	1	2	3	4	5
Cells	1					

 b. Write an equation for the number of cells c at t hours.

 c. Graph the data from your table, showing times up to 4 hours on the horizontal axis.

26. Match these 12 expressions to create six pairs of expressions in which one expression is a simplified version of the other.

 a. $\dfrac{4}{9} - \dfrac{x+1}{9}$ **b.** $\dfrac{5}{6x} - \dfrac{1}{4x}$ **c.** $\dfrac{2}{9x} - \dfrac{5}{6x}$

 d. $\dfrac{4+x}{-2x-8}$ **e.** $-\dfrac{11}{18x}$ **f.** $\dfrac{x}{4x} - \dfrac{x}{3x}$

 g. $\dfrac{7}{12x}$ **h.** $\dfrac{-x+3}{9}$ **i.** $-\dfrac{1}{12}$

 j. $\dfrac{1}{(x-2)^2}$ **k.** $\dfrac{1}{x(x-4)+4}$ **l.** -0.5

Geometry The area of a sector of a circle is found by multiplying the ratio of the sector's angle measure to 360 by the circle's area:

$$\text{area of sector} = \frac{\text{sector's angle measure (in degrees)}}{360°} \cdot \pi r^2$$

Find the area of each sector.

27.
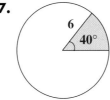

28.
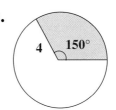

29.
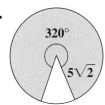

Tell whether each relationship is a function.

30.
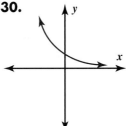

31.
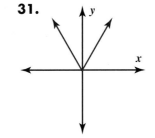

32.
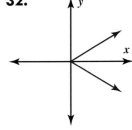

Probability Distributions

Suppose you are planning a camping trip. If you think it is likely to rain during the trip, you will probably decide to bring along a raincoat, boots, a tarp, or other rain gear.

Understanding how likely certain events are can help you make decisions and predictions. Sometimes you can use common sense or logic to decide whether one event is more likely than another.

In some situations, you can determine what is most likely to happen only after more careful analysis. In this lesson, you will analyze situations and games to determine how likely certain events are.

Think & Discuss

List all the possible outcomes from rolling two dice, one after the other.

Are these outcomes all equally likely?

What is the probability you will roll doubles (both dice having the same number)?

What is the probability that at least one die will show a 3?

Which is more likely, rolling at least one 3 or rolling doubles? Could you have answered this question without counting all the combinations for each? Explain.

Investigation ▶ 1 ▶ Comparing Probabilities of Events

You will now look at probabilities involving the roll of two dice and determine which of two events is more likely.

MATERIALS

2 dice

Problem Set A

Tamika rolled two dice 15 times. On each roll, she multiplied the two numbers. Based on her findings, she conjectured that rolling an even product is more likely than rolling an odd product.

1. Roll a pair of dice 15 times, and record whether each product is even or odd. Do your results support Tamika's conjecture?

To figure out whether an even product or an odd product is more likely, you could find the products for all 36 possible dice rolls and count how many are even and how many are odd.

An easier way to analyze Tamika's conjecture is to use what you know about multiplying even and odd factors. On each die, the probability of rolling an odd number is the same as the probability of rolling an even number, so you can simply figure whether each possible combination is odd or even:

even × even   even × odd   odd × even   odd × odd

2. Copy and complete the table to show whether the product of each combination of even and odd factors is even or odd.

Die 1

×	Even	Odd
Even	even	
Odd		

Die 2

3. Which is more likely to occur: an even product or an odd product?

4. Complete these probability statements.

   a. The probability that the product of two dice will be even is ____ out of 4.

   b. The probability that the product of two dice will be odd is ____ out of 4.

5. Suppose that, instead of rolling dice to determine the factors to multiply, Tamika spins these spinners.

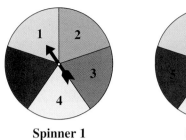

Spinner 1   **Spinner 2**

a. Predict whether an even product or an odd product is more likely.

b. Can you use your table from Problem 2 to determine whether an odd or an even product is more likely? Why or why not?

c. Systematically determine the probability of getting an even product and the probability of getting an odd product. You might find it helpful to complete a multiplication table like the one below.

Spinner 1

	×	1	2	3	4	5
	1					
	2					
Spinner 2	3					
	4					
	5					

Some games are based entirely on skill, some entirely on chance, and some on a combination of the two.

Problem Set B

Tamika was also curious about what would happen if she rolled two dice and considered one number to be the base and the other to be its exponent. Would the result more likely be even or odd?

1. Conduct an experiment to predict whether an even or an odd result is more likely. Before you begin, assign one die as the base and the other as the exponent. Roll the dice 15 times, and record whether the result, $die1^{die2}$, is even or odd. Make a prediction based on your results.

2. Because an odd number is just as likely to be rolled as an even number, you can analyze the possible combinations as you did in Problem Set A. Complete the table to indicate whether raising the given base to the given exponent has an odd or an even result.

	Even Exponent	**Odd Exponent**
Even Base		
Odd Base		

3. Which is more likely to occur: an even result or an odd result?

4. Complete these probability statements.

    a. The probability that $die1^{die2}$ will be even is ____.

    b. The probability that $die1^{die2}$ will be odd is ____.

Share & Summarize

Suppose you spun these two spinners and added the results. Describe two ways you could determine whether an odd sum or an even sum is more likely.

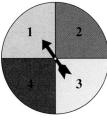

Spinner Y

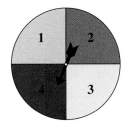

Spinner Z

Investigation ▶2▶ The Rolling Differences Game

Many games involve chance. Figuring out the probabilities of the various events involved in a game can help you determine whether the game is fair.

MATERIALS
• 2 dice
• graph paper

Problem Set C

Héctor and Mikayla are playing a game called *Rolling Differences*. On each round, they each roll a die and find the difference between the numbers. When the numbers are different, they subtract the lesser number from the greater number.

• Player 1 scores 1 point if the difference is 0, 1, or 2.

• Player 2 scores 1 point if the difference is 3, 4, or 5.

Each game consists of 10 rounds (dice rolls).

1. Who do you think is more likely to win a round, Player 1 or Player 2?

2. With a partner, play 10 rounds of *Rolling Differences*. Record who wins. Then, without switching sides—that is, score the second game the same as the first game—play another 10 rounds. Did the same player win both games?

3. You can analyze this game by making a table of differences for all possible rolls. Complete this table to show the difference for each possible roll.

Player 2's Roll

−	1	2	3	4	5	6
1						
2						
3						
4						
5						
6						

Player 1's Roll

4. How many of the 36 dice rolls produce a difference of 5? Which rolls are they?

5. Complete the table to show the probability of rolling each difference.

Difference	0	1	2	3	4	5
Probability						

6. Make a bar graph to show the probability of each difference. Your completed bar graph will show the *probability distribution*—how the probabilities are distributed among the possible differences.

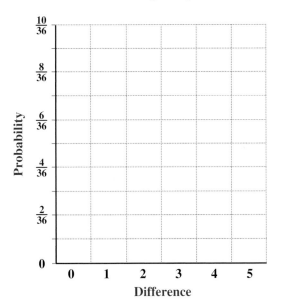

7. Recall that Player 1 scores a point if the difference is 0, 1, or 2. What is the probability that Player 1 will score a point in a round?

8. Use your answer from Problem 7 to calculate the probability that Player 2 will score a point in a round. Does your table confirm your answer?

9. *Rolling Differences* is unfair because one player has a greater probability of winning than the other. Try assigning points in a different way to make the game fair.

Share & Summarize

When two dice are rolled, the 36 possible pairs are equally likely. Why, then, are the chances of rolling numbers with a product of 25 different from the chances of rolling numbers with a product of 12?

Investigation ▶3 The Land Grab Game

In this investigation, you will play a game called *Land Grab.* Analyzing the probabilities involved in the game can help you devise a winning strategy.

MATERIALS
- 2 dice
- game markers
- graph paper

Problem Set D

Land Grab is played by two players. The game board shows three plots of land, each divided into six numbered sections.

Plot A	
12	9
6	4
8	20

Plot B	
30	16
25	1
36	2

Plot C	
3	10
24	15
5	18

Here are the rules for the game.

- Each player selects a different plot of land.
- Each player rolls one die, and the numbers on the two dice are multiplied.
- If the product appears in a section of the plot the player has selected (and it is not already covered), he or she puts a marker on that section.
- The first player to cover all the sections in his or her plot wins.

The challenge is to choose a plot that gives you the best chance of winning.

1. Play *Land Grab* with your partner. For this first game, one of you should choose Plot B and the other should choose Plot C. Before you start, make a prediction about who has a better chance of winning.

Now play two or three more games with your partner. Choose a different plot of land for each game so you get a sense of how easy or difficult it is to win with the various plots.

2. On the basis of the games you played, which plot seems easiest to win with? Which seems most difficult to win with?

Some sections of land are easier to cover than others. For example, a product of 25 can be rolled only one way (5 and 5), while a product of 12 can be rolled four ways (3 and 4, 4 and 3, 6 and 2, and 2 and 6), so the section labeled 12 is easier to cover than the section labeled 25.

3. Determine the probability of covering each section of land. Record your results in a systematic way.

4. Make a bar graph to show the probability distribution for the products.

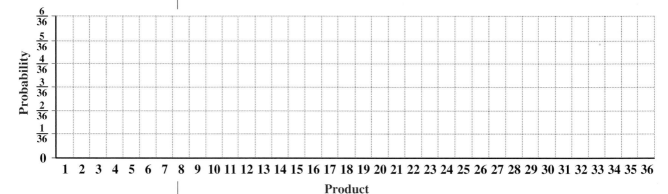

5. Use your results from Problems 3 and 4 to determine which plot of land—A, B or C—is easiest to win the game with. Explain how you decided.

Share & Summarize

You can list all the probabilities for a situation in a table, or you can make a bar graph. Discuss some of the advantages of showing a probability distribution in a bar graph.

On Your Own Exercises

Practice & Apply

1. Two dice are rolled and the two numbers are added.

  a. Copy and complete the table to show all the possible sums.

Die 1

+	1	2	3	4	5	6
1						
2						
3						
4						
5						
6						

Die 2

  b. How many different sums are there? What are they?

  c. Which sum occurs most often? List all the ways it can be created.

  d. Complete the table to indicate whether the sum of each combination of even and odd numbers is even or odd.

Die 1

+	Even	Odd
Even	even	
Odd		

Die 2

  e. What is the probability that the sum of two dice will be odd?

  f. Suppose you spin these spinners and add the results. Can you use your table from Part d to find the probability that the sum will be odd? Explain.

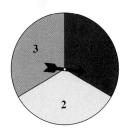

2. Suppose you spin these spinners and multiply the results.

  a. Predict whether an odd product or an even product is more likely.

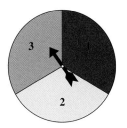
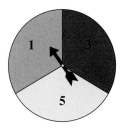

  b. Determine the probability of spinning an odd product and of spinning an even product. Do your results agree with your prediction?

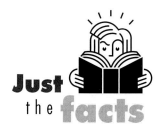
3. Suppose you roll two 8-sided dice with faces numbered 1 to 8.

a. How many possible number pairs can you roll?

b. Complete the table to show all the possible sums of two 8-sided dice.

Die 1

+	1	2	3	4	5	6	7	8
1								
2								
3								
4								
5								
6								
7								
8								

Die 2

c. What is the probability that a sum will be odd?

d. What is the probability that a sum will be a prime number?

4. Suppose you roll two 12-sided dice with faces numbered 1 to 12.

a. How many possible number pairs can you roll?

b. What is the greatest sum possible from a roll of two 12-sided dice?

c. What sum is most likely? What is the probability of this sum?

5. Suppose you roll two dice: a 12-sided die with faces numbered 1 to 12, and an 8-sided die with faces numbered 1 to 8.

a. How many possible number pairs can you roll?

b. What sum is most likely? What is the probability of this sum?

c. Is an odd sum or an even sum more likely?

6. Suppose two people play *Rolling Differences* with one 6-sided die, and one 8-sided die with faces numbered 1 to 8. The players follow these rules:

• Player 1 scores 1 point if the difference is 0, 1, or 2.

• Player 2 scores 1 point if the difference is 3, 4, 5, 6, or 7.

a. Make a table showing the probability of each difference being rolled.

b. Is this game fair? Explain.

c. Make a bar graph to show the probability distribution for this game.

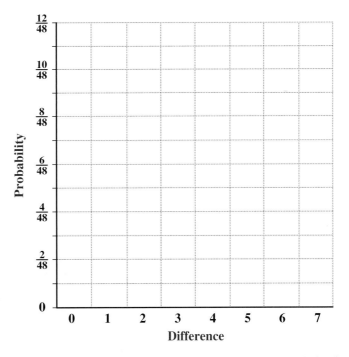

d. Compare the rules for this game to the rules for the original *Rolling Differences* game (which involves two ordinary dice) described in Problem Set C. In which game does Player 2 have a better chance of winning? Explain.

7. Suppose two people play *Rolling Differences* with two 8-sided dice numbered 1 to 8. The players follow these rules:

- Player 1 scores 1 point if the difference is 0, 1, or 2.

- Player 2 scores 1 point if the difference is 3, 4, 5, 6, or 7.

Which player has the advantage? Explain your answer.

8. *Mixing Colors* is a two-player game that uses the spinners and game cards shown below.

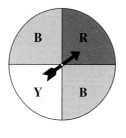
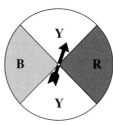

The letters on the spinners and game cards represent the colors red, blue, yellow, green, orange, and purple. Here are the rules.

- Each player selects a different game card.

- Each player spins a spinner. The colors on the two spinners are "mixed" to give a mixture color, as follows:

  RR = R   BB = B   YY = Y   RY = O   RB = P   BY = G

- If the mixture color is on a player's game card (and is not already covered), he or she puts a marker on that section.

- The first player to cover all the sections on his or her game card wins the game.

a. Make a prediction about which game card gives the best chance of winning the game.

b. Determine the probability of spinning each mixture color. Record your results in a systematic way.

c. Make a bar graph to show the probability of spinning each mixture color.

d. Use your results from Parts b and c to determine which card gives the best chance of winning. Explain how you decided.

9. Imagine rolling three regular dice and multiplying all three numbers.

a. How many number triples are possible when you roll three dice?

b. *Without* finding the products of every possible roll, describe a way you could determine whether an odd product or an even product is more likely.

c. Use your method from Part b to determine whether an even product or an odd product is more likely.

10. Imagine rolling five regular dice and looking for outcomes when all five dice match.

a. How many different outcomes are possible on a roll of five dice? Explain.

b. In how many of the possible outcomes do all five dice match?

c. What is the probability of getting all five dice to match on a single roll?

d. Suppose Tamika is given three rolls to get five matching dice. On the second and third rolls, she may roll some or all of the five dice again.

On her first roll, Tamika gets three 3s, a 2, and a 6. She picks up the dice showing 2 and 6 and rolls them again. What is the probability that she will get two more 3s on this roll?

11. Two players each roll a six-sided die and find the difference of the numbers.

• Player 2 receives 2 points each time the difference is 3, 4, or 5.

• Player 1 receives 1 point each time the difference is 0, 1, or 2.

Which player has the advantage in this game? Explain your answer.

12. Kai wants to create a fair dice game for two people, using divisibility by 3. He decided on these rules:

• Each player rolls one die.

• The players find the sum of the two numbers.

• Player 1 scores if the sum is divisible by 3. Player 2 scores if the sum is *not* divisible by 3.

Kai isn't sure how many points each player should score each time.

a. Make a table showing the probability of each sum being rolled.

b. What is the probability that Player 1 scores on a given roll?

c. Kai decided to give 2 points to Player 2 when the sum isn't divisible by 3. For the game to be fair, how many points should Player 1 score each time the sum *is* divisible by 3?

13. *Mountain Climbing* is a two-player game involving two dice and this game board.

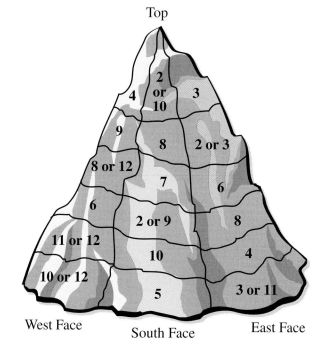

Top

2 or 10

4

3

9

8

2 or 3

8 or 12

7

6

6

2 or 9

8

11 or 12

10

4

10 or 12

5

3 or 11

West Face

South Face

East Face

Here are the rules for *Mountain Climbing.*

• Each player selects a different face of the mountain to climb step by step.

• Each player rolls one die, and the results are added.

• If the sum appears as the next step on the face the player has selected, he or she puts a marker on that step. Players must proceed one step at a time from bottom to top.

• The first player to reach the top step wins.

The challenge is to choose a face that gives the best chance of winning.

a. Copy and complete these tables, which show the probability for reaching each step along each face of the *Mountain Climbing* game board.

West Face

Step	1	2	3	4	5	6
Probability	4/36					

South Face

Step	1	2	3	4	5	6
Probability						

East Face

Step	1	2	3	4	5	6
Probability						

b. Which face gives the greatest probability of winning?

c. Which face gives the least probability of winning?

14. In Investigation 3, you played the game *Land Grab.* Design a new six-section plot that is easier to win with than Plots A, B or C. Explain your reasoning.

15. *Hidden Dice* is a two-player game in which opponents take turns guessing one or both numbers on a pair of dice.

Players take turns rolling the dice and concealing the result. The rolling player announces that the sum of the two dice is "greater than or equal to 7" or "less than or equal to 7."

If the guessing player can correctly identify one of the dice, he or she scores 1 point. The guessing player may then keep the point or give it up for the opportunity to make a guess at the second die. The winner is the first player to score 10 points or to guess both dice correctly in the same turn.

a. Complete the table, which shows all the possible ways you can get a sum that is greater than or equal to 7.

Sum	Die 1	Die 2
12	6	6
11	5	6
	6	5
10	5	5
	4	6
	6	4
9		

Sum	Die 1	Die 2
8		
7		

b. Make another table showing all the possible ways you can roll a sum that is equal to or less than 7.

c. Consider the possible outcomes for a sum of greater than or equal to 7. Which number appears in the greatest number of outcomes? Give the probability that *that* number is on at least one die when the rolling player reports the sum is "greater than or equal to 7."

d. Consider the possible outcomes for a sum of less than or equal to 7. Which number appears in the greatest number of outcomes? Give the probability that *that* number is on at least one die when the rolling player reports the sum is "less than or equal to 7."

e. Suppose the rolling player announces that the sum is greater than or equal to 7, and you correctly guess that a 2 appears on one of the dice. How many possibilities are there for the remaining die, and what are they?

f. Suppose the rolling player announces that the sum is greater than or equal to 7, and you correctly guess that a 1 appears on one of the dice. Explain why you should continue and give up 1 point for a guess at the other die.

g. What is the disadvantage of guessing 1 on your first guess when the sum is greater than or equal to 7?

h. What are the advantages and disadvantages of guessing 6 on your first guess when the sum is greater than or equal to 7?

Mixed Review

16. Match each equation to a graph.

a. $y = {}^-2x^2$

b. $\frac{y}{2} = x^2$

c. $y + 3 = 0.5x^2$

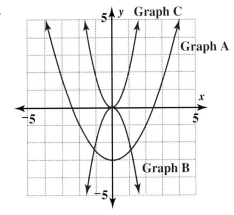

17. Write an equation to represent the value of K in terms of n.

n	0	1	2	3	4
K	1	0.4	0.16	0.064	0.0256

18. Order these numbers from least to greatest.

$$\sqrt[9]{{}^-4} \qquad \sqrt[3]{{}^-4} \qquad \sqrt[11]{{}^-4} \qquad \sqrt[5]{{}^-4} \qquad \sqrt[7]{{}^-4}$$

19. Draw a graph to estimate the solution of this system of equations. Check your answer by substitution.

$$y = 2x + 7$$

$$4x + 3y = 31$$

20. A line segment with length 8 and slope $\frac{1}{4}$ is scaled by a factor of 3. What are the length and the slope of the new segment?

Expand and simplify each expression.

21. ${}^-3a(2a - 3)$

22. $(4k - 7)({}^-2k + 3)$

Simplify each expression.

23. $\dfrac{m - 3}{m(2m - 6)}$

24. $\dfrac{7}{k - 2} - \dfrac{5}{2(k - 2)}$

9.3 Probability Investigations

Many people play state-run lottery games. There is an important social issue associated with these games: they often appear easier to win than they really are! If players knew their real chances of winning, they might make wiser decisions about whether and how often to play.

Think & Discuss

In one state's lottery game, participants choose six different numbers from 1 to 49. To win the grand prize, all six numbers must match those selected in a random drawing. The order of the numbers doesn't matter. For example, 3-32-16-13-48-41 is considered the same as 3-13-16-32-41-48.

This game has a drawing twice a week. Suppose you selected one group of six numbers for every drawing. Make a guess about how often you could expect to win the grand prize.

By the end of this lesson, you will be able to find the exact answer to this problem.

Investigation 1 Analyzing Lottery Games

Let's start by analyzing some simple lottery games. Consider a lottery game in which you must match two out of six numbers.

Think & Discuss

To win the *2-of-6* lottery game, you must match two different numbers from 1 to 6 with those selected in a random drawing. How often do you think you could expect to win this game? Answer this question by completing this probability statement:

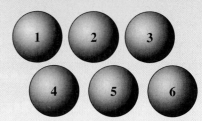

I estimate that I would win the 2-of-6 game once every _____ games.

Record the responses for the entire class on the board.

Problem Set A

With a partner, follow these steps to model the *2-of-6* lottery game.

• Write the whole numbers from 1 to 6 on separate slips of paper. Place the slips in a container, and shake them.

• One partner should write down a two-number lottery pick.

• The other partner should select two numbers from the container.

• Record "win" if the selected number pair matches the recorded number pair and "lose" if it doesn't match. (Both numbers must match in order for you to win.)

• Return the numbers to the container, and shake the container.

• Keeping the same numbers for the lottery pick, repeat the process—drawing two numbers, recording "win" or "lose," and returning them to the container—until you have made 10 selections.

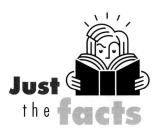

Just the facts

An arrangement or listing of objects in which order is *not* important is called a combination.

1. How many times out of 10 did you win the game?

2. Combine your results with those from your classmates. Use the combined results to complete this statement:

In our class experiment, there was 1 winner every _____ games.

3. List all the possible pairs that can be drawn in the *2-of-6* lottery game. Remember, the order of the numbers doesn't matter; for example, 3-4 is the same as 4-3. How many pairs are there?

4. Instead of listing all the possibilities, you can use a shortcut to find the number of pairs.

  a. How many ways are there to select the first number?

  b. Once the first number is chosen, how many ways are there to select the second number?

  c. If you multiply your answers from Parts a and b, you will have the number of pairs if order *did* matter. To see that this is true, draw a tree diagram showing the different combinations. How many different pairs would there be if order mattered?

  d. Explain why you can just multiply the answers from Parts a and b to find the number of pairs.

  e. Any pair of numbers can be arranged two ways; for example, 1-2 and 2-1. So, your total from Part c counts each number pair twice. To find the actual number of pairs, you need to divide your result from Part c by 2.

  What is the total number of possible pairs for the *2-of-6* lottery game? How does this number compare to the number of possible pairs in the list you made in Problem 3?

5. If you choose just one pair in the *2-of-6* lottery game, what are your chances of winning? Record your answer by completing this probability statement:

The chances of winning are 1 out of _____.

6. Compare your answer for Problem 5 to your class predictions from the Think & Discuss at the bottom of page 582. Do the data support the idea that the game looks easier than it really is?

As you probably would have guessed, the chances of winning a lottery game change as the number of choices and the number of matches required to win change.

Think & Discuss

In the *3-of-7* lottery game, players must match three different numbers from 1 to 7 with those selected in a random drawing. So, players have one more number to choose from, and they must match three numbers instead of two.

Think about how these changes would affect your chances of winning. Then complete this statement:

I estimate that I would win the 3-of-7 game once every _____ games.

Record the responses for the entire class on the board.

Problem Set B

1. You can use the counting strategy from Problem Set A to calculate the number of possible triples for the *3-of-7* game.

  a. How many possibilities are there for the first number selected?

  b. Once the first number is chosen, how many possibilities are there for the second number?

  c. Once the first two numbers are chosen, how many possibilities are there for the third number?

  d. Multiplying your results from Parts a–c will give you the number of triples if order *did* matter—if, for example, 1-2-3 was considered different from 2-3-1. How many different triples would there be if order mattered?

In your answer for Part d, each triple is included more than once. To find how many times each triple is counted, think about how many ways any triple of numbers can be arranged. For example, consider the numbers 1, 2, and 3.

  e. How many possibilities are there for the first number?

  f. Once the first number is chosen, how many possibilities are there for the second number?

  g. Once the first two numbers are chosen, how many possibilities are there for the third number?

  h. How many times is each triple counted in your answer to Part d? That is, how many ways are there to arrange three numbers?

  i. To find the actual number of triples, divide your result from Part d by your answer to Part h. What is the total number of possible triples in the *3-of-7* lottery game?

2. What are your chances of winning the *3-of-7* game?

3. Compare your answer to Problem 2 to your class predictions for the Think & Discuss on page 584. Does the game look easier than it actually is?

You can use the techniques you developed in Problem Sets A and B to calculate the probability of winning other lottery games.

Calculate the probability of winning the *4-of-12* game. To do this, you need to find the number of possible ways 4 numbers can be selected from 12 numbers.

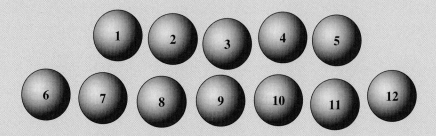

First think about the number of possibilities if order did matter. There would be 12 possibilities for the first number, 11 for the second, 10 for the third, and 9 for the fourth. So, if order mattered, the number of possibilities would be $12 \cdot 11 \cdot 10 \cdot 9$, or 11,880.

Any four numbers can be arranged in $4 \cdot 3 \cdot 2 \cdot 1$, or 24, ways. So, the product $12 \cdot 11 \cdot 10 \cdot 9$ counts each set of four numbers 24 times. To find the number of possible lottery selections, you must divide that result by 24:

$$\frac{12 \cdot 11 \cdot 10 \cdot 9}{4 \cdot 3 \cdot 2 \cdot 1} = \frac{11{,}880}{24} = 495$$

So, there are 495 ways four numbers can be selected. The probability of winning the game with a single ticket is $\frac{1}{495}$.

Share & Summarize

1. Calculate the probability of winning the lottery game described in the Think & Discuss on the top of page 582. Remember, this is a *6-of-49* game. Show your calculation.

2. If you select one group of six numbers for both drawings every week, how often could you expect to win? Show your calculation.

Investigation Analyzing Sports Playoffs

Many sports leagues end their seasons with some type of playoff. Playoff organizers must decide how many teams to include and how to structure the series of games. The organizers try to make the structure as fair as possible for all teams.

MATERIALS

- 3-team playoff charts
- 3 identical slips of paper
- 2 dice

Problem Set C

You will now analyze two playoff structures, *Top Two* and *Top Three*.

Top Two: Teams A and B have the two best records at the end of the season. The two teams play, and the winner is declared the champion.

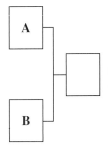

1. Assume the teams have the same chances of winning the game. What is the probability that Team A will win the championship? What is the probability that Team B will win?

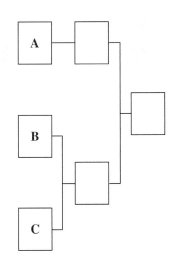

Top Three: At the end of the season, Teams A, B, and C have the three best records. In the playoffs, Team B plays Team C, and the winner plays Team A for the championship.

You can work in a group of three to model the *Top Three* series. In your model, assume that each team has the same chances of winning each game.

To decide who will represent each team, label three identical slips of paper A, B, and C. Put them face down, mix them up, and have each player choose one.

On a copy of the chart above, put the slips of paper on their starting positions. For each game, the two students representing the participating teams each roll a die. The student rolling the higher number wins the game and moves his or her slip of paper on to the next round. If there is a tie, the students roll again.

2. Play the complete playoff series eight times, keeping a tally of the champion teams.

Team	Number of Championships
A	
B	
C	

3. Which team won the most championships? Why do you think this is?

4. If you assume the teams are equally likely to win each game, you can find the probability that each team will win the championship.

   a. If this playoff series were played eight times, how many times would you expect Team B to beat Team C in the first game?

   b. Of the number of times Team B beats Team C, how many times would you expect it to beat Team A in the final game?

   c. Use your results to determine the probability that Team B will win the championship.

   d. Use a similar method to determine the probabilities that Team A and Team C will win the championship. That is, imagine that the series is played eight times, and figure out how many times you can expect each team to win the championship.

5. Suppose a sports league with several teams is planning a special playoff. The organizers are considering including either the top two teams, using the *Top Two* structure, or the top three teams, using the *Top Three* structure. Discuss whether you think one playoff structure would be more appropriate than the other.

Next you will compare playoff structures involving four teams.

MATERIALS
- 4-team playoff charts
- 4 identical slips of paper
- 2 dice

Problem Set D

At the end of the season, Teams A, B, C, and D are the top four teams. Here is one possible playoff structure.

1. Working in a group of four, play through the complete playoff series five times, keeping a record of the champion teams.

2. Combine your results with those from the other groups. Summarize what you find.

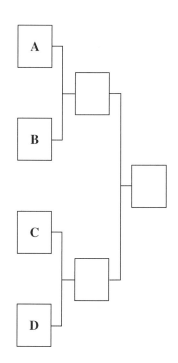

Here is another way to structure a four-team playoff.

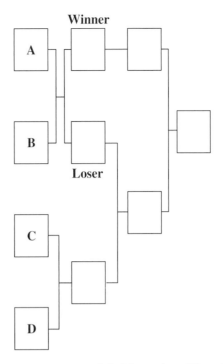

3. Work with your group to model this series. Play the complete series five times, keeping a tally of your results.

4. Combine your results with those from the other groups. Summarize what you find.

5. Discuss any circumstances under which you think one playoff structure would be more appropriate than another.

6. Challenge For each structure in this problem set, determine the probability that each team will win the championship. Assume both teams are equally likely to win each game. (Hint: Imagine that a series is played eight times, and determine how many times you can expect each team to win the championship.)

Share & Summarize

In some of the playoff structures you examined, all the teams had an equal chance of winning the championship. In others, at least one team had a better chance than some of the other teams. What is the difference in the structures that allows one team to have a better chance?

On Your Own Exercises

Practice & Apply

1. In the *Orange-and-White* game, three white marbles and one orange marble are placed in a bag. A player randomly draws two marbles. If the marbles are different colors, the player wins a prize.

   a. List all the possible pairs in the sample space. (Hint: Label the marbles W1, W2, W3, and O.)

   b. What is the probability of winning a prize?

2. To win the *3-of-10* lottery game, players must match three numbers from 1 to 10 with those selected in a random drawing. Remember that order doesn't matter.

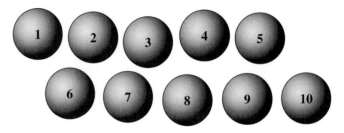

   a. How many possible triples are there in the *3-of-10* lottery game?

   b. What are your chances of winning the *3-of-10* lottery game?

3. To win the grand prize in the *Match 5* lottery game, players must match five numbers from 1 to 30 with those selected in a random drawing. Remember, order doesn't matter.

   a. How many different groups of five numbers are possible?

   b. If you bought a single ticket with five numbers, what is the probability you would win the grand prize?

   c. The *Match 5* lottery game has a drawing three times a week. If you bought a ticket for each drawing—that is, if you played three times a week—how often could you expect to win the grand prize?

   d. Suppose you bought 100 tickets for each drawing (with a different group of five numbers on each ticket). How often could you expect to win the grand prize?

impactmath.com/self_check_quiz

4. Sports A volleyball league has two divisions. At the end of the season, Team A is in first place in Division 1, followed by Team B and Team C. Team D is in first place in Division 2, followed by Team E and Team F. League organizers have structured the playoffs as shown below.

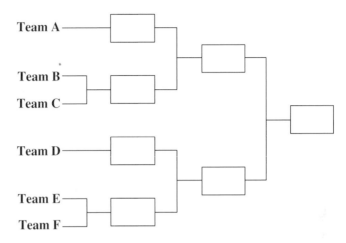

Team A
Team B
Team C
Team D
Team E
Team F

a. Model this series six times, keeping a tally of the champion teams. Flip a coin or roll a die to determine the winner of each game.

b. Which team won the most championships?

c. Determine the probability that each team will win the championship. Assume both teams in each game are equally likely to win. (Hint: Imagine that a series is played eight times, and determine how many times you can expect each team to win the championship.)

5. Design a championship structure for six teams in which all teams play the first round. Assuming each team is equally likely to win a single game, find the probability that each team will win the championship.

Connect & Extend

6. Automobile license plates in one state consist of three different letters followed by three different digits. The state does not use vowels or the letter *Y*, which prevents slang words from accidentally appearing on license plates.

a. Each letter can appear only once on a given license plate. How many different sets of three letters are possible?

b. Each digit can appear only once on a given license plate. How many different sets of three digits are possible?

c. Altogether, how many different license plates with three letters followed by three digits are possible for this state?

7. Suppose the *2-of-6* lottery game was modified so that after the first number was selected, that number was placed back into the group before the next number was selected. In this way, a number could be repeated, meaning pairs such as 2-2 and 3-3 would be possible.

a. Would your chances of winning be better or worse for this modified game? Explain.

b. How many possible pairs are there for this modified game, assuming that order *does* matter? Explain.

c. List all of the possible pairs from part b.

d. Since order really doesn't matter in this game, how many *different* pairs are there? (Remember, if order doesn't matter, 1-2 is the same as 2-1.)

e. Are all of the pairs considered in part d equally likely? Explain.

f. If you choose one number pair for this modified game, what is the probability you will win. (Hint: There are two cases to consider.)

8. Suppose the *3-of-7* lottery game was modified so that after each number was selected, that number was placed back into the group before the next number was selected. In this way, a number could be repeated, meaning triples such as 1-2-2 and 3-3-3 would be possible.

a. How many possible pairs are there for this modified game, assuming that order *does* matter? Explain.

b. Since order really doesn't matter in this game, 1-1-2, 1-2-1, and 2-1-1 are all the same triple. So there are only 84 possible *different* triples. Are all of these different triples equally likely? Explain.

c. If you choose one number triple for this modified game, what is the probability you will win. (Hint: There are three cases to consider.)

In your **own words**

Describe how you would make a given lottery game more difficult to win.

9. Challenge In the *4-of-7* lottery game, players must match four numbers from 1 to 7 with those selected in a random drawing.

a. If you select one group of four numbers, what is the probability that you will win this game?

b. How does the probability of winning the *4-of-7* lottery game compare to the probability of winning the *3-of-7* lottery game (see your work in Problem Set B)?

c. Calculate and compare the probabilities of winning the *5-of-7* lottery game (match five numbers from 1 to 7) and the *2-of-7* lottery game (match two numbers from 1 to 7).

d. Look back over the Example on page 586. Use similar calculations to explain what you discovered in Parts b and c.

e. What game has the same probability of winning as the *4-of-6* game? Explain.

10. Mr. Wegman takes a drive each Sunday afternoon. One Sunday, he decides to let probability determine his destination. The diagram shows the network of roads he is driving. He starts at Point A and drives toward Point B. Each time he reaches a fork in the road, he flips a coin to decide which path to take. If he flips heads, he goes to the left. If he flips tails, he goes to the right.

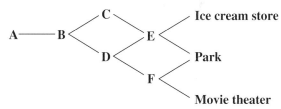

a. What is the probability Mr. Wegman will end at the movie theater? (Hint: Imagine he takes this drive eight times. How many times would you expect him to end at the movie theater?)

b. What is the probability he will end at the park?

c. What is the probability he will end at the ice cream store?

11. Sports In a particular sports league, five teams qualify for the final playoff.

a. Design two ways of structuring the playoff series.

b. Describe a situation in which one playoff structure would be more appropriate than another.

12. **Challenge** In Investigation 2, you always assumed that two teams have the same chances of winning a single game. For this exercise, assume that Team A has a 60% chance of defeating Team B in every game they play against each other.

a. Suppose there is a one-game tournament between the teams and the winner of the game wins the tournament. What is the probability that Team A will win? That Team B will win?

Now suppose you have a "best two out of three" tournament. That means the teams play until one of them wins two games.

b. Use a tree diagram to show all the possibilities for the tournament. For example, in the first game, there are two branches:

A wins or B wins. (Hint: If A wins the first two games, is a third game played?)

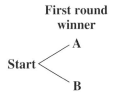

c. Suppose the teams played 1,000 tournaments. In how many tournaments would you expect Team A to win the first game? In how many of those tournaments would you expect Team A to also win the second game?

d. For each combination in your tree diagram, use similar reasoning to find the number of tournaments out of 1,000 you would expect to go that way. For example, one combination should be ABB; in how many tournaments out of 1,000 would you expect the winner to be A, then B, and then B? (Hint: Check your answers by adding them; they should total to 1,000.)

e. Find the total number of tournaments out of 1,000 in which each team wins the tournament. What is the probability that Team A wins a tournament?

f. Which tournament, *one-game* or *best-two-out-of-three,* is better for Team B?

13. Consider the line $y = {}^-5x - 7$.

a. A second line is parallel to this line. What do you know about the equation of the second line?

b. Write an equation for the line parallel to $y = {}^-5x - 7$ that passes through the origin.

c. Write an equation for the line parallel to $y = {}^-5x - 7$ that crosses the y-axis at the point $(0, {}^-2)$.

d. Write an equation for the line parallel to $y = {}^-5x - 7$ that passes through the point $(3, 0)$.

14. Life Science The data in the table represent how a certain population of bacteria grows over time. Write an equation for the relationship, assuming the growth is exponential.

Hours from Start, h	Population, p
0	1
1	5
2	25
3	125

15. Copy this picture. Create a design with rotation symmetry by rotating the figure several times about the point using a 30° angle of rotation.

16. Solve the inequality $0 > 5(7 - x) + 12x$. Draw a graph of the solution.

Factor each expression.

17. $4h^2 - 2h$

18. ${}^-6a^2 + ab + b^2$

19. ${}^-4k^2 - 5kj - j^2$

20. $2m^2 - 9 + 3m$

Use the quadratic formula to solve each equation.

21. $3h^2 - 2h + {}^-6 = 0$

22. ${}^-6a^2 + 3a = {}^-4$

23. $4k^2 = 5k + 2$

24. $2m^2 - 12 = 6m$

Find the x-intercepts for the graph of each equation.

25. $(x - 5)^2 - 49 = 0$

26. $6x^2 + 36 - 30x = 0$

Chapter Summary

VOCABULARY
sample space

In this chapter, you found the sizes of *sample spaces* for several situations. Sometimes this required counting all the possible ways a group of items could be selected from a larger group. Other times you had to figure out the number of ways a group of things could be ordered.

You can find the sample space for a situation by systematically listing all the possibilities. You learned that you can sometimes discover a pattern to help you determine the size of the sample space without listing all the outcomes.

You also found the probabilities of events for various situations, and you determined whether one event was more likely than another. You saw that sometimes finding probabilities can help you make decisions or devise game-winning strategies.

Strategies and Applications

The questions in this section will help you review and apply the important ideas and strategies developed in this chapter.

Making a systematic list of every possible outcome

1. Ally, Brevin, Carol, Doug, and Evan are setting up a chess tournament among themselves that will be a round-robin tournament—that is, every participant will play every other participant once. How many games will there be? Make a systematic list of every possible tournament pairing.

2. In science class, Ally, Brevin, Carol, and Doug are assigned to sit next to one another in the first row.

   a. List all the arrangements in which Ally sits in the first seat, then all the arrangements in which Brevin sits in the first seat, and so on.

   b. How many different arrangements are there?

A First	B First	C First	D First
ABCD			
ABDC			

impactmath.com/chapter_test

Using a pattern or shortcut to find the size of a sample space without listing every outcome

3. The manager of a baseball team is responsible for assigning the nine players to a batting order.

a. Explain how you can find the number of different batting orders *without* listing all the possibilities. How many batting orders are possible?

b. How many different batting orders are possible if one of the nine players, the pitcher, always bats ninth?

4. Suppose a basketball coach has 12 players. *Without* listing all the possibilities, explain how you can find the number of different 5-player teams the coach could create. How many teams are possible?

Determining the probability of an event

5. In Question 1, you determined all the possible pairings for a round-robin chess tournament. Imagine that each pairing is written on a slip of paper. The slips are placed into a container and mixed. One slip is then chosen at random.

a. What is the probability that the chosen names are Doug and Evan?

b. What is the probability that Doug or Evan (or both) is included in the chosen pairing?

c. What is the probability that the chosen names contain neither Doug or Evan?

6. In Question 2, you listed all the ways four students could be arranged in a row. Use your list to answer these questions.

a. If seats are assigned randomly to the four students, what is the probability that Brevin and Carol will sit next to each other?

b. If seats are assigned randomly to the four students, what is the probability that Doug will sit at one end of the row?

Using probability to determine whether a game is fair

7. Héctor is trying to invent a two-player game that involves rolling two dice and adding the results. He is considering several scoring rules.

   a. Suppose Player 1 scores 1 point for a sum of 2, 3, 4, 9, 10, 11, or 12, and Player 2 scores 1 point for any other sum. Is this game fair? Explain.

   b. Suppose Player 1 scores 1 point when the sum is 6, 7, or 8. Otherwise, Player 2 scores 1 point. Is this game fair? Explain.

   c. Suppose Player 1 scores 3 points when the sum is 11 or 12, and Player 2 scores 1 point when the sum is 3, 4, or 5. For other sums, neither player scores. Is this a fair game? Explain.

Using probability to make decisions

8. *Higher or Lower?* is a two-player game in which opponents take turns guessing whether the sum of the numbers on a pair of rolled dice will be higher than 7 or lower than 7. The play of one turn proceeds as follows:

   • Each player rolls one die. The guessing player for the roll makes his or her roll in full view; the nonguessing player conceals his or her roll.

   • The guessing player then guesses "higher than 7" or "lower than 7." The guess is made with the knowledge of one die but not the other.

   • The guessing player scores 1 point if he or she is correct. The nonguessing player scores 1 point if the guess is incorrect. If the sum is 7, no points are awarded.

   Suppose you are the player guessing higher or lower.

   a. The first die shows a 2. Should you guess the sum will be "higher than 7" or "lower than 7"? What is the probability you will be correct?

   b. The first die shows a 4. Should you guess the sum will be "higher than 7" or "lower than 7"? What is the probability you will be correct?

Demonstrating Skills

9. Suppose you have seven chairs in a row. How many different seating orders are possible for seven people?

10. At Barak's Burgers you can order a cheeseburger with up to seven different condiments: mustard, catsup, pickles, onions, tomatoes, lettuce, and mayonnaise. How many different cheeseburgers are possible, including a plain cheeseburger with no condiments?

11. The *21-and-5* lottery game requires players to match 5 numbers drawn randomly from 21. How many different groups of 5 numbers are possible?

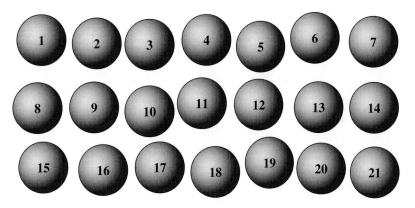

CHAPTER 10

Modeling with Data

Real-Life Math

It Makes Good Census The United States Constitution mandates that a census of the U.S. population be conducted every 10 years. This complicated process requires that every U.S. citizen be counted.

Detailed information about each citizen is collected and analyzed as part of the census. For example, age, race, religion, and education data are collected and analyzed. These analyses enable us to make generalizations and draw conclusions about the U.S. population.

Think About It What would you like to know about the U.S. population? Write a few questions, and then use an almanac, the Internet, or another resource to find the answers.

Family Letter

Dear Student and Family Members,

Throughout the past year, we have seen mathematics at work in several situations. In this chapter, we will look more closely at how to work with data, including data from many real-world contexts. We will use tables to organize data, look for trends, and draw conclusions. We will use visual displays, including a map describing a historic cholera outbreak in London. Using the map, we will gather evidence to determine the source of the epidemic.

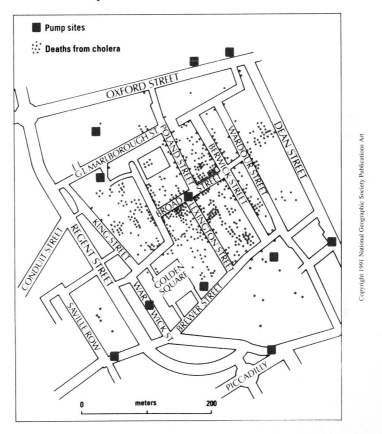

In the second half of the chapter, we will use *mathematical models*. For example, we will use equations to model population growth and make predictions. We will also analyze a business using tables, and use a model population to explore how left-handedness might change from one generation to another.

What can you do at home?

You might ask your student about the London cholera outbreak or the other situations he or she is learning to analyze. You might also explore information in the newspaper together and look for patterns: Are there patterns in the temperatures displayed on the weather map?

Data Patterns in Tables and Graphs

The times in which you are living have been called the *information age* because the amount of information available is increasing faster than ever before. Through the Internet, you can find an enormous amount of data, on almost any topic you can think of, in a matter of minutes.

Just having data won't help you, however, if you can't interpret the data. By organizing and analyzing data, you can sometimes discover trends and connections that will help you understand the information better.

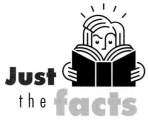

Just **the facts**

If you have access to the Internet, you may be interested in *fedstats.gov*, where you can find some of the data collected by the U.S. government.

Think & Discuss

What kinds of data do you think your school might have about you? How might the school use these data?

In math class, you have often collected and drawn conclusions from data. Consider some investigations in which you collected numerical data earlier this year and then formulated a conjecture. How did you use mathematics to make sense of the collected data?

Investigation Analyzing Data Using Tables

Schools often collect data about test scores of groups of students and use them to evaluate the performance of the students and the schools.

One high school draws its students from two small towns, Northtown and Southtown, each with one middle school. At the beginning of the school year, students in algebra classes take a test to determine how well prepared they are. The results of the algebra pretest for two classes are given on the opposite page.

The designation "S" in the "Town" column indicates that the student was from Southtown, and "N" indicates the student was from Northtown. The letter "M" in the "Class" column represents the morning algebra class, and "A" represents the afternoon class.

Algebra Pretest Results

Student	Score	Town	Class	Student	Score	Town	Class
1	100	S	A	23	54	N	A
2	81	N	M	24	72	N	A
3	55	S	M	25	100	S	M
4	74	N	A	26	66	S	A
5	58	N	A	27	90	S	M
6	59	S	A	28	84	N	A
7	94	N	M	29	68	N	M
8	72	N	A	30	73	N	M
9	100	S	M	31	44	S	M
10	100	S	M	32	82	N	A
11	77	N	A	33	60	S	A
12	94	S	A	34	79	S	M
13	66	N	M	35	94	S	A
14	85	N	M	36	89	N	A
15	63	S	A	37	69	N	A
16	74	S	M	38	69	N	M
17	90	N	M	39	62	S	M
18	66	N	A	40	87	N	M
19	59	S	M	41	76	N	A
20	92	S	A	42	70	S	M
21	73	S	M	43	100	S	A
22	81	N	M	44	88	S	A

The vice principal used the scores to compare the mathematics preparation of students in the two towns. He first calculated the mean score for all 44 students. He then created a new table that would help him compare the performance of students in the two towns.

Problem Set A

1. Design a table that could help the vice principal make a town-to-town comparison of the scores.

2. Find the mean score of the entire group of 44 students and the mean score of the students from each town. Compare these statistics for the two towns.

3. Find the median score of the entire group of 44 students and the median score of the students from each town. Compare these statistics for the two towns.

4. Assuming the test gives an accurate indication of each student's preparation, consider what these scores might mean about the preparation of students from the two towns.

 a. Which town has a wider range of student scores?

 b. Do the towns have the same number of students above the overall median? If not, which has more students above the median?

 c. Five students scored 100 on the test. Which town is each of these top scorers from?

 d. Which town has the lowest test score? How many points' difference is there between the lowest scores for the two towns?

 e. Based on these data, which town's middle school do you think prepares students better? Explain your reasoning.

Problem Set B

The same data are often used for quite different purposes. The vice principal was comparing the performance of students from two *towns*. The teacher of the two classes wondered whether either *class* was better prepared.

 1. Construct a table to make the teacher's analysis easier. That is, find a way to display the data so you can more easily differentiate the scores for the two classes, morning and afternoon.

 2. Do you think the data support a conclusion that one class is better prepared than the other? Justify your answer. You may want to consider mean, median, mode, range, or other factors.

Share & Summarize

In this investigation, you used a table of data to create two new tables to help you analyze the information in the original table.

 1. For each new table, how did you decide how to reorganize the original table?

 2. Consider the table on page 603. Give an advantage and a disadvantage to presenting the data in the single table, rather than the two individual tables you created in Problem Sets A and B.

 3. What things did you look at when trying to form conclusions about the differences between students from the two towns or the two classes?

Investigation 2 ▶ Organizing Data

Lydia wants her parents to buy a new car. When her parents argued that new cars are expensive, Lydia decided to try to convince them that they should buy a new car now (rather than wait a year or two until she was ready to drive and would undoubtedly be asking for the keys!).

Although she would have preferred a sports car, Lydia knew her parents would be more likely to listen if she talked about a midsize model like the one they currently owned. She collected the data below about the base price of their model of car each year.

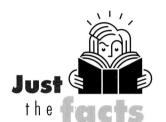

Just the facts

The *base price* for a particular model car is the price without any extra options, such as a CD player and tinted windows.

Price of New Car

Year	Price	Year	Price
1985	$8,449	1995	$15,775
1987	10,598	1997	17,150
1989	11,808	1999	18,922
1991	12,698	2001	19,940
1993	14,198	2003	21,319

Think & Discuss

Describe the trend in the data.

What might cause the price to change the way it does?

MATERIALS

graphing calculator

Problem Set C

Plot the data from Lydia's table on your calculator. It will help if you treat 1981 as Year 1.

1. Use your graph to decide whether the relationship between year and price appears to be linear, quadratic, exponential, or some other type of function. If it appears to be linear, find an equation for a line that seems to be a good fit for the data.

2. Assume the price continues to increase in the same way. Use your equation to find out how much Lydia's parents would save by buying this year's model rather than waiting two years.

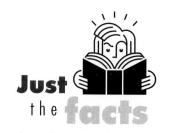

Just the facts

An odometer measures the distance a vehicle has traveled.

Lydia's parents weren't convinced, so Lydia decided to find more information. Her father kept a monthly record of odometer readings and the amount of gas bought for the family car. Lydia studied the records, hoping the data might suggest that the car was using so much gas that it would save money to replace it with a model with better fuel economy.

She prepared this record for 12 months, starting with the gasoline bought in June 2003. The second column shows the odometer reading (rounded to the nearest mile) on the last day of the month. The third column shows the gallons of gasoline bought during that month.

Month	Odometer Reading (mi)	Gasoline Bought (gal)
May	119,982	
Jun	121,142	42.8
Jul	122,564	36.6
Aug	126,354	139.7
Sep	127,459	42.0
Oct	128,106	26.5
Nov	128,919	34.7
Dec	129,939	41.5
Jan	131,052	44.6
Feb	131,695	27.2
Mar	132,430	29.6
Apr	134,114	60.0
May	135,135	35.3

Problem Set **D**

1. How many miles did the car travel during the year? What was the average number of miles driven per month?

2. Fuel economy is often measured in miles per gallon. Calculate this measure for July and November to the nearest tenth.

3. Construct a table that shows the miles driven and the fuel economy (in miles per gallon) for each month.

4. Which four months show the worst fuel economy? List them in order, starting with the worst.

FUEL GAUGE

Problem Set E

It is natural to wonder what caused the fuel economy to be better in some months than others. Lydia suspects that it's related to temperature; she thinks the car gets fewer miles per gallon in cold weather.

To test her theory, she researched the average temperatures in her city for the months during which she had data. From the Internet, she obtained the average of the daily mean temperatures, in °F, for each month.

Average Daily Mean Temperatures

Jun	Jul	Aug	Sep	Oct	Nov	Dec	Jan	Feb	Mar	Apr	May
64.6	74.3	72.5	66.3	54.4	44.6	39.1	29.5	33.6	39.4	49.2	58.2

1. Are any of the monthly temperature records surprising? Why?

2. Use your calculator to plot the (*temperature, fuel economy*) data. (The fuel economy data, in mpg, can be found in your answer to Problem 3 of Problem Set D.) Adjust the window settings to fill the screen with the data, as much as possible. Does there seem to be a connection between these two variables? Explain.

3. There should be one point on your graph that looks far from the others. Make a new graph without including that point. Adjust the window settings to fill the screen with the remaining data points. Now do you think there is a connection between temperature and fuel economy? Explain.

Problem Set F

Lydia's father said that he thought the car got fewer miles per gallon in months when they didn't drive much. Lydia decided to use her data to test his theory.

1. Construct a table that could be used to more easily see whether the fuel economy is worse in low-mile months and better in high-mile months. Can you see any evidence in your table to support Lydia's father's theory?

2. Plot the (*miles driven, fuel economy*) data on your calculator. Does there seem to be a connection between these two variables? Explain.

Share & Summarize

1. In Problem Set C, you plotted the data from a given table. In Problem Sets E and F, however, you had to do something to the given data before plotting. In each case, what data *did* you graph, and why couldn't you graph the given data?

2. Compare the temperature graph (from Problem Set E) that you feel is most helpful to the graph of miles driven (from Problem Set F). Which graph suggests a stronger connection between the variables?

3. Lydia's brother Kyle observed that the family usually took their longer trips in the summer. He reasoned that fuel economy would be better on longer trips, since he had read that highway driving gives better fuel economy than city driving. Do the data fit this observation? Explain.

Investigation ▶ 3 Looking for Trends

The Social Security Administration has collected information on the names given to newborn boys and girls in the United States.

The two tables show the three most common names for newborn boys and girls in the United States for each year from 1900 through 2002. Each table lists the years in which each name was in first, second, and third place. To conserve space, the tables don't show the "19" part of the years, so the entry "87, 88, 90–93" represents the years 1987, 1988, 1990, 1991, 1992, and 1993.*

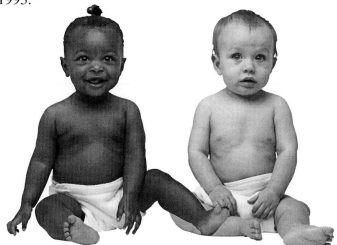

*Source: These data are derived from information at the Social Security Administration Web site, *www.ssa.gov.*

Boys' Names

Name	First Place	Second Place	Third Place
Christopher		72, 73, 75, 81–93	74, 76–80, 94
David	60, 61, 63	58, 59, 66–70	56, 62, 64, 71–72
Jacob	99, 2000–2002	95, 97, 98	96
James	35, 40, 42, 43, 45–47, 49, 52	27, 30, 31, 33, 34, 36–39, 44, 48, 50, 51, 55, 57, 65	00–17, 19–22, 24, 25, 28, 29, 32, 41, 53, 59, 67, 68, 73
Jason		74, 76–80	75, 81
John	00–25, 50	26, 28, 29, 32, 41, 47, 62–64, 71	27, 30, 31, 33–40, 42–46, 48, 49, 51, 52, 54, 60, 61, 65, 66, 69, 70
Joshua			89, 90, 92, 93, 2000
Matthew		94, 96	82–88, 91, 95, 97–99, 2000–2001
Michael	53, 55–59, 62, 64–98	54, 60, 61, 99, 2000–2002	63
Robert	26–34, 36–39, 41, 44, 48, 51, 54	21–25, 35, 40, 42, 43, 45, 46, 49, 52, 53, 56	18, 47, 50, 55, 57, 58
William		00–20	23, 26

Girls' Names

Name	First Place	Second Place	Third Place
Alexis			99
Amanda			79–82, 86–88, 91
Amy		74–77	73
Anna			00–02
Ashley	88, 91	85–87, 89, 90, 92–95	83, 84, 96
Barbara		37–44	32, 33, 35, 36, 45
Betty		27–34	25
Brittany			89, 90
Deborah		54, 55	53
Debra			56
Dorothy		20–26	12–15, 17–19, 27–31
Emily	95–99, 2000–2002		94
Hannah		98, 99, 2000	2001–2002
Helen		00–02, 04–19	03, 20–24, 26
Jennifer	70–84	69	85
Jessica	85–87, 89, 90, 92–94	78, 80–84, 88, 91, 96	95
Karen			65
Kimberly			67, 68, 70
Linda	47–52	45, 46, 53, 58	54, 55, 57, 59
Lisa	62–69	61, 70	71, 72
Madison		2001–2002	2000
Margaret		03	04–11, 16
Maria		65–67	60, 62–64
Mary	00–46, 53–61	47–52, 62–64	66
Melissa		79	76–78
Michelle		68, 71–73	69, 74, 75
Patricia			37–44, 46–52
Samantha			98
Sarah		97	92, 93
Shirley		35, 36	34
Susan		56, 57, 59, 60	58, 61
Taylor			97

Problem Set G

Michael and Mary were discussing the information in the tables on page 609. Looking at the first table, Michael said, "Hey, I've got the most popular name of the century!"

"Not true," said Mary, who was looking at the second table. "My name is the most popular."

1. Consider how you might decide which name is the most popular for the years listed in the table.

   a. Using the total number of years in first place as your measure, determine the two most popular boys' names and the two most popular girls' names.

   b. Considering the total number of years each name was in first place, who would you say was right, Michael or Mary? Defend your answer.

   c. Lee suggested a different measure: assign 3 points for each first-place appearance, 2 for each second, and 1 for each third. Scan the list to pick out four likely top boys' names and four likely top girls' names, using Lee's measure, and calculate the scores for those eight names. Which two boys' names and two girls' names are rated most popular this way?

2. Which do you think changes more frequently: the popularity of particular boys' names or the popularity of particular girls' names? Write a sentence or two defending your answer.

3. Which names have stayed in the top three for the most *consecutive* years? Explain.

4. Which names made sudden appearances into the top three for only 1 to 3 years and then disappeared? Do you have any explanation for the popularity of these particular names?

Data can be organized in many ways. For example, on the opposite page are two new tables for the girls' names.

   • The first table combines the names into 10-year intervals. The number in parentheses after a name tells how many years in that interval that name was in the given place.

   • The second table gives the number of years each name was in first, second, and third place. It also lists the maximum number of consecutive years each name was in each place.

Interval	First Place		Second Place		Third Place	
1900–1909	Mary (10)		Helen (9)	Margaret (1)	Anna (3) Margaret (6)	Helen (1)
1910–1919	Mary (10)		Helen (10)		Margaret (3)	Dorothy (7)
1920–1929	Mary (10)		Dorothy (7)	Betty (3)	Helen (6) Dorothy (3)	Betty (1)
1930–1939	Mary (10)		Betty (5) Barbara (3)	Shirley (2)	Barbara (4) Patricia (3)	Dorothy (2) Shirley (1)
1940–1949	Mary (7)	Linda (3)	Barbara (5) Mary (3)	Linda (2)	Patricia (9)	Barbara (1)
1950–1959	Linda (3)	Mary (7)	Mary (3) Deborah (2)	Linda (2) Susan (3)	Patricia (3) Linda (4) Susan (1)	Deborah (1) Debra (1)
1960–1969	Mary (2)	Lisa (8)	Susan (1) Mary (3) Michelle (1)	Lisa (1) Maria (3) Jennifer (1)	Maria (4) Karen (1) Kimberly (2)	Susan (1) Mary (1) Michelle (1)
1970–1979	Jennifer (10)		Lisa (1) Amy (4) Melissa (1)	Michelle (3) Jessica (1)	Kimberly (1) Amy (1) Melissa (3)	Lisa (2) Michelle (2) Amanda (1)
1980–1989	Jennifer (5) Ashley (1)	Jessica (4)	Jessica (6)	Ashley (4)	Amanda (6) Jennifer (1)	Ashley (2) Brittany (1)
1990–1999	Jessica (4) Emily (5)	Ashley (1)	Ashley (5) Sarah (1)	Jessica (2) Hannah (1)	Brittany (1) Sarah (2) Jessica (1) Taylor (1)	Amanda (1) Emily (1) Ashley (1) Samantha (1) Alexis (1)
2000–2002	Emily (3)		Hannah (1)	Madison (2)	Madison (1)	Hannah (2)

Name	Total Years			Maximum Consecutive Years		
	1st Place	2nd Place	3rd Place	1st Place	2nd Place	3rd Place
Alexis	0	0	1	0	0	1
Amanda	0	0	8	0	0	4
Amy	0	4	1	0	4	1
Anna	0	0	3	0	0	3
Ashley	2	9	3	1	4	2
Barbara	0	8	5	0	8	2
Betty	0	8	1	0	8	1
Brittany	0	0	2	0	0	2
Deborah	0	2	1	0	2	1
Debra	0	0	1	0	0	1
Dorothy	0	7	12	0	7	5
Emily	8	0	1	8	0	1
Hannah	0	3	2	0	3	2
Helen	0	19	7	0	16	5
Jennifer	15	1	1	15	1	1
Jessica	8	9	1	3	5	1
Karen	0	0	1	0	0	1
Kimberly	0	0	3	0	0	2
Linda	6	4	4	6	2	2
Lisa	8	2	2	8	1	2
Madison	0	2	1	0	2	1
Margaret	0	1	9	0	1	8
Maria	0	3	4	0	3	3
Mary	56	9	1	47	6	1
Melissa	0	1	3	0	1	3
Michelle	0	4	3	0	3	2
Patricia	0	0	15	0	0	8
Samantha	0	0	1	0	0	1
Sarah	0	1	2	0	1	2
Shirley	0	2	1	0	2	1
Susan	0	4	2	0	2	1
Taylor	0	0	1	0	0	1

5. Using the first table on page 611, find the answers to Problem 1 (Parts a and c), Problem 3, and Problem 4 again, for just the girls' names. Then find the answers one more time, using the second table. Describe how each arrangement of the data makes it easier or harder to answer the questions.

6. Describe any major trends you see over the course of the century. For example, can you see any significant differences in popular names toward the end of the century as opposed to the beginning or the middle of the century?

7. Suppose you are designing wall hangings with babies' names on them that you plan to sell through a toy company that distributes nationwide. You want to sell as many as possible, but the toy company has agreed to carry only six boys' names and six girls' names at first, to see how well customers like the product. Predict the most popular names for the current year, and support your prediction with an explanation.

8. Predict the most-used names for 1899 (the year before the data shown in the table). Support your prediction with an explanation.

Share & Summarize

1. Compare the advantages and disadvantages of the three tables in this investigation.

2. Write a question about the most popular names that you would like to answer but don't have enough data to do so.

Investigation ▶4▶ From Maps to Graphs

As you've seen, a large table filled with data can be hard to analyze. A visual display, such as a graph, of the information in a table sometimes makes essential patterns apparent immediately. A map is sometimes the best visual display of data.

Think & Discuss

What sort of geographic data is usually presented on a map?

Have you seen maps that present other types of information? What sort of information?

In 1854, the Broad Street area of London suffered a severe outbreak of *cholera,* a serious bacterial infection of the small intestine. A London physician, John Snow, was determined to stop the spread of this disease.

Dr. Snow recorded the locations of the victims' homes on a map of this section of the city. At the time, London did not have pipes bringing the public water supply directly into homes and business. The map shows the locations of the public pumps that residents used as their water source.

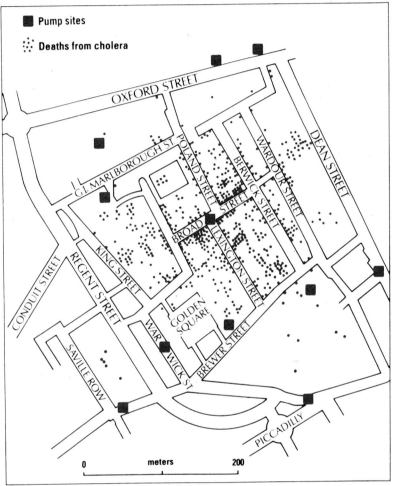

■ Pump sites

Deaths from cholera

MATERIALS

- Snow's map of London
- ruler
- compass
- graph paper

Problem Set H

1. What pattern, if any, do you observe in the deaths?

As Dr. Snow studied the pattern of deaths on the map, he noticed that they seemed to be concentrated in certain locations. Even though many people lived around Oxford Street, Regent Street, and Piccadilly, there were few cholera deaths in those areas. The visual display of deaths on the map convinced him that drinking water was the likely cause of the epidemic—and that the Broad Street pump, where the deaths clustered most densely, might be the source.

Dr. Snow was interested in how far each cholera death was from the Broad Street pump. A length representing 200 meters is given on the map. It will be convenient to have a much smaller distance—such as 25 meters—to help you analyze what was happening.

2. Divide the 200-meter length on your map in half to make 100-meter intervals. Divide each part in half again to make 50-meter intervals. Finally, divide each 50-meter segment in half to make 25-meter intervals.

3. Draw several circles on your map, using the Broad Street pump as the center. Give the circles radii of 25, 50, 75, . . ., and 200 meters. This will create eight rings of 25-meter width about the pump, something like this:

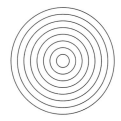

4. Consider the innermost circle to be Ring 1, and the area between the innermost circle and the next circle to be Ring 2. Count the number of victims who lived within Ring 3. Describe the distance of the residences of this ring from the Broad Street pump.

5. Count the number of victims who lived in each of the other rings, and list them in a table. You will be adding two more columns to your table later, so leave space.

6. Construct a histogram with the distance from the Broad Street pump on the horizontal axis and the number of victims on the vertical axis. Give each bar a width of 25 meters.

7. Do your table and histogram support the conclusion that there are more deaths closer to the pump? Explain.

8. Kai used his own table to calculate the median, mode, and mean for the distance of the victims' residences from the pump. To make things simpler, he rounded distances according to the number of the ring in which the victim lived. For example, for all people in Ring 1, he used the distance 12.5 m; for Ring 2, he used 37.5 m; and for Ring 3, he used 62.5 m.

Kai found a median of about 87.5 m, a mode of about 62.5 m, and a mean of about 78.3 m. Consider what these measurements might mean in terms of the cholera epidemic. What useful information, if anything, do these statistics reveal about the situation?

Just the facts

The circles you draw in Problem 3 are *concentric* because they have the same center.

Remember

The area of a circle with radius r is πr^2.

9. As you counted dots on the map, you might have noticed that each ring covers more area than the next smaller ring. For example, Ring 2 has 3 times the area of Ring 1!

a. Ring 1 represents a circular piece of land with radius 25 m. What is the area, in square meters, covered by that ring?

To calculate the area of each ring, you have to find the area of two circles and subtract. For example, the area of the shaded ring to the right is the area of the large circle minus the area of the small circle.

b. Find the area of each ring, and add a column to your table from Problem 5 to list these new data.

c. Now, by dividing the number of victims in a ring by the area of that ring, find the number of victims per square meter for each ring. Add another column to your table to record this information, called the *population density* of the victims, to the nearest ten-thousandth.

d. Does the pattern in the population density support the conclusion that there are more deaths closer to the pump? Explain.

10. Make a plot of the (*ring number, population density*) data.

Dr. Snow's presentation, complete with maps and tables, convinced the city administration that water from the Broad Street pump was to blame for the cholera epidemic. When the handle was removed so people could no longer get water from that pump, the epidemic subsided.

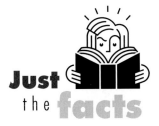

Just the facts

Dr. Snow had to work hard to convince people of his conclusions. In the mid–nineteenth century, not much was known about the spread of disease. His work with cholera is a classic example in *epidemiology*, the study of diseases that spread rapidly through a population.

Share & Summarize

1. Consider the map; the table; the mean, median, and mode; the histogram; and the plot showing population density versus ring number. Explain why each is or is not useful in supporting Dr. Snow's conclusion about the relationship between the number of deaths and the distances from the victims' residences to the Broad Street pump.

2. Which display or statistic provides the clearest support for such a connection?

On Your Own Exercises

In Exercises 1–3, use this information:

Economics Ben decided to write an article for the school newspaper comparing compact disk (CD) prices in several stores. He found the prices of five current releases at three local stores and at an Internet music site and listed them in a table.

CD Prices

Store	Artist	Price
Castle	A K Mango	$12.19
Castle	Screaming Screamers	12.50
Castle	Front Street Girls	13.09
InstantMusic	A K Mango	13.25
InstantMusic	Front Street Girls	13.25
InstantMusic	Screaming Screamers	13.49
InstantMusic	Out of Sync	13.59
GLU Sounds	Front Street Girls	13.95
Castle	Out of Sync	14.29
GLU Sounds	A K Mango	14.50
Pineapples	A K Mango	14.99
InstantMusic	Aviva	15.00
GLU Sounds	Out of Sync	15.49
GLU Sounds	Screaming Screamers	15.50
Pineapples	Screaming Screamers	15.99
Pineapples	Front Street Girls	16.00
Pineapples	Out of Sync	16.25
GLU Sounds	Aviva	16.75
Pineapples	Aviva	18.89
Castle	Aviva	19.00

1. Study Ben's table, and write a brief paragraph describing the most noticeable differences in CD prices.

2. Consider how CD prices vary from store to store.

    a. Reorganize the table to make it easier to compare the stores' prices for the various CDs. (Abbreviate the store and artist names if you want to.)

    b. What is the mean price of the CDs in each store?

    c. Which store would you recommend for the best price in general? Explain your choice.

3. Consider the prices for each artist's CD.

    a. Reorganize the table to make it easier to compare the prices for a particular CD.

    b. What is the mean price of each artist's CD?

4. Ecology You can find data about air pollution and other environmental issues on the U.S. Environmental Protection Agency (EPA) web site, *www.epa.gov.*

**Garbage-Disposal Methods, 1960–2000
(in millions of tons per year)**

	1960	1970	1980	1990	2000
Recycled or Composted	5.6	8.0	14.5	33.2	67.7
Combustion (Burning)	27.0	25.1	13.7	31.9	33.7
Discarded in Landfills	55.5	87.9	123.4	140.1	130.6
Total	88.1	121.1	151.6	205.2	232.0

Source: *Characterization of Municipal Solid Waste in the U.S.: 2001 Update,* U.S. Environmental Protection Agency, Washington, D.C.

Remember

The *percentage increase* is the difference in two quantities divided by the original quantity and expressed as a percentage.

Ellis and Gabriela want to use the garbage-disposal data to help decide whether people are recycling more or less than they used to.

a. What is the percentage increase in tons of waste recycled or composted in 2000 compared to 1960?

b. Graph the amount of waste recycled or composted over the years 1960–2000.

c. Ellis argued that the answer to Part a and the graph from Part b show that recycling has improved a great deal in the last few decades. Do you agree? Why or why not?

d. Gabriela argued that they should compute a ratio for each year: the amount of waste recycled or composted to the total amount of waste generated. Make a table showing this ratio, as a decimal, for each year in the original table. Then graph the ratios over the 1960–2000 period.

e. What does your graph from Part d tell you about whether people are recycling more or less than they used to?

5. Social Studies In Investigation 3, you considered tables listing the most common names for newborns in the United States. When a name falls from first place, you might expect it to descend slowly, in which case it would appear in the second or third position for a year or two. For which years does this actually happen? Are there years when it doesn't happen (meaning the first-place name must have fallen to fourth place or lower)?

Social Studies In Exercises 6–8, use the table on the opposite page, which gives the years in which each party's presidential candidates received the most votes in each state from 1948 to 2000.

There were 14 elections during those years. In the table, the two major parties—Democrats and Republicans—have their own categories. All other parties are included in the "Other" category. Note that Alaska (AK) and Hawaii (HI) did not become states until 1959. Also, the District of Columbia (DC), although not a state, received voting status similar to that of each state beginning in 1964.

6. In a presidential election, the candidate receiving the most votes in a state wins that state. Each state counts for a certain number of points, called *electoral votes,* with some states worth more than others because of population differences. For Parts a and b, assume each state is worth the same number of electoral votes.

   a. Just from looking at the table, try to guess whether more Democrats or more Republicans were elected president in the years listed.

   b. Can you think of a way to reorganize the table that would make it easier for you to guess the answer to Part a? If so, describe it; you may want to give a sample table with just a row or two to illustrate. If not, explain why you think the given table is best.

   Now consider how you might try to make a more accurate guess using each state's actual number of electoral votes. For example, Alaska—and many other states—has only 3 electoral votes, while California has 54.

   c. How might you modify the table to reflect this information?

   d. **Challenge** To determine who actually won each election, you should add the electoral votes of the states that each candidate won. A candidate must have at least 270 electoral votes to win.

   How would you use the number of electoral votes for each state to help guess which party has had more candidates win elections, without actually calculating the winner each year?

Years in Which Each Party's Presidential Candidate Received the Most Votes

	Democrat	Republican	Other
AL	52, 56, 60, 76	64, 72, 80, 84, 88, 92, 96, 2000	48, 68
AK	64	60, 68, 72, 76, 80, 84, 88, 92, 96, 2000	
AZ	48, 96	52, 56, 60, 64, 68, 72, 76, 80, 84, 88, 92, 2000	
AR	48, 52, 56, 60, 64, 76, 92, 96	68, 72, 80, 84, 88, 2000	
CA	48, 64, 92, 96, 2000	52, 56, 60, 68, 72, 76, 80, 84, 88	
CO	48, 64, 92	52, 56, 60, 68, 72, 76, 80, 84, 88, 96, 2000	
CT	60, 64, 68, 92, 96, 2000	48, 52, 56, 72, 76, 80, 82, 88	
DE	60, 64, 76, 92, 96, 2000	48, 52, 56, 68, 72, 80, 84, 88	
DC	64, 68, 72, 76, 80, 84, 88, 92, 96, 2000		
FL	48, 64, 76, 96	52, 56, 60, 68, 72, 80, 84, 88, 92, 2000	
GA	48, 52, 56, 60, 76, 80, 92	64, 72, 84, 88, 96, 2000	68
HI	60, 64, 68, 76, 80, 88, 92, 96, 2,000	72, 84	
ID	48, 64	52, 56, 60, 68, 72, 76, 80, 84, 88, 92, 96, 2000	
IL	48, 60, 64, 92, 96, 2000	52, 56, 68, 72, 76, 80, 84, 88	
IN	64	48, 52, 56, 60, 68, 72, 76, 80, 84, 88, 92, 96, 2000	
IA	48, 64, 88, 92, 96, 2000	52, 56, 60, 68, 72, 76, 80, 84	
KS	64	48, 52, 56, 60, 68, 72, 76, 80, 84, 88, 92, 96, 2000	
KY	48, 52, 64, 76, 92, 96	56, 60, 68, 72, 80, 84, 88, 2000	
LA	52, 60, 76, 92, 96	56, 64, 72, 80, 84, 88, 2000	48, 68
ME	64, 68, 92, 96, 2000	48, 52, 56, 60, 72, 76, 80, 84, 88	
MD	60, 64, 68, 76, 80, 92, 96, 2000	48, 52, 56, 72, 84, 88	
MA	48, 60, 64, 68, 72, 76, 88, 92, 96, 2000	52, 56, 80, 84	
MI	60, 64, 68, 92, 96, 2000	48, 52, 56, 72, 76, 80, 84, 88	
MN	48, 60, 64, 68, 76, 80, 84, 88, 92, 96, 2000	52, 56, 72	
MS	52, 56	64, 72, 76, 80, 84, 88, 92, 96, 2000	48, 60, 68
MO	48, 56, 60, 64, 76, 92, 96	52, 68, 72, 80, 84, 88, 2000	
MT	48, 64, 92, 96	52, 56, 60, 68, 72, 76, 80, 84, 88, 2000	
NE	64	48, 52, 56, 60, 68, 72, 76, 80, 84, 88, 92, 96, 2000	
NV	48, 60, 64, 92, 96	52, 56, 68, 72, 76, 80, 84, 88, 2000	
NH	64, 92, 96	48, 52, 56, 60, 68, 72, 76, 80, 84, 88, 2000	
NJ	60, 64, 92, 96, 2000	48, 52, 56, 68, 72, 76, 80, 84, 88	
NM	48, 60, 64, 92, 96, 2000	52, 56, 68, 72, 76, 80, 84, 88	
NY	60, 65, 68, 76, 88, 92, 96, 2000	48, 52, 56, 72, 80, 84	
NC	48, 52, 56, 60, 64, 76	68, 72, 80, 84, 88, 92, 96, 2000	
ND	64	48, 52, 56, 60, 68, 72, 76, 80, 84, 88, 92, 96, 2000	
OH	48, 76, 92, 96	52, 56, 60, 65, 68, 72, 80, 84, 88, 2000	
OK	48, 64	52, 56, 60, 68, 72, 76, 80, 84, 88, 92, 96, 2000	
OR	64, 88, 92, 96, 2000	48, 52, 56, 60, 68, 72, 76, 80, 84	
PA	60, 64, 68, 76, 92, 96, 2000	48, 52, 56, 72, 80, 84, 88	
RI	48, 60, 64, 68, 76, 80, 88, 92, 96, 2000	52, 56, 72, 84	
SC	52, 56, 60, 76	64, 68, 72, 80, 84, 88, 92, 96, 2000	48
SD	64	48, 52, 56, 60, 68, 72, 76, 80, 84, 88, 92, 96, 2000	
TN	48, 64, 76, 92, 96	52, 56, 60, 68, 72, 80, 84, 88, 2000	
TX	48, 60, 64, 68, 76	52, 56, 72, 80, 84, 88, 92, 96, 2000	
UT	48, 64	52, 56, 60, 68, 72, 76, 80, 84, 88, 92, 96, 2000	
VT	64, 92, 96, 2000	48, 52, 56, 60, 68, 72, 76, 80, 84, 88	
VA	48, 64	52, 56, 60, 68, 72, 76, 80, 84, 88, 92, 96, 2000	
WA	48, 64, 68, 88, 92, 96, 2000	52, 56, 60, 72, 76, 80, 84	
WV	48, 52, 60, 64, 68, 76, 80, 88, 92, 96	56, 72, 84, 2000	
WI	48, 64, 68, 76, 88, 92, 96, 2000	52, 56, 60, 72, 80, 84	
WY	48, 64	52, 56, 60, 68, 72, 76, 80, 84, 88, 92, 96, 2000	

In Exercises 7 and 8, refer to the information on pages 618 and 619.

7. Suppose that if the Democratic candidates have lost in a state no more than three times from 1948 to 2000, that state is considered to be "largely Democrat" during this period. Similarly, a state in which the Republican candidates have lost no more than three times is considered "largely Republican" during this period.

a. Name all the states that are largely Democrat or largely Republican, by this definition.

b. For the states that are largely Republican, in which years were most of them won by the Democratic candidate? For the states that are largely Democratic, in which years were most of them won by the Republican candidate?

c. Can you think of a way to reorganize the table that would make it easier to find the answers to Parts a and b? If so, describe it; you may want to give a sample table with just a row or two to illustrate. If not, explain why you think the given table is best.

8. There were a few years in which a candidate other than the Republican or Democratic candidate won a state.

a. List the states and the years.

b. What do you notice about the states? (Hint: Locating them on a map may be helpful.)

c. Can you think of a way to reorganize the table that would make it easier to determine the answers to Parts a and b? If so, describe it; you may want to give a sample table with just a row or two to illustrate. If not, explain why you think the given table is best.

In Exercises 9 and 10, use this information:

Social Studies Aftermath Inc. offers after-school math classes for seventh and eighth graders in Massachusetts. During one year, 603 students were enrolled in the program. Use the following data, which show the number of students in Aftermath, the total population of seventh and eighth graders, and the percentage of the seventh- and eighth-grade population enrolled.

Students Enrolled in Aftermath

City or Town	Number Enrolled	Population (Grades 7–8)	Percent Enrolled
Boston	98	8,873	1.104
Braintree	40	828	4.831
Brockton	2	2,505	0.080
Brookline	52	899	5.784
Cambridge	3	1,061	0.283
Canton	35	433	8.083
Cohasset	5	191	2.618
Dedham	29	457	6.346
Duxbury	2	461	0.434
Easton	26	537	4.842
Hanover	2	390	0.513
Hingham	45	511	8.806
Mansfield	18	535	3.364
Marshfield	2	688	0.291
Medfield	2	347	0.576
Milton	77	630	12.222
Needham	5	605	0.826
Newton	7	1664	0.421
Norwell	17	286	5.944
Quincy	69	1,290	5.349
Randolph	4	646	0.619
Westwood	52	330	15.758

Source: Adapted from information at the Massachusetts Department of Education Web site, *www.doe.mass.edu.*

9. Luna suspects that the town or city sending the *most* students must be where Aftermath is located. Orlando thinks the town or city sending the greatest *percentage* of students must be where Aftermath is located.

a. Which town or city sends the most students to Aftermath?

b. Which town or city sends the greatest percentage of its seventh and eighth graders to Aftermath?

c. Who do you think is more likely to be correct, Luna or Orlando? Explain your reasoning.

10. Geography For this exercise, refer to the information on page 621.

This is a map of Eastern Massachusetts.

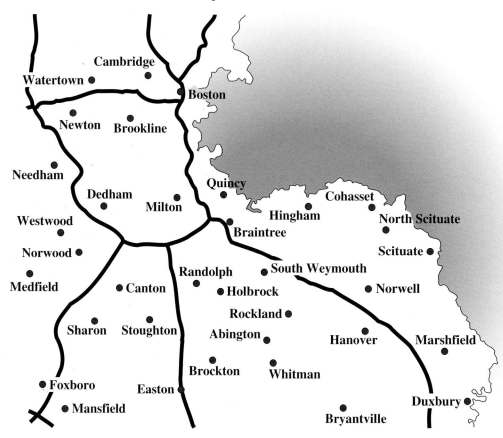

a. Of all the cities and towns that send any number of students to Aftermath, which is farthest north? Farthest south? Farthest west? Farthest east?

b. Lydia suspects that Aftermath is located somewhere in the middle of all the towns and cities listed in the table. Where would that be?

c. Consider only cities and towns sending three or more students. How does that change your answers to Parts a and b?

d. Explain why it might be reasonable to exclude towns with fewer than three students from your analysis.

e. Make your best guess about the location of the Aftermath program, and explain your reasoning.

In Exercises 11 and 12, use this information:

An important branch of mathematics is *cryptology,* the study of making and breaking codes. Its uses range from deciphering intercepted enemy messages in wartime to encrypting credit card information over the Internet.

A simple—but not very secure!—method of sending secret messages is the use of a substitution table. This method is used in cryptogram puzzles in your daily newspaper. To send a message, it is helpful to present the entries in a substitution table in alphabetical order, like this:

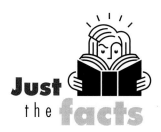

Input	a	b	c	d	e	f	g	h	i	j	k	l	m	n	o	p	q	r	s	t	u	v	w	x	y	z
Output	C	F	I	L	O	R	U	X	A	D	G	J	M	P	S	V	Y	B	E	H	K	N	Q	T	W	Z

Remember

This table represents a function since it provides a unique output for each input.

To avoid confusion, it's useful to use lower-case letters for your original (English) message and upper-case letters for your secret message, as the table shows. To write a message, look up each letter in the top row and change it to the corresponding letter in the bottom row. For example, to write "come here," you would send ISMO XOBO.

11. Using the original table as it is sorted is a nuisance when you are *deciphering* a message.

    a. Suppose you receive the message BOHKBP AMMOLACHOJW. Translate it into plain English.

    b. Make a new table in which the *outputs* are in alphabetical order.

    c. Use your new table to decipher the message WSK CBO ISBBOIH.

Just the facts

Cryptology played a part in the victory of the Allies in World War II. Mathematician Alan Turing, among others, had a critical role in breaking the Enigma code used by the Germans.

a	7.3
b	0.9
c	3.0
d	4.4
e	13.0
f	2.8
g	1.6
h	3.5
i	7.4
j	0.2
k	0.3
l	3.5
m	2.5
n	7.8
o	7.4
p	2.7
q	0.3
r	7.7
s	6.3
t	9.3
u	2.7
v	1.3
w	1.6
x	0.5
y	1.9
z	0.1

Source: Trinity College Web site, *www.trincoll.edu/depts/cpsc/cryptography/caesar.html.*

12. Challenge Refer to the information on page 623.

If you intercept someone else's encrypted message and don't know the table that was used to construct it, you have a problem to solve: breaking the code.

To solve a code, it is helpful to have some idea of which letters are most frequent in English and what combinations of letters are typical. In the short encoded message in Part a of Exercise 11, for instance, the letter "O" occurs three times; since "e" is the most common letter in English, you might guess (correctly) that "O" represents "e" in this message.

The table at left shows one estimate of the frequencies of letters in English text, as percentages.

a. Use the letter frequencies as a guide in attempting to translate the message below. It will help to start with a blank table in which the inputs are in alphabetical order. Fill in your guesses about the outputs (in pencil!) as you go along, and soon a pattern will emerge.

b. Explain how you solved Part a.

The table above comes from a Web site. You can find letter-frequency tables from many sources that differ in the percentages assigned to the letters of the alphabet.

c. Explain why sources might have apparently contradictory tables for frequencies of letters in English.

d. Explain why a particular message, like the one above, might *not* have the same letter frequencies as those in the table.

Just the facts

The construction of secret messages is cryptography. Breaking codes is cryptanalysis. Both are part of cryptology.

In Exercises 13–15, use this information:

Ecology Automobiles are an essential mode of transportation at this point in U.S. history. Unfortunately, they are also a major source of air pollution. In recent years, a series of technological improvements has reduced the amount of pollution emitted per mile driven, but people are driving more miles.

The following figure includes a line graph and a bar graph. The line graph displays the average per-vehicle emissions (estimated) from 1960 through 2015. The bar graph shows the vehicle miles traveled (in billions) for these years. Certain assumptions have been made for the estimates of future emissions, such as no increase in regulations, no cutbacks in driving, and no unexpected improvements in technology.

Comparison of the Number of Miles Driven and Emissions per Vehicle in the U.S.

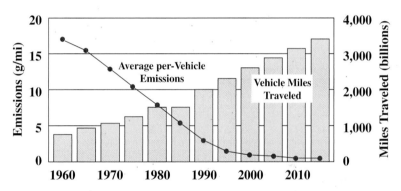

Source: "Automobiles and Ozone," Fact Sheet OMS-4 of the Office of Mobile Sources, the U.S. Environmental Protection Agency, Jan 1993.

13. At first glance, do the improvements (as indicated by the line graph) seem to be keeping up with the increase in miles driven (as indicated by the bar graph)? Explain.

14. Paul claimed that the total amount of pollutants produced by cars in 2010 will be less than that in 1960 due to technological improvements. Sara claimed that the total amount of pollutants from cars will be greater in 2010 than in 1960. Who do you think is correct? Explain your answer, using the graphs to justify your reasoning.

In your
own
words

Find a magazine article, Internet web site, or television newscast that gives information in a table or visual display of some sort. Describe the display, and explain how it is useful for the information presented.

15. Use the graphs on page 625 to create a new graph by following these directions.

a. Complete the following table, giving an approximation of the average amount of emissions from cars each year.

Year	Average per-Vehicle Emissions (grams of hydrocarbon per mile)	Vehicle Miles Traveled (billions)	Total Emissions (billions of grams of hydrocarbon)
1960	17	750	12,750
1965			
1970			
1975			
1980			
1985			
1990			
1995			
2000			
2005			
2010			
2015			

b. Draw a graph displaying the average number of grams of hydrocarbon produced per year for each year on the original graphs.

c. Describe how the amount of hydrocarbon produced by cars changed over the observed and predicted years.

d. Use the graphs to predict the average per-vehicle emissions and vehicle miles traveled in 2030.

e. Calculate the estimated grams of hydrocarbon that cars will produce in 2030, and add these data to your graph.

16. Earth Science Diana decided to do her science project on the frequency of earthquakes. She already knew that an earthquake's magnitude is sometimes measured on a scale of positive numbers called the *Richter scale.* This scale is not linear: a quake measuring 7 is very severe—10 times as severe as one measuring 6; a quake measuring 8 is devastating.

Diana conjectured that there were more earthquakes of higher magnitude than lower. She explained her hypothesis by saying that she often reads about major earthquake disasters but rarely hears about small quakes. Do you agree with Diana's hypothesis? If not, why not?

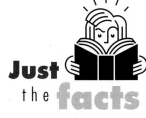

Just the facts

Earthquake data are collected and exchanged by countries throughout the world. Teams of scientists examine the data in hopes of discovering patterns that might allow them to predict future earthquakes.

In Exercises 17 and 18, use this information:

Earth Science Diana (see Exercise 16) browsed through earthquake pages on the Internet and found the U.S. Geological Survey National Earthquake Information Center. The following table was on one of their Web pages.

Table 1

Frequency of Occurrence of Earthquakes

Description	Magnitude	Average Number Annually
Great	8 and higher	1
Major	7–7.9	17
Strong	6–6.9	134
Moderate	5–5.9	1,319
Light	4–4.9	13,000 (estimated)
Minor	3–3.9	130,000 (estimated)
Very Minor	2–2.9	1,300,000 (estimated)

Source: U.S. Geological Survey National Earthquake Information Center Web site, *wwwneic.cr.usgs.gov.*

17. Look carefully at Table 1.

    a. Do the data support Diana's hypothesis (in Exercise 16)?

    b. Use the data from Table 1 to construct a histogram that shows the average annual number of earthquakes at each magnitude level. Give each bar a width of 1 unit of Richter magnitude.

    c. What significant difficulties did you have when constructing your histogram?

    d. Does your histogram emphasize any patterns that connect frequency with magnitude?

18. Diana (see Exercises 16 and 17) had a second hypothesis: that the frequency of earthquakes increased during the twentieth century. Again, she referred to the coverage of earthquakes in the news media: "Every year we hear about more of them!"

She then found a table that appeared to trace the magnitude and frequency of quakes from the years 1992 to 2003.

Table 2

Magnitude	1992	1993	1994	1995	1996	1997	1998	1999	2000	2001	2002	2003
8.0–9.9	0	1	2	3	1	0	1	0	1	1	0	1
7.0–7.9	23	15	13	22	21	16	11	18	14	15	13	13
6.0–6.9	104	141	161	185	160	129	117	128	158	126	132	128
5.0–5.9	1,541	1,449	1,542	1,327	1,223	1,118	979	1,106	1,345	1,243	1,198	954
4.0–4.9	5,196	5,034	4,544	8,140	8,794	7,938	7,303	7,042	8,045	8,084	8,603	7,121
3.0–3.9	4,643	4,263	5,000	5,002	4,869	4,467	5,945	5,521	4,784	6,151	7,004	6,524
2.0–2.9	3,068	5,390	5,369	3,838	2,388	2,397	4,091	4,201	3,758	4,162	6,420	6,652
1.0–1.9	887	1,177	779	645	295	388	805	751	1,026	944	1,137	2,101
0.1–0.9	2	9	17	19	1	4	10	5	5	1	10	107

Source: U.S. Geological Survey National Earthquake Information Center Web site, *wwwneic.cr.usgs.gov.*

a. Look carefully at Table 2. Do the data seem to support Diana's hypothesis at any or all magnitude levels?

b. In which years from 1992 to 2003 did the number of earthquakes with magnitude 7.0–7.9 reach or exceed the long-term average given in Table 1 on page 627?

c. If you compare the two tables, something seems very wrong. For which magnitudes of earthquakes are the data in Table 2 *so* different from the data in Table 1 that you just can't believe they are correct? Explain.

d. Returning to the Internet, Diana examined Table 2 more closely. She discovered that this table includes only those earthquakes whose locations could be determined by the Geological Survey National Earthquake Information Center. How does this information explain the surprising numbers in Table 2?

e. Describe at least two ways to use the data in Table 2 to better analyze the patterns in the table. (Ignore the rows for magnitudes less than 4.)

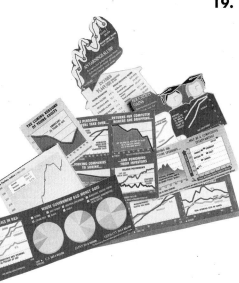

19. Look through some newspapers to find an example of data presented using maps or displays other than pie charts, bar graphs or histograms, line graphs, or tables. Write a report about your example, including the following information:

* description of the display

* description of the data presented

* advantages to using that type of display for the type of data presented

* disadvantages to using that type of display for the type of data presented

* other types of displays, if any, that might reasonably be used for the data, and disadvantages to using these displays over the one in the newspaper

Mixed Review

Factor each expression.

20. $2n^2 - 6n$

21. $4a^2 - 1$

22. $3x^2 - 9x - 30$

Expand and simplify each expression.

23. $2(g + 3)(2g - 7) + g(2g^2 + 3)$

24. $x + 2 - (x + 1)(3x + 4) + 2(8 - x)$

25. $0.5t + 3t - 1.5(t + 1)(7t + 1)$

Make a rough sketch showing the general shape and location of the graph of each equation.

26. $y = x^2 - 3x - 4$

27. $y = \frac{3}{x - 1} + 2$

28. $y = x^3 + 3$

Evaluate or simplify without using a calculator.

29. $\left(\frac{1}{7}\right)^3$

30. $2\sqrt[4]{81a^6}$, where a is nonnegative

31. 3^{-3}

32. $\sqrt[3]{\left(\frac{8}{125}\right)^2}$

33. Statistics Surveys are often used to estimate a characteristic of a large group of people based on a relatively small group of people.

For example, a recent study found that about 1,050 of the 1,500 American adults surveyed said they found news helpful when making practical decisions, but only about 795 trusted what their local television anchors told them. Network anchors were believed by about 675 people, newspaper reporters by about 465, and radio talk-show hosts by only 210 people.

There were about 200 million American adults in the year of the survey. As you answer the following questions, assume the survey's sample was representative of all American adults at that time. That is, assume the proportions in the sample are the same as the proportions for all American adults.

a. Set up and solve a proportion to estimate how many American adults found news helpful when making practical decisions. What percentage is this?

Set up and solve a proportion to estimate how many American adults trust each of the following news sources. Then find the estimated percentage of American adults who trust each news source.

b. local television anchors

c. network anchors

d. newspaper reporters

e. radio talk-show hosts

Models, Data, and Decisions

In Lesson 10.1, you drew conclusions directly from data. In this lesson, you will construct mathematical models to help you interpret data sets and make predictions from them.

Think & Discuss

Every detail about a rocket launch is controlled by computers. Suppose that for one launch, a computer calculated the height of the rocket every 2 seconds and produced the following data.

Time (s)	Height (m)	Time (s)	Height (m)
0	0	14	1,344
2	48	16	1,704
4	144	18	2,064
6	288	20	2,424
8	480	22	2,784
10	720	24	3,144
12	1,008	26	3,504

- Describe what is happening to the *rate* at which the rocket's height changes over the 26 seconds that are recorded.

- At what time does the nature of the rocket's motion change?

- Do you think the computed data are reasonable? Why or why not?

Investigation Data and Algebra

If you can write equations that describe the rocket-launch data given on page 631, you will be able to predict the rocket's height at times not given in the table.

M A T E R I A L S
- graph paper
- graphing calculator

Problem Set **A**

1. Draw a graph by plotting the rocket-launch data. Use the horizontal axis for time since launch (in seconds) and the vertical axis for height (in meters). Connect the points using line segments.

2. The point at which the nature of the data changes is called a *break point*. Consider just the data points up to the break point you identified in the Think & Discuss.

   a. Calculate the first and second differences between the height values. What type of relationship does there seem to be between time and height?

   In Chapter 1, you used a graphing calculator to find a *line of best fit* for data that seemed approximately linear. In the same way, you can use your calculator to fit curves to data that show nonlinear trends.

   b. Enter the data for the points before the break point. Then use your calculator to find an equation of the type you answered in Part a that describes the relationship between time t and height h.

3. Now consider just the data beginning with the break point and continuing after that point.

   a. What type of relationship does there appear to be between time and height for these data?

   b. Write an equation that describes the relationship between time t and height h for these data.

4. For what times should you use the equation you wrote in Problem 2 to estimate the rocket's height? For what times should you use your equation from Problem 3?

5. Predict the rocket's height at 3 seconds, 15 seconds, and 23 seconds.

Even though equations can help model a rocket's motion, many things will affect an actual flight, such as shifting winds and flaws in the rocket's construction.

Human activities can be even more difficult to describe with mathematics. For example, population growth is often exponential in nature, so we might expect to be able to find an exponential equation that approximately fits population data over time. However, events often alter the expected pattern of population growth.

MATERIALS

graphing calculator

Problem Set B

The table gives population data for the state of California from 1900 through 1990.

Enter the data into your calculator. Use *years after 1900* as the input; for example, 1910 is $t = 10$ and 1970 is $t = 70$.

California Population

Year	Population
1900	1,485,053
1910	2,377,549
1920	3,426,861
1930	5,677,251
1940	6,907,387
1950	10,586,223
1960	15,717,204
1970	19,971,069
1980	23,667,764
1990	29,760,021
2000	33,871,648

Source: *World Almanac and Book of Facts 2003.* Copyright © 2003 World Almanac Education Group, Inc.

1. Use your calculator to graph the population of California over time, connecting the points with line segments. Make a sketch of your graph. Does the graph look roughly exponential? Explain.

2. Use the calculator's curve-fitting features to find an exponential function C for which the input is the years after 1900, t, and the output is the (approximate) population of California in that year. Write the base of the exponent to the nearest ten thousandth.

3. Now enter your equation into the calculator and graph it. Plot the data on the same graph, without connecting the points. Do you think the equation does a good job of fitting the data? Explain.

4. Use the equation to predict the population of California in the year 1890 and in the year 2010.

5. The actual population of California in 1890 was 1,213,398. How does this compare to your prediction in Problem 4?

Investigation Modeling a Simple Economic Problem

Equations are constructed not only by aerospace engineers and population scientists, but also by economists and business analysts. In this investigation, you will analyze a business and create a simple model for it.

MATERIALS
- graphing calculator
- graph paper (optional)

Problem Set C

Tamika lives in a neighborhood with lots of young children and not very many teenagers, so there is a great demand for baby-sitters. She and four of her friends decided to pool their efforts and start a baby-sitting business.

The five friends began by charging $3 per hour. They quickly had lots of business—almost more than they could handle! In fact, they did a combined total of 120 hours of baby-sitting in each of the first few months.

1. How many hours per month, on average, did *each* friend work?

2. How much income per month, on average, did *each* friend earn from baby-sitting?

Adam suggested they could make more money by tripling their hourly rate. Unfortunately, most people didn't want to pay $9 per hour, so they went elsewhere for baby-sitters. Business dropped to a total of 60 hours over the next 2 months—less than what the group was used to getting in a *single* month.

3. How many hours per month, on average, did each friend work during these two months?

4. How much income per month, on average, did each friend earn from baby-sitting? How did this compare with their income when they were charging only $3?

Rebecca then suggested lowering the hourly rate to $7. "We might not get as much business as we had the first few months," she reasoned, "but we will get more than in the past two!" They received calls for 60 hours of baby-sitting during the next month alone.

5. How much income, on the average, did each friend get from baby-sitting for this month? How did this compare with their average monthly income when they were charging $9 and $3?

Finally, Hilda proposed that they graph all these results. "We could find an equation for a curve that fits the data points," she said, "and then determine how much we should charge to earn the maximum possible income."

6. Make a table with hourly rate in the first column and average monthly income per baby-sitter in the second. Fill the table using the information from Problems 1–5.

7. Graph the data in your table. What kind of curve might describe the relationship between hourly rate and average monthly income?

When you create a model to fit data, you must sometimes make assumptions to help decide what kind of relationship to use. In this case, Hilda proposed that the connection between the rate and the hours of baby-sitting they get might be linear—that is, for every dollar they raise their rate, they lose a certain number of hours.

8. Using Hilda's assumption, the number of hours each sitter works, on average, is a linear expression like $a - br$, where a and b are constants and r is the hourly rate.

   a. What does the value of b represent?

   b. What does the value of a represent? (Hint: What happens when r is 0?)

   c. Each baby-sitter's income is the rate multiplied by the number of hours worked, or $r(a - br)$. What kind of expression is $r(a - br)$? Does this support your answer to Problem 7?

9. Use your calculator to fit an equation for income i of the type you answered in Part c of Problem 8 to the data in your table from Problem 6.

10. Use your equation to determine the hourly rate that results in the greatest average monthly income. Use the Trace feature on your calculator, if needed. What is the rate, and what is the income?

Share & Summarize

Compare your work in Investigations 1 and 2.

1. How are the approaches used to find models for the data similar in the investigations?

2. How are the approaches different?

Investigation 3 Starting a Business

Evita and Tariq noticed that there were quite a few broken bicycles—and lots of discarded bicycle *parts*—lying by the side of the road in their town. They would see a frame in one place, a rear wheel in another, and a bike missing its wheels in a third. They decided to form a business recovering old bicycle parts and rebuilding good bicycles from ones that no longer worked. They couldn't resist calling their business *Re-cycle*.

Problem Set D

Before starting their business, Evita and Tariq researched and collected statistics from others who had been successful at this kind of business. They decided to construct a mathematical model of how their Re-cycle project might go for 5 years. The model required several assumptions:

• They would need some tools. Evita estimated their cost to be $600.

• Tariq predicted they could rebuild 12 bicycles the first year. He thought that after the first year, they would have a better idea about what they could find where. With greater experience, each year they could rebuild 10 more bicycles than the previous year.

• Tariq also predicted they could sell 7 bicycles the first year, and that in each subsequent year they could sell 50% more bicycles than the preceding year.

• They would need to buy some bicycle parts, since they couldn't be sure they would find working versions of everything they needed. Evita estimated they would have to spend about $1,000 on parts in each of the first two years. She also thought that, in each year after the first two, they would have to spend $200 more on parts than the preceding year.

• Evita predicted they could sell a rebuilt bicycle for an average price of $250 this year, and that for each of the next 4 years the price would increase. Tariq and Evita agreed that a yearly increase equal to 10% of the previous year's price seemed reasonable.

1. Calculate the expected costs that Evita and Tariq will incur each year. (Tools don't have to be bought after the first year.)

2. Calculate the expected number of bicycles they will sell each year. Decide what to do when the model predicts a fractional number of bicycles, and justify your decision.

3. Now find the number of bicycles the friends will build each year. How many extra bicycles will Re-cycle have on hand at the end of each year? (Remember to include the leftovers from the previous year.)

4. Calculate the price to be charged per bicycle each year.

5. Copy the table, and complete it with information from Problems 1–4 and by calculating the values for the last two columns. *Gross income* is simply the amount of money taken in. *Net profit* is the amount of money left after all costs have been paid.

Year	Costs	Bikes Sold	Price per Bike	Gross Income	Net Profit
1	$1,600	7	$250.00	$1,750.00	$150.00
2	1,000				
3					
4					
5					

6. What is Re-cycle's predicted net profit over the 5-year period?

Share & Summarize

1. In the business model for Re-cycle, many assumptions had to be made. Although you don't know the conditions in their town, which of their assumptions would you ask Evita and Tariq to reconsider, and why?

2. Evita and Tariq planned on advertising to help sell their bikes, but they neglected to consider that advertising costs money!

   a. Revise your table with an added assumption of spending $500 per year on advertising. What is the net profit for each year?

   b. What is the new predicted net profit over the 5-year period?

Investigation ▶4▶ Model Populations

Scientists are increasingly interested in why some people are left-handed, because left-handed patients may recover faster from some ailments than right-handed patients do. They may also be less likely to develop certain disorders in the first place.

Estimates vary, but one estimate is that 12.6% of males and 9.9% of females are left-handed. To make things easier in the problems that follow, assume that approximately 10% of women and 13% of men are left-handed.

Just the facts

Some famous left-handed people are Julius Caesar, Ludwig van Beethoven, H. G. Wells, Babe Ruth, Pablo Picasso, and John F. Kennedy.

Think & Discuss

• Does it follow from the estimates that in a group of 100 women, selected randomly, exactly 10 will be left-handed? Explain.

• Is it possible that all 100 of a group of 100 randomly selected women will be left-handed?

When you imagine a fictitious group of 100 women and say that 10 are left-handed, you have created a *model population* that fits the data exactly. Creating such a model is often helpful in answering complex questions about characteristics in the real population.

MATERIALS

graph paper or protractor

Problem Set E

An imaginary population of 500 men and 500 women can be used as a model of left- and right-handedness.

1. Draw a graph of the model—either a bar chart or a pie chart—that visually shows the number of right- and left-handed people in the group.

2. How many adults (male or female) in your model are left-handed? What percentage of the total population is that?

Left-handedness seems to be genetic: left-handed adults are more likely to have left-handed children. One pair of scientists concluded from a study that if both your parents are right-handed, you have a 9.5% chance of being left-handed. Your chances rise to 19.5% if one of your parents is left-handed and to 26.1% if *both* are left-handed.

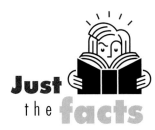
Suppose the 500 men in the model each marry one of the 500 women. Assume that handedness of a potential spouse is not a factor in who marries whom. That is, assume that in each couple, the chances that the man is left-handed are 13% and that the woman is left-handed are 10%.

3. How many of the couples are likely to consist of two right-handed people? Explain how you found your answer.

4. How many of the couples are likely to consist of two left-handed people? Explain how you found your answer.

5. How many of the couples are likely to consist of one right-handed person and one left-handed person? Explain.

6. Suppose each couple has exactly two children.

   a. Consider the children of couples in which both parents are right-handed. About how many of those children would you expect to be left-handed? Show how you found your answer.

   b. Now consider the children of couples in which both parents are left-handed. How many of those children would you expect to be left-handed? Show how you found your answer.

   c. Finally, consider the children of couples in which one parent is left-handed and the other is right-handed. How many of those children would you expect to be left-handed? Show how you found your answer.

   d. The 500 couples have a total of 1,000 children. How many of those children would you expect to be left-handed? What percentage of all the children is that?

7. Compare the percentage of left-handed adults to the percentage of left-handed children in this group. What do you think this might mean about left-handedness far into the future?

Sometimes a mathematical model can help us understand data that are hard to believe or seem contradictory. A situation that appears to be contradictory is known as a *paradox*. In Problem Set F, you will consider one paradox and discover why it can actually be true.

Problem Set F

Evan's younger twin sisters, Hannah and Elana, attend a small elementary school. There are only two fifth-grade classrooms in the school; Hannah is in the Green Room and Elana is in the Blue Room. The twins told Evan the following things:

- The average height of the boys in the Blue Room is greater than that of the boys in the Green Room.

- The average height of the girls in the Blue Room is greater than that of the girls in the Green Room.

- The combined average height of everyone in the Blue Room is *less* than the combined average height of everyone in the Green Room!

Evan was skeptical. "If *both* the boys *and* the girls in the Blue Room are taller than their Green Room classmates," he reasoned, "it seems the Blue Room should, in general, have taller students."

The twins insisted this wasn't the case. Evan decided to try to create a model population for which this odd situation could have occurred. Then at least he would know it was possible.

He reasoned that the paradox couldn't occur unless the two classes had different ratios of girls to boys. To create an extreme example, Evan distributed 30 girls very unevenly—27 in one room, 3 in the other—and 30 boys in the opposite way.

Green Room

G G G G G G G
B G B G
G G G G G G
G G G G G G
G G G B G G G

Blue Room

B B B B B B B
B G B B
B B B B B B
B B B B B G
G B B B B B B

By adjusting the average heights of the girls and the boys, Evan tried to see whether it was possible to match the situation at the school. He started by assigning heights to the boys and girls in the Green Room. Then he needed to create a Blue Room class that had a lower average overall height than the Green Room class, even though both the boys and the girls were taller than those in the Green Room.

Below are the tables Evan began. Try to complete the tables in a way that meets the conditions of the problem.

Green Room

	Girls	Boys	Total
Students	27	3	30
Average Height (in.)	56	54	

Blue Room

	Girls	Boys	Total
Students	3	27	30
Average Height (in.)			

Share & Summarize

In Problem Set E, you used a model to predict how the percentage of left-handers might change from one generation to the next.

1. Consider trying to predict this change by working with percentages rather than with a model population. What advantage does working with the model have?

2. When you worked with the model, you were asked to make several assumptions.

   a. List the assumptions. Decide whether each assumption is reasonable for a real population, like that of the United States.

   b. How do you think it might change your prediction if the assumptions you listed are not true?

3. **Challenge** Think about the paradox in Problem Set F. Can you see the key to this paradox? That is, can you see why it sounds impossible but actually isn't?

On Your Own Exercises

1. **Physical Science** The table shows the distance a toy car has traveled from its starting point at various times. The car is moving in a straight line.

   a. Make a graph of the data.

   b. Describe the car's motion.

   c. When is the break point? That is, at what time does the car's motion change?

   d. Write an equation for the distance d the car has traveled, in feet, from time $t = 0$ seconds to the break point.

   e. Does the motion after the break point seem to be quadratic or exponential? Explain why you think so.

   f. Use a graphing calculator to find an equation for the distance d the car has traveled for times after the break point.

Time (s)	Distance (ft)
0	0
1	2
2	4
3	6
4	8
5	11
6	16
7	23
8	32
9	43

2. **Physical Science** Crystal once heard that liquids cool faster when they are hotter. A friend said that meant you should use hot water—not cold—to make ice cubes quickly. This didn't make much sense to Crystal, so she decided to conduct an experiment to see just how hot water cools.

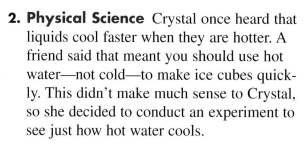

   She boiled some water and put a thermometer in the liquid. When the mercury stopped rising, she recorded the temperature and then put the water in her freezer. Every 2 minutes she checked the temperature, recording her results in a table.

   a. Plot the data from her table.

   b. What kind of relationship does there appear to be between these variables?

   c. Consider the time it would take water at 50°C to cool to 15°C in Crystal's freezer. Then consider the time it would take water to cool from 100°C to 15°C. Do you think her friend was right—that she should use hot water to make ice cubes? If not, what does it mean to say that hot water cools faster? Support your answer.

   d. Use a graphing calculator to find an equation relating the water temperature T to the time t after Crystal put the water in the freezer.

Time (min)	Temp (°C)
0	120
2	97
4	79
6	64
8	52
10	42
12	34
14	27
16	22
18	18
20	15

impactmath.com/self_check_quiz

3. **Economics** David designs and makes jewelry boxes to sell at crafts fairs. At a special week-long fair, he introduced a new design. He priced the boxes at $30 each, but they weren't as popular as he had hoped: he didn't sell any the first day. The next day he lowered the price to $20 and sold 10 boxes. He thought he might do even better if he lowered the price to $10; he sold 20 at that price.

a. Find David's *revenue* (the amount of money he received) for these boxes on each of the first three days of the fair.

b. Assume that the number of boxes sold is related to their price by a linear relationship. The revenue is then the price p times a linear expression involving the price, such as $a - bp$. What type of relationship is there between the price and the revenue?

c. Suppose David wants to take in the most money possible and that he has enough jewelry boxes to fill whatever demand there is for them. What price should he try for the rest of the fair? Explain your answer. (Hint: Consider the symmetry in graphs of the type of relationship you answered for Part b, and use the revenues for the first three days.)

4. **Economics** Aysha makes handblown glass ornaments. She found that, for one style of ornament, the amount of money she received for selling them depended on the price she asked. She created this graph to estimate her revenue (money received) on a single day for any price.

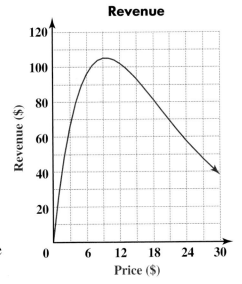

a. Approximately what price gives the greatest revenue for a single day? What is the corresponding revenue?

b. What price or prices would give a revenue of about $89?

c. Aysha considers more than just the amount of money she receives when she prices her ornaments. She also considers the cost of creating each ornament, both in the time she spends and the cost of the materials.

If Aysha had to choose between the two prices you found in Part b, which she should choose? Explain why.

5. Suppose Evita and Tariq were overly optimistic about the cost of parts in their Re-cycle business of Problem Set D. Suppose they will really spend $2,000 the first year, and in each subsequent year they will spend $400 over the previous year. If you include $500 per year for advertising, what is the new profit each year? Over the 5 years?

6. Consider Evita and Tariq's Re-cycle business in Problem Set D. Make appropriate changes in their assumptions so that the business has a negative net profit over the 5-year period. Defend each of your changes by explaining why it is plausible.

7. Life Science Medical tests sometimes give incorrect results. Suppose that one disease occurs in 4% of the population and that a test that screens for this disease misdiagnoses a healthy person 5% of the time and an ill person 10% of the time.

a. Suppose you take the test and the result is positive, suggesting that you have the disease. What do you estimate is the probability that you really have it?

Now consider a model population of 1,000 people, with 4% of those people afflicted by the disease.

b. How many people in the model have the disease? How many of these are expected to be reported by the test as *not* having the disease—that is, how many are expected to incorrectly test negative?

c. How many people in the model do not have the disease? How many of these will be expected to incorrectly test positive?

d. Copy and complete the table using your results from Parts b and c.

	Population	Tests Positive	Tests Negative
Has Disease			
Doesn't Have Disease			
Total	1,000		

e. What is the probability that a person who tests positive actually has the disease?

f. What is the probability that a person who tests negative doesn't have the disease?

8. Social Studies The table gives population data for the United States from 1900 through 2000.

a. Here is a plot of these data. What kind of relationship—for example, linear, quadratic, or exponential—do you think best describes these data?

U.S. Population

Year	Population
1900	76,212,168
1910	92,228,496
1920	106,021,537
1930	123,202,624
1940	132,164,569
1950	151,325,798
1960	179,323,175
1970	203,302,031
1980	226,542,203
1990	248,709,873
2000	281,421,906

Source: *World Almanac and Book of Facts 1999.* Copyright © 2003 World Almanac Education Group, Inc.

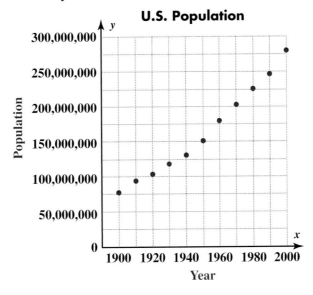

b. The graph is repeated twice below. On one, an exponential curve of best fit, about $y = 0.0022(1.0129^x)$, is included. On the other, a line of best fit, about $y = 2,019,000x - 3,772,000,000$, is included. Identify which is the line and which is the exponential curve.

i.
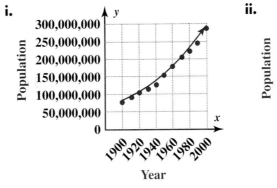

ii.
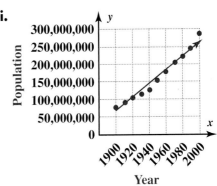

c. Which of the two equations appears to fit the data better? Does this agree with your answer to Part a?

d. Use the equation that fits better to estimate the population of the United States in the years 1890 and 2010.

e. The actual population of the United States in 1890 was 62,979,766. How does this compare to your estimate in Part d?

9. Challenge For Parts a–c, choose the graph below that you think will best fit the data.

a. burning rate for light to heavy fabric

Density of Fabric (g/m²)	50	100	150	200	250	300	350	400
Speed of Burning (cm/s)	15.4	7.9	5.2	3.9	3.1	2.6		2.0

b. a braking car on dry concrete

Speed of Car (mph)	20	25	30	35	40	50	60	70
Stopping Distance (ft)	16	25	36		64	100	144	196

c. regular polygon angle

interior angle

Sides of the Polygon	3	4	5	10	15	20	25	30
Interior Angle (degrees)	60	90	108	144	156	162		

Quadratic Functions

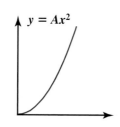

$y = Ax^2$

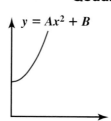

$y = Ax^2 + B$

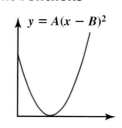

$y = A(x - B)^2$

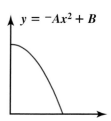

$y = {}^-Ax^2 + B$

Exponential Functions

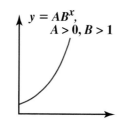

$y = AB^x,\ A > 0, B > 1$

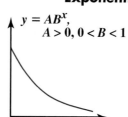

$y = AB^x,\ A > 0, 0 < B < 1$

Reciprocal Functions

$y = \dfrac{A}{x}$

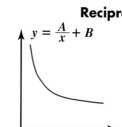

$y = \dfrac{A}{x} + B$

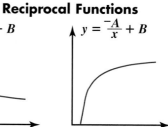

$y = \dfrac{-A}{x} + B$

d. Choose *one* table above. Use the general equation given with the graph to help find a specific equation relating the two variables.

e. Use your equation to find the missing value in your chosen table.

10. Economics The manager of a small fast-food company was reviewing the restaurant's profits. The company had been operating only 5 years. The first year saw no profit, but profits had been positive each year since. Profits had increased at first and then decreased some when the company was preparing for an expansion. The current year's sales suggested the profits that year would be only $5,000. However, the manager projected major profit increases, predicting that the company's profit would reach $35,000 in 2 years.

a. Assuming the current year is Year 0, which of these graphs best represents the information above?

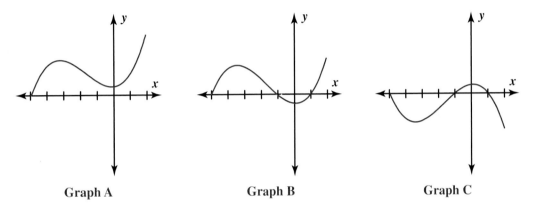

Graph A   Graph B   Graph C

b. Match each function to a graph from Part a.

i. $P(x) = 1,000(x + 5)(1 - x^2)$

ii. $P(x) = 1,000(x + 5)(x^2 + 1)$

iii. $P(x) = 1,000(x + 5)(x^2 - 1)$

c. Use the function that matches the graph you chose in Part a to check that the profit in the current year (that is, Year 0) is $5,000. Calculate the profit from 4 years ago and the profit in 2 years.

d. Choose one of the other functions, and assume it is the profit function for a rival fast-food company. Write a paragraph like the one at the beginning of this exercise, describing the company's profit over the last 5 years and the predictions for the next few years. Use your imagination to include some reasons for changes!

In your
own
words

In what ways can mathematical models be helpful to someone who is starting a business? What advice would you give about using such models?

For the situations described in Exercises 11–14, answer Parts a–d.

a. What two variable quantities are related in the situation?

b. What assumptions were made about how the quantities varied?

c. What type of function—linear, quadratic, exponential, or reciprocal—would best model the situation if the assumptions were correct?

d. Do you find the assumptions reasonable in the context? Why?

11. Héctor can run 100 meters in 12 seconds, so it will take him about 2 minutes to run 1 kilometer.

12. Lydia and her brother go to the movies every week, and together it costs $15 to get in. If their sister were to go with them, it would cost $22.50 to get in.

13. An 8-inch-diameter pizza will feed one hungry student quite well, so a 16-inch pizza should feed four students.

14. Ben's grandmother put $1,000 in a savings account. By the end of a year, the amount had increased to $1,050, so by the end of the second year it will have increased to $1,100.

15. Life Science A test for one fairly common medical condition has false-positive results of 30% (meaning 30% of healthy persons are wrongly reported to have the condition) and false-negative results of 10% (10% of people with the condition are wrongly reported to be healthy). The disease is present in about 10% of the population.

a. Create a sample population that represents the results of testing for the condition, and display the results in a table like the one below.

	Population	Tests Positive	Tests Negative
Doesn't Have Condition			
Has Condition			
Total			

b. What is the probability that someone who tests negative doesn't have the condition?

c. What is the probability that someone who tests positive actually has the condition?

A new test has been created for the condition. The designers have found that the new test gives false-positive results 10% of the time and false-negative results 30% of the time.

d. Using your population size from Part a, create a table that displays the results of testing for the condition using the new test.

e. What is the probability that someone who tests negative with the new test doesn't have the condition?

f. What is the probability that someone who tests positive with the new test actually has the condition?

g. Compare your answers to Parts b and e, and compare your answers to Parts c and f. Which test do you think is better, the original test (with a relatively high percentage of false positives) or the new test (with a high percentage of false negatives)? Explain.

Mixed Review

Expand each expression.

16. $\frac{4}{7}\left(\frac{1}{2}t + 12\right)$

17. $x(4 - 13x)$

18. $0.2(72v - 3)$

Find the value of n in each equation.

19. $3.582 \times 10^n = 3{,}582{,}000$

20. $n \times 10^7 = 34{,}001$

21. $82.882 \times 10^3 = n$

22. $28.1 \times \frac{1}{10^3} = n$

23. Graph this inequality on a number line, and give three values that satisfy it.

$$^-3 \le y < 7$$

24. In how many ways can the letters on the sign be ordered?

Find an equation for each line described.

25. passing through the point $(3,\ ^-14)$ and parallel to the line $y = 16x - 2$

26. passing through the points $(^-8, 9)$ and $(^-1, 3)$

27. with slope 0 and passing through the point $(2, 0.5)$

28. Technology Computers use a *binary* number system, a system that has 0 and 1 as its only digits. Each digit (0 or 1) is called a *bit*.

Computers translate everything—including letters—into series of 0s and 1s. Using only one bit, there are two possible series: 0 and 1. Using two bits, there are four possible series: 00, 01, 10, and 11. Because of the use of binary numbers, powers of 2 show up in many ways when you analyze how computers work.

a. To be able to distinguish 26 lower-case letters and 26 upper-case letters, a computer needs at least 52 different series. What is the least number of bits that will give at least 52 series? How many series will that many bits give?

b. In actuality, a single letter or symbol, called a *character,* is identified using a *byte,* which is 8 bits. (Notice that 8 is a power of 2.) How many different characters can be identified using a single byte?

c. A *kilobyte* is not 1,000 bytes, as you might think. It's actually 1,024 bytes, because 1,024 is a power of 2. What power of 2 is 1,024?

d. How many *bits* (not bytes) are in 1 kilobyte? Express your answer as a power of 2.

Chapter Summary

In this chapter, you worked with various kinds of displays for organizing and analyzing data. You reorganized tables to make it easier to see trends in the variables. In some cases, you made calculations using the given data in order to produce the data you really needed.

You also worked with models of different types. You fit curves to data, using the shape of the data points or assumptions about the situation to select the type of curve to use. Sometimes, though, an algebraic equation wasn't helpful, so you used tables and *model populations* to analyze situations and predict outcomes.

M A T E R I A L S

graphing calculator

Strategies and Applications

The questions in this section will help you review and apply the important ideas and strategies developed in this chapter.

Analyzing data presented in tables

The table on page 653 gives 2003 statistics for all the Major League baseball teams: total number of runs, percentage of games won, and earned run average. The earned run average (ERA) gives the average number of runs the *opposing* teams earned per inning. The teams all played approximately the same number of games.

1. Evita and Marcus were arguing about which league had better teams, the American League or the National League.

   a. Use the table to compare the number of runs each team scored over the year. Based on just this one statistic, which league would you think has better teams? Support your answer.

   b. How could you reorganize the table to make it easier to answer Part a?

2. Evita and Marcus want to compare the teams in each region rather than in each league. How could they reorganize the table to make this task easier?

impactmath.com/chapter_test

2003 Major League Baseball Statistics

Team	Total Runs	Games Won (%)	ERA	League	Region
Los Angeles	574	52.5	3.16	National	West
Detroit	591	26.5	5.30	American	Central
New York Mets	642	41.0	4.48	National	East
San Diego	678	39.5	4.87	National	West
Cincinnati	694	42.6	5.09	National	Central
Cleveland	699	42.0	4.21	American	Central
Montreal	711	51.2	4.01	National	East
Milwaukee	714	42.0	5.02	National	Central
Tampa Bay	715	38.9	4.93	American	East
Arizona	717	51.9	3.84	National	West
Chicago Cubs	724	54.3	3.83	National	Central
Anaheim	736	47.5	4.28	American	West
Baltimore	743	43.6	4.76	American	East
Florida	751	56.2	4.04	National	East
Pittsburgh	753	46.3	4.64	National	Central
San Francisco	755	62.1	3.73	National	West
Oakland	768	59.3	3.63	American	West
Philadelphia	791	53.1	4.04	National	East
Chicago White Sox	791	53.1	4.17	American	Central
Seattle	795	57.4	3.76	American	West
Minnesota	801	55.6	4.41	American	Central
Houston	805	53.7	3.86	National	Central
Texas	826	43.8	5.67	American	West
Kansas City	836	51.2	5.05	American	Central
Colorado	853	45.7	5.20	National	West
St.Louis	876	52.5	4.60	National	Central
New York Yankees	877	62.0	4.02	American	East
Toronto	894	53.1	4.69	American	East
Atlanta	907	62.3	4.10	National	East
Boston	961	58.6	4.48	American	East

Source: MLB.com

Using visual displays to identify trends

3. The baseball statistics from page 653 are plotted below.

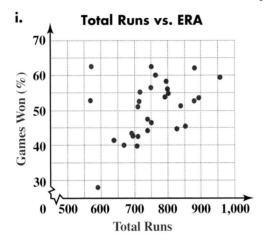

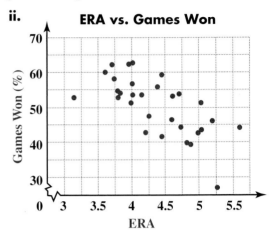

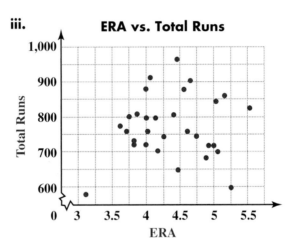

a. Which plot or plots do you think show some connection between the given variables, even a weak one? Which do you think show no connection at all? Explain.

b. Which plot do you think shows the strongest connection between the variables?

4. On the opposite page are two displays of the United States.

a. Describe the pattern in the first map, which categorizes the states by amount of farmland, in millions of acres. Using that map, which states do you think may be the most important farming states?

b. Describe the pattern in the second map, which categorizes the states by the portion of their land used for farming. Using that map, which do you think may be the most important farming states?

c. Which map do you think gives a better sense of how important farming is to each state? Explain.

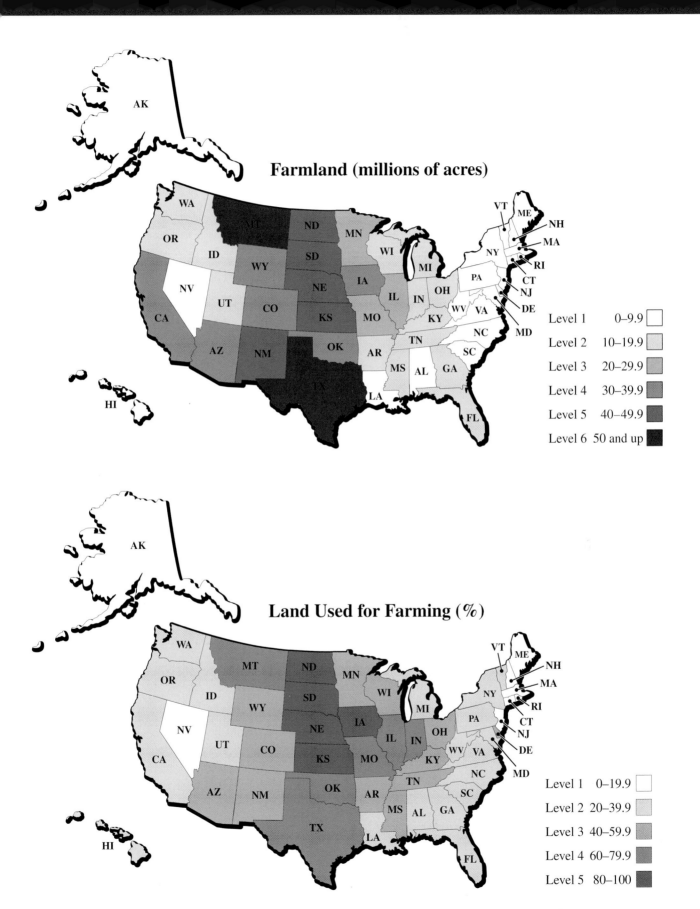

Farmland (millions of acres)

Level 1   0–9.9
Level 2   10–19.9
Level 3   20–29.9
Level 4   30–39.9
Level 5   40–49.9
Level 6   50 and up

Land Used for Farming (%)

Level 1   0–19.9
Level 2   20–39.9
Level 3   40–59.9
Level 4   60–79.9
Level 5   80–100

Creating and using models

5. Have your ears ever "popped" as you changed elevation, perhaps when driving up or down a mountain or landing or taking off in an airplane? The popping is caused by changes in *air pressure*. As your altitude or elevation increases, the pressure exerted by the air on your ears lessens.

A *barometer* is a device that measures air pressure. Mercury in the barometer rises and falls depending on the air pressure. The table shows the average barometer reading of air pressure, in inches of mercury, at various altitudes.

Altitude (ft)	Average Air Pressure (in. of mercury)
0 (sea level)	29.92
5,000	24.90
10,000	20.58
20,000	13.76
30,000	8.90
40,000	5.56
50,000	3.44
60,000	2.14
70,000	1.32
80,000	0.82
90,000	0.51
100,000	0.33

Source: *New York Public Library Science Desk Reference.* New York: Macmillan, 1995.

a. Plot the (*altitude, average air pressure*) data on your calculator. What kind of relationship appears to exist between the two variables?

b. Use your calculator to find an equation that models the relationship between these two variables.

c. The highest mountain in the United States is Mount McKinley in Alaska, at 20,320 ft. Use your model to estimate the average air pressure at the top of Mount McKinley.

d. The highest altitude attained by an airplane in a horizontal flight was 85,068.997 ft, by U.S. Air Force captain Robert C. Helt on July 28, 1976. The highest altitude ever attained by an airplane was 123,523.58 ft, by Alexander Fedotov of the USSR on August 31, 1977. Use your model to estimate the average air pressures at these two altitudes.

6. Hernando, an accountant, is reviewing the Algora Corporation's taxes. The company owns stock in three other companies, giving Algora the following percentage ownership in each.

Company	Ownership Percentage
Binomi	20%
The Co-Efficiency Company	40%
Diagon Inc.	30%

In examining these other companies, Hernando found that they each own stock in Algora! Binomi owns 15%, Co-Efficiency owns 25%, and Diagon owns 10% of Algora.

a. Create a model by supposing that there are 1,000 shares, or equal-sized pieces, of the Algora Corporation. Binomi owns 15% of them, or 150 shares. How many of the 1,000 shares of Algora are owned by Co-Efficiency? By Diagon?

b. Since Algora owns 20% of Binomi, you might consider 20% of Binomi's 150 shares of Algora to be owned by Algora *through* Binomi. That is, Algora owns 30 shares through Binomi. Find the number of shares Algora owns through Co-Efficiency and through Diagon.

c. Together the three other companies own 50% of Algora. Suppose Algora itself still owns the remaining 50% of the 1,000 shares. How many shares does Algora own in total, including the shares owned through other companies?

d. What percentage of Algora is owned by the company itself?

Demonstrating Skills

7. The table lists the federal minimum hourly wage from 1978 to 2003. Suppose a person works 40 hours each week at minimum wage, for 50 weeks each year. Create a new table giving the person's annual (yearly) income for each rate.

Federal Minimum Hourly Wage, 1978–2003

Effective Date	Minimum Wage
Jan 1, 1978	$2.65
Jan 1, 1979	2.90
Jan 1, 1980	3.10
Jan 1, 1981	3.35
Apr 1, 1990	3.80
Apr 1, 1991	4.25
Oct 1, 1996	4.75
Sep 1, 1997	5.15

8. Match each graph with the type of relationship—linear, quadratic, exponential growth, exponential decay, or inverse variation—that best describes it. Use each type of relationship once.

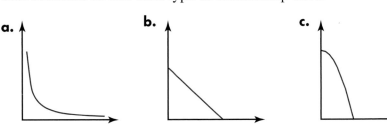

a.   b.   c.

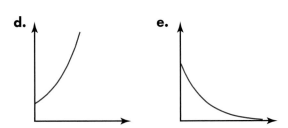

d.   e.

APPENDIX

Trigonometry

Trigonometric Ratios

There are many ways to look at linear relationships. In Chapter 1 you looked at direct variation and other linear relationships in words, tables, graphs, and equations. You also looked at the slopes of lines on graphs and in equations. Let's look at lines from a different perspective.

Remember

In a right triangle, the longest side is called the *hypotenuse*. The other two sides are called *legs*.

Think & Discuss

Look at the right triangle in this drawing. What are the coordinates of the end points of the hypotenuse?

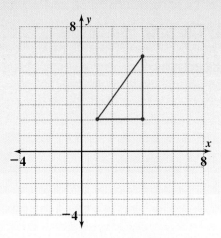

How would you find the slope of the line connecting these points?

In this lesson, you will learn a way to find the slope of a line by thinking of that line as the hypotenuse of a right triangle. You will be able to find the slope by using the angles in the triangle even if you don't know the coordinates of the endpoints of the hypotenuse. But first, you will review some of your work with similar triangles.

Investigation Special Ratios

You can determine the height of a tree if you know the length of its shadow and the height and the length of a shadow of some other object, such as a ruler. To do this you can use the length of the tree's shadow, the height of the ruler, and the length of the ruler's shadow to set up a proportion.

$$\frac{\text{ruler } (R)}{\text{ruler shadow } (S_R)} = \frac{\text{tree } (T)}{\text{tree shadow } (S_T)}$$

When using proportions like these, you only need to know three of the measurements to determine the fourth measurement.

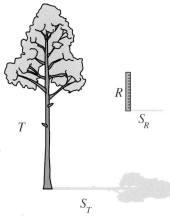

But some of these measurements are not really important by themselves. For example, you could use rulers of different lengths casting their different length shadows, and you *still* could figure out the height of the tree. Can you explain why? If the *actual length* of the ruler is not really important, then what is?

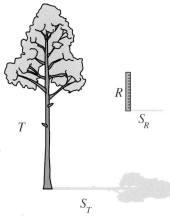

MATERIALS

protractors

rulers

Explore

Each person in your group should do the following:

• Draw a right triangle with one angle that measures 40°. Make sure your triangle is not congruent to anyone else's in your group.

• Measure the length of each leg of your right triangle.

• Make a table like the one below and record the lengths of the legs of your triangle. Express the ratio of the shorter leg to the longer leg as $\frac{S}{L}$. It is not necessary to reduce this fraction. Finally, convert that fraction to a decimal.

Shorter leg length (S)	Longer leg length (L)	Ratio $\frac{S}{L}$ (fraction)	Ratio $\frac{S}{L}$ (decimal)

• Collect the measurements and ratios from the other members of your group and record them in your table so you have all the information.

• Compare the ratios from your group's measurements. What do you notice about the ratios?

• Compare the ratios of the measurements of your entire class. Is the same thing still true?

• Why do you think this is happening? Why *should* this be true?

Problem Set **A**

1. Consider all the triangles that you and your classmates drew for the Explore activity. Are the triangles all similar? Explain.

2. Imagine you have another right triangle with a 40° angle.

   a. What is the ratio, to the nearest hundredth, of its shorter leg to its longer leg?

   b. Suppose the longer leg is 14 inches. How long is the shorter leg?

3. Given the two similar triangles shown, fill in the missing parts of each proportion.

   a. $\frac{c}{b} = \frac{?}{?}$

   b. $\frac{b}{e} = \frac{?}{?}$ and $\frac{?}{?}$

   c. $\frac{f}{d} = \frac{?}{?}$

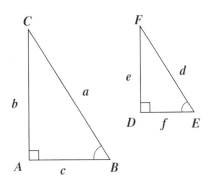

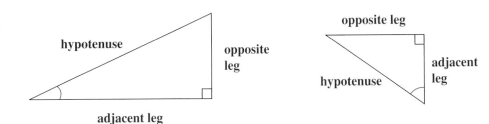

In each of the triangles above, one of the acute angles is marked. One of the legs is **opposite** this angle, or across from it. The other leg is **adjacent**, or next to, the acute angle.

As you work on the next Problem Set, pay attention to the relationship of an angle to the legs that are opposite and adjacent to it.

Problem Set **B**

For each problem, draw a right triangle such that the ratio of the lengths of the leg opposite ∠A and the leg adjacent to ∠A is the same as the ratio given. As you are drawing, be sure that you have labeled one of the *acute* angles ∠A. Measure and draw the legs of the triangle to fit the ratio. Then find the measure of ∠A.

1. $\frac{\text{leg opposite } \angle A}{\text{leg adjacent } \angle A} = \frac{1}{4}$ or 0.25

2. $\frac{\text{leg opposite } \angle A}{\text{leg adjacent } \angle A} = \frac{1}{2}$ or 0.5

3. $\dfrac{\text{leg opposite } \angle A}{\text{leg adjacent } \angle A} = \dfrac{4}{3}$ or 1.3333 . . .

4. $\dfrac{\text{leg opposite } \angle A}{\text{leg adjacent } \angle A} = \dfrac{5}{2}$ or 2.5

5. What happens to the size of $\angle A$ as the ratio increases? Why is this happening?

Share & Summarize

1. Look back at the triangles and ratios you found in the Explore section at the beginning of Investigation 1. The last question in the Explore asked you to explain why you got the results you did. What do you think now?

2. How large and how small can the ratios of the legs of a right triangle be?

3. In Problem Set A, Problem 2, you knew only the length of one leg of a right triangle. Still, you were able to figure out the length of the other leg using a ratio. What feature of the triangle does that ratio depend on? Its size? Its angles? Its position?

Investigation ▶2 The Tangent of an Angle

VOCABULARY
tangent

The special ratio you looked at in the previous investigation has a name. The **tangent** of an acute angle $\angle A$ is the ratio of the length of the leg opposite $\angle A$ to the length of the leg adjacent to $\angle A$ in a right triangle. The abbreviation for tangent is *tan*.

$$\tan \angle A = \dfrac{\text{leg opposite } \angle A}{\text{leg adjacent } \angle A}$$

You can use the tangent to calculate the length of a side in a right triangle that you cannot measure or, if the triangle is graphed on the coordinate plane, whose endpoints you do not know.

Think & Discuss

Suppose you have several right triangles, each with a 20° angle. Why would the tangent ratio of the 20° angle be constant in all of these triangles, no matter the triangle's size? How can you be sure of this?

Since a 20° angle has the same tangent no matter what right triangle it is in, you can refer to the tangent of 20°, or tan 20°, in general terms. How could you find tan 20° using only a ruler and a protractor?

You have used the tangent ratio to calculate the lengths of legs in a right triangle. In the following problems, you will use this technique in a variety of situations.

MATERIALS

graphing calculator

EXAMPLE

Find the length of the leg opposite the 28° angle in the triangle.

$$\tan 28° = \frac{\text{opposite leg}}{\text{adjacent leg}}$$

$$\tan 28° = \frac{? \text{ mi}}{110 \text{ mi}}$$

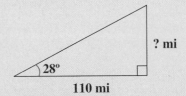

To find the tan 28°, you could certainly draw a right triangle that has an acute angle measuring 28°, measure the lengths of its legs, and calculate the ratio $\frac{\text{opposite leg}}{\text{adjacent leg}}$. Fortunately, you can also find tan 28° using a scientific or graphing calculator. On a graphing calculator, press [MODE] [↓] [↓] [→] [ENTER] to be sure that your calculator is in degree mode. Then press [TAN] 28. The calculator should display 0.531709432 which is about 0.5317.

Let x = the unknown length in miles.

Now finish the equation.

$$\tan 28° = \frac{x \text{ mi}}{110 \text{ mi}}$$

$$0.5317 = \frac{x \text{ mi}}{110 \text{ mi}}$$

$$0.5317 \cdot 110 \text{ mi} = x \text{ mi}$$
$$58.487 = x$$

So the length of the leg opposite the 28° angle is about 58.49 miles.

Problem Set C will give you some practice using the tangent to find lengths of sides in right triangles.

Problem Set C

Find the value of x in each of the triangles shown below. Round your answers to the nearest hundredth.

1.

5.2 ft
40°
x ft

2.

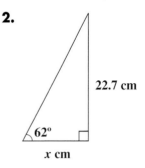

22.7 cm
62°
x cm

3.

72°
3 m
18°
x m

4.

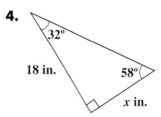

32°
18 in.
58°
x in.

5. For safety and comfort, a landing airplane must approach a runway at an angle of about 3°. At what altitude is a plane that it is still 1,000 horizontal feet away from touching down on a runway? Round to the nearest foot.

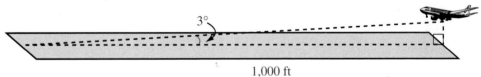

3°

1,000 ft

6. Caity is trying to estimate the height of a flagpole. Using her protractor, she estimates the angle measure from her eye up to the top of the flagpole to be about 34°.

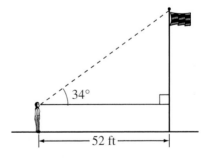

34°

52 ft

a. Caity knows she is 52 feet away from the flagpole. She uses those numbers to calculate the opposite leg of the right triangle shown in the drawing above to the nearest foot. What answer does she get?

b. She submits that number for her answer, but her teacher tells her she is not correct. What did she forget to do?

7. Two hikers come to a deep gorge with steep, almost vertical, cliffs. Since they have no way to cross it, they need to find out how much rope they will need to climb down the side of the gorge to the path at the bottom. They decide to approximate the height of the cliff by taking some measurements.

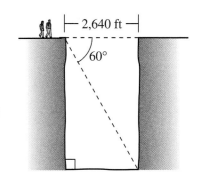

They are able to estimate that the angle from where they are standing to the bottom of the opposite cliff is about 60°. They know from their map that the width of the gorge is about $\frac{1}{2}$ mile, or about 2,640 feet. About how deep is the gorge?

8. Erica has to clean out the leaves from the gutters on her house. She knows that to be safe she should put her ladder at an angle of about 75° to the ground. She wants to figure out how far from the house to put the ladder and how long a ladder she will need to reach the roof.

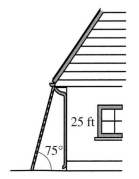

a. If her house is about 25 feet to the roof, how far out from the house should she put the foot of her ladder?

b. How long of a ladder will she need to the nearest foot?

9. Challenge The students in Mr. Claus's math class are trying to determine the height of a tree outside the school.

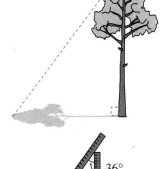

Luisa places her ruler vertical to the ground as shown. She then places one end of a yardstick at the end of the ruler's shadow and leans the yardstick on the ruler so that it forms a right triangle.

She then measures the angle formed by the ruler and the yardstick and finds that it is about 36°. Using this information and the length of the tree's shadow, Luisa is able to determine the height of the tree.

a. Explain how Luisa determines the tree's height.

b. If she knows that the length of the tree's shadow is 42.5 feet, what is the height of the tree?

Investigation ▶ 3 Connections Between Tangent, Slope, Rate, and Proportionality

In this investigation, you will see how rate (or speed) may be connected to the slope of a graph and the tangent of an angle.

Think & Discuss

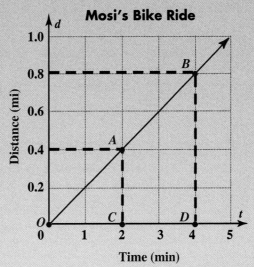

Mosi rode his bike from home to school at a constant speed. The graph above shows the distance Mosi traveled. Two minutes after he started, he was 0.4 mile away from his home. Two minutes later, he was 0.8 mile away. Point A has coordinates $(2, 0.4)$; point B coordinates $(4, 0.8)$; point O coordinates $(0, 0)$; point C coordinates $(2, 0)$; and point D coordinates $(4, 0)$.

What is Mosi's speed in miles per minute?

What is the slope of $\overrightarrow{OA}$? of $\overrightarrow{OB}$?

What is the tangent of $\angle AOC$?

What do you notice about the speed of Mosi's bike, the slope of the line, and the tangent of the angle?

Problem Set D

Look back at the graph on page 667.

1. Compare the ways you calculated the slope of the $\overrightarrow{OA}$ and the tangent of $\angle AOC$. How are they similar or different?

2. Express in words the meaning of the ratio $\frac{AC}{OC}$. Express in words the meaning of the ratio $\frac{BD}{OD}$.

3. Compare the ratios $\frac{AC}{OC}$ and $\frac{BD}{OD}$. Can you make a proportion with these ratios? Why or why not?

4. Is the relationship on the graph one of direct variation (is it proportional)? Why or why not?

5. Are $\triangle AOC$ and $\triangle BOD$ similar? Explain.

Problem Set E

Sancha started to bike with Mosi and rode beside him for two minutes. Then she became tired and went more slowly. Four minutes after they started, she was 0.6 mile away from home. Here is a graph showing both Mosi's and Sancha's progress.

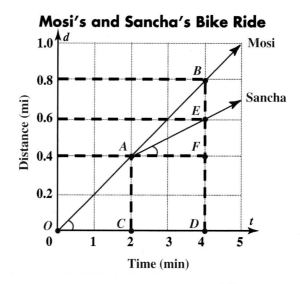

1. Express in words the meaning of the ratio $\frac{AC}{OC}$. Express in words the meaning of the ratio $\frac{ED}{OD}$.

2. Compare the ratios $\frac{AC}{OC}$ and $\frac{ED}{OD}$ for Sancha. Can you make a proportion with these ratios? Why or why not?

3. Is the graphed relationship for Sancha one of direct variation (is it proportional)? Why or why not?

4. a. What is Sancha's speed during the first two minutes of her ride?

b. What is her speed during the second two minutes?

5. What is the tangent of $\angle AOC$? What is the tangent of $\angle EAF$?

6. Does the graph of Sancha's bike ride have a constant slope over the first 4 minutes? Explain.

7. What should be true about the speed in order for the ratios $\frac{AC}{OC}$ and $\frac{ED}{OD}$ to be equal?

Problem Set F

Hannah works as a Homework Helper for an elementary school. She is paid a fixed amount of $20 for being available each weekend, even if she is not needed. If she is needed, she is paid an additional $5 for each hour she works. The graph below shows how Hannah's pay on weekends depends on the number of hours worked.

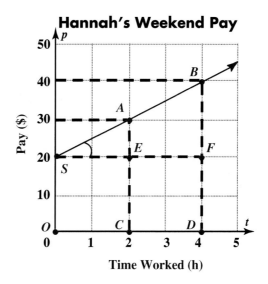

1. What is Hannah's hourly rate of pay, not including the fixed amount?

2. What is the slope of the line on the graph?

3. Compare the hourly rate and the slope. How are they similar or different?

4. Express in words the meaning of the ratio of $\frac{AC}{OC}$. Express in words the meaning of the ratio $\frac{BD}{OD}$.

5. Compare the ratios $\frac{AC}{OC}$ and $\frac{BD}{OD}$ for Hannah. Can you form a proportion with these ratios? Why or why not?

6. Is the relationship on the graph a proportional one? Why or why not?

Share & Summarize

1. Consider the hourly rate of pay and the slope of Hannah's graph. Which angle's tangent is equal to these?

2. This graph shows a line making an angle of 40° with the *x*-axis.

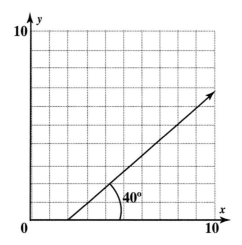

   a. Find the slope of the line using the tangent ratio.

   b. Find two points on the line and carefully measure their coordinates. Then find the slope of the line from your two points.

   c. Compare your answers to Part a and Part b.

Investigation 4 ▶ The Sine and Cosine

In the previous investigations, you worked with the tangent ratio by constructing right triangles. You selected one of the acute angles, measured the legs opposite and adjacent to that angle, and then formed the ratio of those legs.

Think & Discuss

Given right triangle *ABC* with sides *a*, *b*, and *c*, how many different ratios of sides can you form? State all of them.

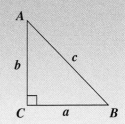

Looking at $\angle A$, you know that the ratio $\frac{a}{b}$ has been given the special name of tangent. In fact, there are names for each of the other ratios as well. These ratios are the focus of a branch of mathematics called *trigonometry*. In this investigation, you will work with two more of these ratios.

Problem Set G

1. Draw three different right triangles that have one angle measuring 65°.

   a. Measure the length of each leg opposite the 65° angle and each hypotenuse.

   b. Make a table with these headings and record each measure. Then calculate the ratio of the opposite leg to the hypotenuse, and write it as a fraction and as a decimal.

Opposite leg	Hypotenuse	Ratio (fraction)	Ratio (decimal)

   c. What is true about this ratio?

   d. Will this ratio be the same for any right triangle with a 65° angle? Explain your answer.

The name given to this ratio is **sine**. The abbreviation for sine is *sin*.

$$\sin \angle A = \frac{\text{leg opposite } \angle A}{\text{hypotenuse}}$$

2. How can you use your triangles to approximate sin 25°? What is your approximation?

Problem Set H

1. Use the three right triangles that you drew for Problem 1 in Problem Set G.

   a. Measure the length of each leg adjacent to the 65° angle.

   b. Make a table with these headings and record each measure. Then calculate the ratio of the adjacent leg to the hypotenuse, and write it as a fraction and as a decimal.

Adjacent leg	Hypotenuse	Ratio (fraction)	Ratio (decimal)

   c. What is true about this ratio?

   d. Will this ratio be the same for any right triangle with a 65° angle? Explain your answer.

The name given to this ratio is **cosine.** The abbreviation for cosine is *cos.*

$$\cos \angle A = \frac{\text{leg adjacent } \angle A}{\text{hypotenuse}}$$

2. Ian said he found a right triangle where the cosine of one of the angles was $\frac{5}{4}$. Marcus said that was impossible. Explain who was correct and why.

3. Challenge How small can the sine and cosine of an angle be in a right triangle? How large can they be?

In Problem Set I, you will use your calculator to solve problems involving lengths in right triangles. You should use the sine and cosine keys, ⌷SIN⌷ and ⌷COS⌷ respectively, in the same way you used the tangent key to find the sines and cosines of angles.

MATERIALS
graphing calculator

EXAMPLE

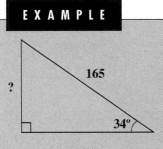

How can you find the length of the leg opposite the 34° angle?

$$\sin 34° = \frac{\text{opposite leg}}{\text{hypotenuse}}$$

$$\sin 34° = \frac{?}{165}$$

Press ⌷SIN⌷ 34. The calculator should display 0.559192903 which is about 0.5592.

$$0.5592 = \frac{x}{165}$$

$$0.5592(165) = x$$

$$92.3 = x$$

MATERIALS
graphing calculator

Problem Set I

1. Find x and y.

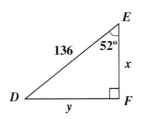

2. A car is traveling on a road that has an incline of 4°. If the car travels 1,500 feet on this road, how much has its elevation changed to the nearest foot?

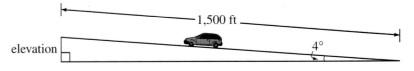

3. Colin found the cosine of one of the angles in a right triangle was 0.2. Find the lengths of all sides in at least two right triangles for which this could be true. How many right triangles are there in all?

4. Challenge Anson claimed that no matter what right triangle he examined, the sine of one of the acute angles was always equal to the cosine of the other acute angle. In other words, in the triangle shown, $\sin A = \cos B$. Do you think Anson is correct? Why or why not?

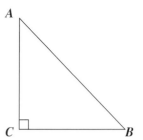

Share & Summarize

1. You are now familiar with the right triangle ratios tangent, sine, and cosine. What is the relationship of these ratios to similar triangles?

2. Which of the ratios—sine, cosine, or tangent—can be greater than 1? Explain your answer.

3. David found sin 30° on his calculator to be 0.5. Explain in words what this means in terms of the legs of a right triangle with a 30° angle.

1. Consider these similar right triangles.

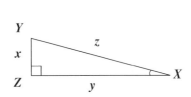

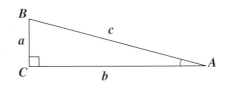

a. $\frac{x}{y} = \frac{1}{5}$, and $b = 2.4$ inches. Find the length of a.

b. What is $\frac{1}{5}$ in decimal form?

c. What is $\frac{a}{b}$ in decimal form?

2. Consider these similar right triangles.

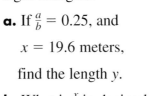 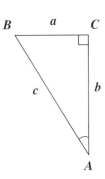

a. If $\frac{a}{b} = 0.25$, and $x = 19.6$ meters, find the length y.

b. What is $\frac{x}{y}$ in decimal form?

Each of the following ratios compares the leg opposite $\angle A$ to the leg adjacent to $\angle A$. Copy the triangle and label $\angle A$.

3. $\frac{4}{3}$

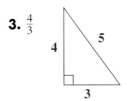

4. $\frac{24}{7}$

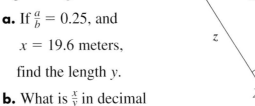

5. $\frac{5}{12}$

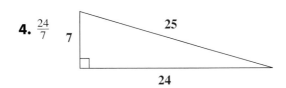

In Exercises 6–9, find the numerical value of sine, cosine, and tangent of the angle marked A.

6.

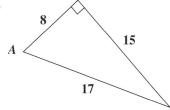

7.

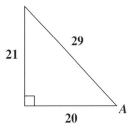

8.

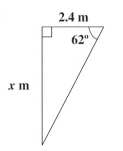

9.

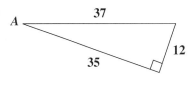

In Exercises 10–19, use a calculator to find the value of x and/or y to the nearest hundredth.

10.

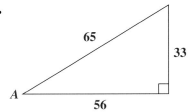

11.

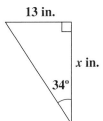

12.

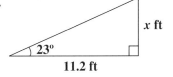

13.

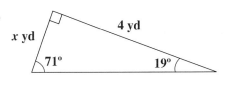

14.

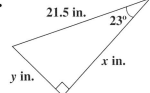

15.

16.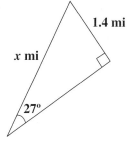

17.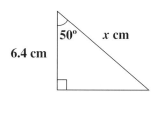

18.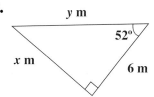

19.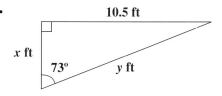

20. At the Ice Cream Club, each ice cream cone costs $1.50.

   a. Graph this relationship with the number of cones from 0 and 15 on the *x*-axis and the cost of cones from $0 to $25 on the *y*-axis.

   b. What is the slope of the line you graphed?

   c. Plot the points (0, 0), (5, 0), and (10, 0) and label them *O, C,* and *D* respectively. Draw a vertical line through point *C* and label the point where this line meets your graph *A*. Draw a vertical line through point *D* and label the point where this line meets your graph *B*.

   d. What is the tangent of ∠*BOD*? Compare it with the slope of the line.

   e. Express in words the meaning of ratio $\frac{AC}{OC}$. Express in words the meaning of ratio $\frac{BD}{OD}$.

   f. Compare ratios $\frac{AC}{OC}$ and $\frac{BD}{OD}$. Can you make a proportion with these ratios? Why or why not?

   g. If you express these two ratios as one number, what does this number represent?

   h. Is the relationship graphed a direct variation? Why or why not?

   i. Are △*AOC* and △*BOD* similar? Why or why not?

21. Look at the graph below.

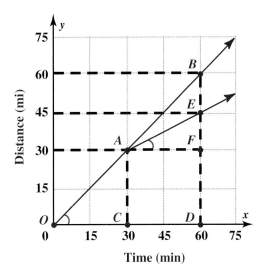

a. Write a problem about a real-life situation that could be represented by this graph.

b. Express in words the meaning of the ratio $\frac{AC}{OC}$. Express in words the meaning of the ratio $\frac{ED}{OD}$.

c. Compare the ratios $\frac{AC}{OC}$ and $\frac{ED}{OD}$. Can you make a proportion with these ratios? Why or why not?

d. Does the graph of *OAE* represent a relationship with direct variation (a proportional relationship)? Why or why not?

e. What is the slope of $\overleftrightarrow{OA}$? What is the slope of the $\overleftrightarrow{AE}$?

f. What is the tangent of $\angle AOC$? What is the tangent of $\angle EAF$?

g. Is the slope of *OAE* constant?

h. If this were a graph of speed, what should be true about the speed in order for the ratios $\frac{AC}{OC}$ and $\frac{ED}{OD}$ to be equal?

Connect & Extend

22. Airports need to know the height of the cloud cover in order to control air traffic landing and taking off. One method the airport employees can use is to beam a light straight up, and measure the angle from a distant location up to the light on the clouds. One night, an airport employee stood 250 feet from the light and measured this angle as 70°. How high was the cloud cover that night?

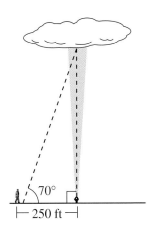

Explain the connection between the slope of a line and the tangent of an angle.

23. A hot air balloon is hovering directly over a 200-foot tower for power lines. The pilot in the balloon is concerned that he has enough clearance to be safe and radios his assistant on the ground. The assistant quickly moves to a place 100 feet from the center of the tower and estimates the measure of the angle up to the bottom of the balloon's basket as 65°. About how much clearance does the balloon have?

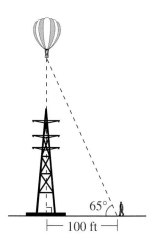

24. As a sailboat passes a river buoy, the crew drops the anchor. Once the anchor hits the bottom of the river, the boat continues past the buoy another 50 feet before the line goes taut and the boat comes to a stop. The captain estimates that the measure of the angle between the anchor line and a straight line down to the bottom of the river is 48°. About how deep is the river at that point?

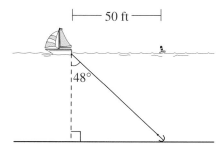

25. Jessica kneels a horizontal distance of 20 feet from the kite she is flying. Holding the end of the kite string to the ground, she estimates the angle the string makes with the ground to be 35°.

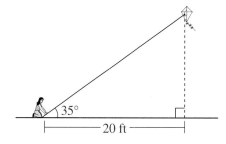

Estimate the length of kite string Jessica has let out to the nearest tenth of a foot.

26. The tailgate of a moving truck is 2 feet above the ground. The incline of the ramp used for loading the truck is 15° as shown. Find the length of the ramp to the nearest tenth of a foot.

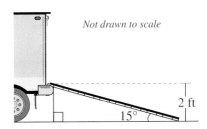

Not drawn to scale

2 ft

15°

27. Look at the drawing of the staircase.

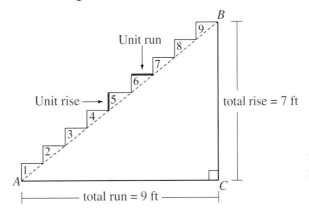

B

Unit run

total rise = 7 ft

Unit rise

A

C

total run = 9 ft

a. Calculate the unit rise and the unit run for the steps of the staircase. Round to the nearest tenth of an inch if necessary.

b. Building manuals have specifications for building staircases. In one such manual it states that the unit rise on a staircase needs to be in the range of 6–8 inches, and that the total of the unit rise plus the unit run needs to be between 17 and 18 inches. Is the staircase pictured above built to code? Why or why not?

c. What is the ratio of the unit rise to the unit run?

d. How is this ratio related to the slope of the staircase, or the slope of $\overleftrightarrow{AB}$?

e. Calculate the ratio of total rise to total run of the staircase. How is this ratio related to the slope of the staircase or $\overleftrightarrow{AB}$?

f. Compare the ratios of the unit rise to the unit run and the total rise to total run. Can you form a proportion with these ratios? Why or why not?

g. How are these ratios related to the tangent of $\angle BAC$? How would you find the tangent of this angle?

h. Draw a graph which shows how the height you reach depends on the number of stairs you climb.

i. What is the slope of the line on the graph? How is the slope related to the slope of the staircase, to the ratio of unit rise to unit run, and to the tangent of $\angle ABC$?

28. Derrick drew a rhombus and then drew in the diagonals of that rhombus as shown below.

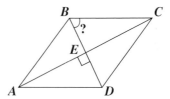

a. When Derrick measured the diagonals, he found that diagonal AC was twice as long diagonal BD. If the length of diagonal BD is x, express the length of diagonal AC in terms of x

b. Derrick is trying to estimate the measure of $\angle CBE$. He knows that the diagonals of a rhombus bisect each other and are perpendicular. He also has the following table of trigonometric values at his disposal.

Trigonometric Table

Degrees	Sine	Cosine	Tangent
0	0.0000	1.0000	0.0000
5	0.0872	0.9962	0.0875
10	0.1736	0.9848	0.1763
15	0.2588	0.9659	0.2679
20	0.3420	0.9397	0.3640
25	0.4226	0.9063	0.4663
30	0.5000	0.8660	0.5774
35	0.5736	0.8192	0.7002
40	0.6428	0.7660	0.8391
45	0.7071	0.7071	1.0000
50	0.7660	0.6428	1.1918
55	0.8192	0.5736	1.4281
60	0.8660	0.5000	1.7321
65	0.9063	0.4226	2.1445
70	0.9397	0.3420	2.7475
75	0.9659	0.2588	3.7321
80	0.9848	0.1736	5.6713
85	0.9962	0.0872	11.4301
90	1.0000	0.0000	....

Explain how Derrick could estimate the measure of $\angle CBE$.

c. If the measure of diagonal BD is 6 inches, estimate the measure of $\angle CBE$.

GLOSSARY / GLOSARIO

English	Español

adjacent leg In a right triangle with acute angle A, the leg next to $\angle A$.

cateto adyacente En un triángulo rectángulo con ángulo agudo A, cateto el lado del $\angle A$.

algebraic expression A combination of numbers, variables, and operation symbols that gives a number when all variables are replaced by numbers. Examples of *algebraic expressions* are $3n + 2$, $x^2 - 2x + 7$, and $p + q$. [page 356]

expresión algebraica Combinación de números, variables y símbolos de operaciones que resulta en un número cuando todas las variables se reemplazan con números. Ejemplos de *expresiones algebraicas* son $3n + 2$, $x^2 - 2x + 7$ y $p + q$.

binomial The sum or difference of two unlike terms. For example, $x + 7$, $x^2 - 3$, and $a + c$ are *binomials*. [page 373]

binomio La suma o diferencia de dos términos no semejantes. Por ejemplo: $x + 7$, $x^2 - 3$ y $a + c$ son *binomios*.

coefficient The numeric multiplier in an algebraic term. For example, in the expression $3x^2 - 2x + 7$, 3 is the coefficient of x^2, and $^{-}2$ is the coefficient of x. [page 31]

coeficiente El multiplicador numérico en un término algebraico. Por ejemplo: en la expresión $3x^2 - 2x + 7$, 3 es el coeficiente de x^2 y $^{-}2$ es el coeficiente de x.

congruent Having the same size and shape. [page 294]

congruente Que tiene el mismo tamaño y la misma forma.

conjecture An educated guess or generalization that you haven't yet proved correct. [page 127]

conjetura Suposición o generalización informada que aun no se ha probado como correcta.

cosine In a right triangle with acute angle A, the cosine of $\angle A = \dfrac{\text{leg adjacent } \angle A}{\text{hypotenuse}}$.

cosena En un triangulo rectangulo con angulo agudo A, el coseno del $\angle A = \dfrac{\text{cateto adyacente al } \angle A}{\text{hypotenusa}}$.

cubic equation An equation that can be written in the form $y = ax^3 + bx^2 + cx + d$, where $a \neq 0$. For example, $y = 2x^3$, $y = 0.5x^3 - x^2 + 4$, and $y = x^3 - x$ are *cubic equations*. [page 93]

ecuación cúbica Ecuación que se puede escribir en la forma $y = ax^3 + bx^2 + cx + d$, *donde* $a \neq 0$. Por ejemplo: $y = 2x^3$, $y = 0.5x^3 - x^2 + 4$ y $y = x^3 - x$ son *ecuaciones cúbicas*.

decay factor In a situation in which a quantity decays exponentially, the *decay factor* is the number by which the quantity is repeatedly multiplied. A *decay factor* is always greater than 0 and less than 1. For example, if the value of a computer decreases by 15% per year, then its value each year is 0.85 times its value the previous year. In this case, the *decay factor* is 0.85. [page 176]

factor de desintegración En una situación en que una cantidad se desintegra exponencialmente, el *factor de desintegración* es el número por el cual se multiplica la cantidad repetidas veces. El *factor de descomposición* siempre es mayor que 0 y menor que 1. Por ejemplo: si el costo de una computadora disminuye en un 15% por año, entonces su valor cada año es 0.85 veces el valor del año anterior. En este caso, el *factor de descomposición* es 0.85.

dilation A transformation that creates a figure similar, but not necessarily congruent, to an original figure. [page 329]

dilación Transformación que crea una figura semejante, pero no necesariamente congruente, a una figura original.

direct variation A relationship in which two variables are directly proportional. The equation for a *direct variation* can be written in the form $y = mx$, where $m \neq 0$. The graph of a *direct variation* is a line through the origin $(0, 0)$. [page 7]

directly proportional Term used to describe a relationship between two variables in which, if the value of one variable is multiplied by a number, the value of the other variable is multiplied by the same number. For example, if Lara earns \$8 per hour, then the variable *hours worked* is *directly proportional* to the variable *dollars earned*. [page 7]

distributive property The *distributive property of multiplication over addition* states that for any numbers n, a, and b, $n(a + b) = na + nb$. The *distributive property of multiplication over subtraction* states that for any numbers n, a, and b, $n(a - b) = na - nb$. [page 358]

domain The set of allowable inputs to a function. For example, the *domain* of $f(x) = \sqrt{x}$ is all non-negative real numbers. The *domain* of $g(t) = \frac{1}{t - 3}$ is all real numbers except 3. [page 495]

elimination A method for solving a system of equations that involves possibly rewriting one or both equations and then adding or subtracting the equations to *eliminate* a variable. For example, you could solve the system $x + 2y = 9$, $3x + y = 7$ by multiplying both sides of the first equation by 3 and then subtracting the second equation from the result. [page 266]

equation A mathematical sentence stating that two quantities are equal. For example, the sentence $3 - 11 = {}^-4 + {}^-4$ and $x^2 - 4 = 0$ are *equations*. [page 226]

expanding Using the distributive property to multiply the factors in an algebraic expression. For example, you can *expand* $x(x + 3)$ to get $x^2 + 3x$. [page 359]

exponent A symbol written above and to the right of a quantity that tells how many times the quantity is multiplied by itself. For example, $t \cdot t \cdot t$ can be written as t^3. [page 146]

exponential decay A decreasing pattern of change in which a quantity is repeatedly multiplied by a number less than 1 and greater than 0. [page 175]

variación directa Relación en que dos variables son directamente proporcionales. La ecuación para una *variación directa* se puede escribir en la forma $y = mx$, donde $m \neq 0$. La gráfica de una *variación directa* es una recta a través del origen $(0, 0)$.

directamente proporcional Término que se usa para describir una relación entre dos variables en el cual, si el valor de una de las variables se multiplica por un número, el valor de la otra variable se multiplica por el mismo número. Por ejemplo: si Lara gana \$8 por hora, entonces la variable *horas trabajadas* es *directamente proporcional* a la variable *dólares ganados*.

propiedad distributiva La *propiedad distributiva de la multiplicación sobre la adición* establece que para todo número n, a y b, $n(a + b) = na + nb$. La *propiedad distributiva de la multiplicación sobre la sustracción* establece que para todo número n, a y b, $n(a - b) = na - nb$.

dominio El conjunto de entradas permitidas para una función. Por ejemplo: el *dominio* de $f(x) = \sqrt{x}$ son todos los números reales no negativos. El *dominio* de $g(t) = \frac{1}{t - 3}$ son todos los números reales excepto 3.

eliminación Método para resolver un sistema de ecuaciones que posiblemente involucra reescribir una o ambas ecuaciones y luego sumar o restar las ecuaciones para *eliminar* una de las variables. Por ejemplo: podrías resolver el sistema $x + 2y = 9$, $3x + y = 7$ al multiplicar ambos lados de la primera ecuación por 3 y luego restar la segunda ecuación del resultado.

ecuación Enunciado matemático que establece la igualdad de dos cantidades. Por ejemplo: los enunciados $3 - 11 = {}^-4 + {}^-4$ y $x^2 - 4 = 0$ son *ecuaciones*.

desarrollar Uso de la propiedad distributiva para multiplicar los factores en una expresión algebraica. Por ejemplo: puedes *desarrollar* $x(x + 3)$ para obtener $x^2 + 3x$.

exponente Símbolo que se escribe más arriba y a la derecha de una cantidad y el cual indica el número de veces que la cantidad se multiplica por sí misma. Por ejemplo: $t \cdot t \cdot t$ se puede escribir como t^3.

desintegración exponencial Patrón decreciente de cambio en que una cantidad se multiplica repetidamente por un número menor que 1 y mayor que 0.

exponential growth An increasing pattern of change in which a quantity is repeatedly multiplied by a number greater than 1. [page 169]

crecimiento exponencial Patrón creciente de cambio en que una cantidad se multiplica repetidamente por un número mayor que 1.

factoring Writing an algebraic expression as a product of factors. For example, $x^2 - x - 6$ can be *factored* to get $(x + 3)(x - 3)$. [page 443]

factorizar Escribir una expresión algebraica como el producto de factores. Por ejemplo: $x^2 - x - 6$ se puede *factorizar* para obtener $(x + 3)(x - 3)$.

function Term used to describe a relationship between an input variable and an output variable in which there is only one output for each input. [page 488]

función Término que se usa para describir la relación entre una variable de entrada y una variable de salida en que sólo hay una salida para cada entrada.

growth factor In a situation in which a quantity grows exponentially, the *growth factor* is the number by which the quantity is repeatedly multiplied. A *growth factor* is always greater than 1. For example, if a population grows by 3% every year, then the population each year is 1.03 times the population the previous year. In this case, the *growth factor* is 1.03. [page 169]

factor de crecimiento En una situación en la cual una cantidad crece exponencialmente, el *factor de crecimiento* es el número por el cual se multiplica la cantidad repetidamente. El *factor de crecimiento* es siempre mayor que 1. Por ejemplo: si una población crece un 3% cada año, entonces cada año la población es 1.03 veces la población del año previo. En este caso, el *factor de crecimiento* es 1.03.

hyperbola The graph of an inverse variation. [page 112]

hipérbola La gráfica de una variación inversa.

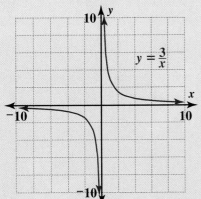

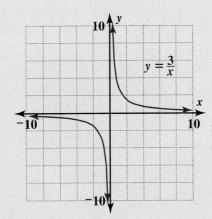

image The figure or point that results from a transformation. [page 292]

imagen Figura o punto que resulta de una transformación.

inequality A mathematical statement that uses one of the symbols $<$, $>$, $\leq$, $\geq$, or $\neq$ to compare quantities. Examples of inequalities are $n - 3 \leq 12$ and $9 - 2 > 1$. [page 226]

desigualdad Enunciado matemático que usa uno de los símbolos $<$, $>$, $\leq$, $\geq$ o $\neq$ para comparar cantidades. Ejemplos de desigualdades son $n - 3 \leq 12$ y $9 - 2 > 1$.

inverse variation A relationship in which two variables are inversely proportional. The equation for an *inverse variation* can be written in the form $xy = c$, or $y = \frac{c}{x}$, where c is a nonzero constant. The graph of an *inverse variation* is a hyperbola. [page 113]

variación inversa Relación en que dos variables son inversamente proporcionales. La ecuación de una *variación inversa* se puede escribir en la forma $xy = c$, o $y = \frac{c}{x}$, donde c es una constante no nula. La gráfica de una *variación inversa* es una hipérbola.

inversely proportional Term used to describe a relationship in which the product of two variables is a nonzero constant. If two variables are *inversely proportional,* then when the value of one variable is multiplied by a number, the value of the other variable is multiplied by the *reciprocal* of that number. For example, the time it takes to travel 50 miles is *inversely proportional* to the average speed traveled. [page 113]

inversamente proporcional Término que se usa para describir una relación en la cual el producto de dos variables es una variable no nula. Si dos variables son *inversamente proporcionales,* entonces cuando el valor de una de las variables se multiplica por un número, el valor de la otra variable se multiplica por el *recíproco* de ese número. Por ejemplo: el tiempo que toma viajar 50 millas es *inversamente proporcional* a la rapidez promedio viajada.

irrational numbers Numbers that cannot be written as ratios of two integers. In decimal form, *irrational numbers* are non-terminating and non-repeating. Examples of *irrational numbers* include π, $\sqrt{17}$, and $3\sqrt{2}$. [page 200]

números irracionales Números que no se pueden escribir como razones de dos enteros. En forma decimal, los *números irracionales* son decimales no terminales y no periódicos. Ejemplos de *números irracionales* incluyen π, $\sqrt{17}$ y $3\sqrt{2}$.

like terms In an algebraic expression, terms with the same variables raised to the same powers. For example, in the expression $x + 3 - 7x + 8x^2 - 2x^2 + 1$, $8x^2$ and $-2x^2$ are *like terms*, x and ^-7x are *like terms,* and 3 and 1 are *like terms.* [page 363]

términos semejantes En una expresión algebraica, los términos con las mismas variables elevadas a las mismas potencias. Por ejemplo: en la expresión $x + 3 - 7x + 8x^2 - 2x^2 + 1$, $8x^2$ y $-2x^2$ son *términos semejantes*, x y ^-7x son *términos semejantes* y 3 y 1 son *términos semejantes.*

line of reflection A *line* over which a figure is *reflected.* In the figure below, the blue K has been *reflected* over the *line of reflection l* to get the orange K. [page 292]

eje de reflexión Un *eje* sobre el cual se *refleja* una figura. En la siguiente figura, la K azul ha sido *reflejada* sobre el *eje de reflexión l* para obtener la K anaranjada.

line of symmetry A line that divides a figure into two mirror-image halves. [page 289]

eje de simetría Recta que divide una figura en dos mitades especulares. [pág. 289]

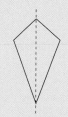

line symmetry See *reflection symmetry.*

simetría lineal Ver *simetría de reflexión.*

English	**Español**

linear relationship A relationship with a graph that is a straight line. Linear relationships are characterized by a constant rate of change—each time the value of one variable changes by a fixed amount, the value of the other variable changes by a fixed amount. The equation for a *linear relationship* can be written in the form $y = mx + b$, where m is the slope of the graph and b is its y-intercept. [page 4]

relación lineal Relación cuya gráfica es una recta. Las relaciones lineales se caracterizan por una tasa constante de cambio: cada vez que el valor de una de las variables cambia por una cantidad fija, el valor de la otra variable cambia por una cantidad fija. La ecuación de una *relación lineal* se puede escribir en la forma $y = mx + b$, donde m es la pendiente de la gráfica y b es su intersección y.

nth root An nth *root* of a number a is a number b, such that $b^n = a$. For example, -3 and 3 are *fourth roots* of 81 because $(-3)^4 = 81$ and $3^4 = 81$. [page 199]

enésima raíz La *enésima raíz* de un número a es un número b, tal que $b^n = a$. Por ejemplo: -3 y 3 son las *cuartas raíces* de 81 porque $(-3)^4 = 81$ y $3^4 = 81$.

opposite leg In a right triangle with a acute angle A, the leg across from $\angle A$.

cateto apuesto En un triangula rectángulo con ángulo aguda A, el cateto al otra lada del $\angle A$.

parabola The graph of a quadratic relationship. [page 71]

parábola La gráfica de una relación cuadrática.

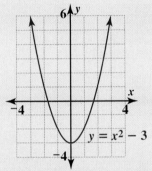

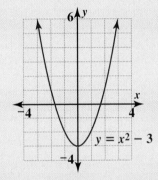

perpendicular bisector A line that intersects a segment at its midpoint and is perpendicular to the segment. [page 294]

mediatriz Recta que interseca un segmento en su punto medio y que es perpendicular al segmento.

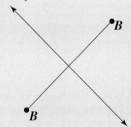

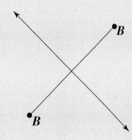

quadratic equation An equation that can be written in the form $y = ax^2 + bx + c$, where $a \neq 0$. For example, $y = x^2$, $y = 3x^2 - x + 4$, and $y = -2x^2 + 1$ are *quadratic equations*. [page 83]

ecuación cuadrática Ecuación que se puede escribir en la forma $y = ax^2 + bx + c$, donde $a \neq 0$. Por ejemplo: $y = x^2$, $y = 3x^2 - x + 4$ y $y = -2x^2 + 1$ son *ecuaciones cuadráticas*.

quadratic expression An expression that can be written in the form $ax^2 + bx + c$, where $a \neq 0$. For example, $x^2 - 4$, $x^2 + 2x + 0.5$, and $-3x^2 + 1$ are *quadratic expressions*. [page 83]

expresión cuadrática Expresión que se puede escribir en la forma $ax^2 + bx + c$, donde $a \neq 0$. Por ejemplo: $x^2 - 4$, $x^2 + 2x + 0.5$ y $-3x^2 + 1$ son *expresiones cuadráticas*.

English / Español

English	Español

radical sign A symbol $\sqrt{}$ used to indicate a root of a number. The symbol $\sqrt{}$ by itself indicates the positive square root. The symbol $\sqrt[n]{}$ indicates the nth root of a number. For example, $\sqrt{25} = 5$ and $\sqrt[3]{-64} = -4$. [page 191]

signo radical Símbolo $\sqrt{}$ que se usa para indicar la raíz de un número. El símbolo $\sqrt{}$ por sí sólo indica la raíz cuadrada positiva. El símbolo $\sqrt[n]{}$ indica la *ené*sima raíz de un número. Por ejemplo: $\sqrt{25} = 5$ y $\sqrt[3]{-64} = -4$.

range All the possible output values for a function. For example, the *range* of $h(x) = x^2 + 2$ is all real numbers greater than or equal to 2. The *range* of $f(x) = -\sqrt{x}$ is all real numbers less than or equal to 0. [page 518]

rango Todos los posibles valores de salida de una función. Por ejemplo: el *rango* de $h(x) = x^2 + 2$ son todos los números reales mayores que o iguales a 2. El *rango* de $f(x) = -\sqrt{x}$ son todos los números reales menores que o iguales a 0.

rational numbers Numbers that can be written as ratios of two integers. In decimal form, *rational numbers* are terminating or repeating. For example, 5, -0.274, and $0.\overline{3}$ are *rational numbers*. [page 200]

números racionales Números que se pueden escribir como razones de dos enteros. En forma decimal, los *números racionales* son números terminales o periódicos. Por ejemplo: 5, -0.274 y $0.\overline{3}$ son *números racionales*.

real numbers The set of rational and irrational numbers. All the numbers that can be located on the number line. [page 200]

números reales El conjunto de números racionales e irracionales. Todos los números que se pueden ubicar en la recta numérica.

reciprocal relationship See *inverse variation*. [page 115]

relación recíproca Ver *variación inversa*.

reflection over a line A transformation that matches each point on a figure to its mirror image over a line. In the figure below the blue curve has been *reflected over the line* to create the orange curve. [page 292]

reflexión sobre una recta Transformación en que cada punto de una figura corresponde con su imagen especular sobre una recta. En la siguiente figura, la curva azul se *reflejó sobre la recta* para crear la curva anaranjada.

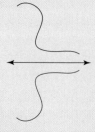 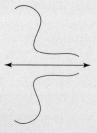

reflection symmetry A figure has *reflection symmetry* (or line symmetry) if you can draw a line that divides the figure into two mirror-image halves. The figures below have reflection symmetry. [page 289]

simetría de reflexión Una figura tiene *simetría de reflexión* (simetría lineal) si puedes dibujar una recta que divida la figura en dos mitades especulares. Las siguientes figuras tienen simetría de reflexión.

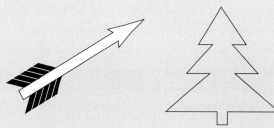 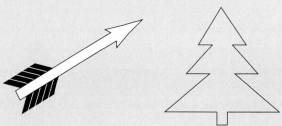

English	Español

rotation A transformation in which a figure is turned about a point. A positive angle of rotation indicates a counterclockwise rotation; a negative angle of rotation indicates a clockwise rotation. For example, the orange triangle at the right was created by *rotating* the blue triangle 90° about point *P*. [page 305]

90° rotation about point *P*

rotación Transformación en que se le da vuelta a una figura alrededor de un punto. Un ángulo de rotación positivo indica una rotación en dirección contraria a las manecillas del reloj; un ángulo de rotación negativo indica una rotación en la dirección de las manecillas del reloj. Por ejemplo: el triángulo anaranjado a la derecha se creó al *rotar* el triángulo azul 90° alrededor del punto *P*.

rotación de 90° alrededor del punto *P*

rotation symmetry A figure has *rotation symmetry* if you can rotate it about a centerpoint *without turning it all the way around,* and find a place where it looks exactly as it did in its original position. The figures below have *rotation symmetry*. [page 303]

simetría de rotación Una figura tiene *simetría de rotación* si se puede rotar alrededor de un punto central *sin voltearla completamente a su alrededor* y se puede hallar un lugar en donde se ve exactamente como se veía en su posición original. Las siguientes figuras tienen *simetría de rotación*.

sample space In a probability situation, the set of all possible outcomes. For example, when two coins are tossed, the sample space consists of head/head, head/tail, tail/head, tail/tail. [page 547]

espacio muestral En una situación de probabilidad, el conjunto de todos los resultados posibles. Por ejemplo: al lanzar dos monedas al aire, el espacio muestral consta de cara/cara, cara/escudo, escudo/cara, escudo/escudo.

scale drawing A drawing that is similar to some original figure. [page 330]

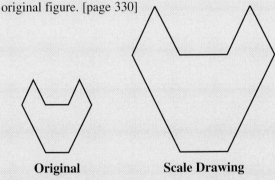

Original   **Scale Drawing**

dibujo a escala Dibujo que es semejante a alguna figura original.

 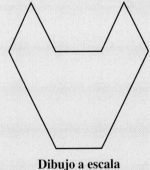

Original   **Dibujo a escala**

scale factor The ratio between corresponding side lengths of similar figures. There are two *scale factors* associated with every pair of non-congruent similar figures. For example, in the figures above, the *scale factor* from the small figure to the large figure is 2, and the *scale factor* from the large figure to the small figure is $\frac{1}{2}$. [page 330]

factor de escala La razón entre las longitudes de lados correspondientes de figuras semejantes. Hay dos *factores de escala* asociados con cada par de figuras semejantes no congruentes. Por ejemplo: en las figuras anteriores, el *factor de escala* de la figura pequeña a la figura grande es 2 y el *factor de escala* de la figura grande a la figura pequeña es $\frac{1}{2}$.

English	Español

scientific notation The method of writing a number in which the number is expressed as the product of a power of 10 and a number greater than or equal to 1 but less than 10. For example, 5,000,000 written in *scientific notation* is 5×10^6. [page 148]

notación científica Método de escribir un número en la cual el número se expresa como el producto de una potencia de 10 y un número mayor que o igual a 1, pero menor que 10. Por ejemplo: 5,000,000 escrito en *notación científica* es 5×10^6.

similar Having the same shape. [page 294]

semejante Que tiene la misma forma.

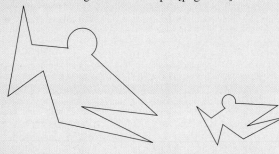

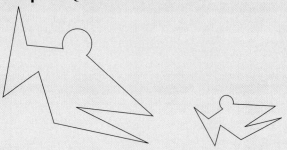

sine In a right triangle with acute angle A, the sine of $\angle A = \frac{\text{leg opposite } \angle A}{\text{hypotenuse}}$.

seno En un triángulo rectángulo con ángulo agudo A, el seno del $\angle A = \frac{\text{cateto opuesto al } \angle A}{\text{hipotenusa}}$.

slope The ratio $\left(\frac{\text{rise}}{\text{run}}\right)$ used to describe the steepness of a non-vertical line. Given the two points on a non-vertical line, you can calculate the *slope* by dividing the difference in the y coordinates by the difference in the x coordinates. (Be sure to subtract the x and y coordinates in the same order.) If a linear equation is written in the form $y = mx + b$, the value m is the *slope* of its graph. For example, the graph of $y = {}^-x - 2$, has *slope* $^-1$. [page 27]

pendiente La razón $\left(\frac{\text{altura}}{\text{carrera}}\right)$ que se usa para describir el grado de inclinación de una recta no vertical. Dados los dos puntos de una recta no vertical, puedes calcular la *pendiente* al dividir la diferencia de las coordenadas y entre la diferencia de las coordenadas x. (Asegúrate de restar las coordenadas x y y en el mismo orden.) Si una ecuación lineal se escribe en la forma $y = mx + b$, el valor m es la *pendiente* de su gráfica. Por ejemplo: la *pendiente* de la gráfica de $y = {}^-x - 2$ es $^-1$.

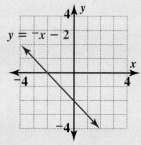

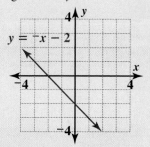

slope-intercept form The form $y = mx + b$ of a linear equation. The graph of an equation of this form has slope m and y-intercept b. For example, the graph of $y = {}^-x - 2$ (shown above) has slope $^-1$ and y-intercept $^-2$. [page 49]

forma pendiente-intersección La forma $y = mx + b$ de una ecuación lineal. La gráfica de una ecuación de esta forma tiene pendiente m e intersección y b. Por ejemplo: la gráfica de $y = {}^-x - 2$ (que se muestra arriba) tiene pendiente $^-1$ e intersección y igual a $^-2$.

square root A *square root* of a number a is a number b, such that $b^2 = a$. For example, $^-9$ and 9 are *square roots* of 81 because $(^-9)^2 = 81$ and $9^2 = 81$. [page 190]

raíz cuadrada La *raíz cuadrada* de un número a es un número b, tal que $b^2 = a$. Por ejemplo: $^-9$ y 9 son *raíces cuadradas* de 81 porque $(^-9)^2 = 81$ y $9^2 = 81$.

English

substitution A method for solving a system of equations that involves using one of the equations to write an expression for one variable in terms of the other variable, and then *substituting* that expression into the other equation. For example, you could solve the system $y = 2x + 1$, $3x + y = 11$ by first *substituting* $2x + 1$ for y in the second equation. [page 264]

system of equations A group of two or more equations with the same variables. [page 257]

term A part of an algebraic expression made up of numbers and/or variables multiplied together. For example, in the expression $5x - 7x^2 + 2$, the terms are $5x$, $^-7x^2$, and 2. [page 363]

transformation A way of creating a figure similar or congruent to an original figure. Reflections, rotations, translations, and dilations are four types of *transformations*. [page 288]

translation A transformation within a plane in which a figure is moved a specific distance in a specific direction. For example, the first figure below was *translated* 1 inch to the right to get the second figure. [page 313]

trigonometric ratios The ratios of the measures of two sides of a right triangle.

trinomial An expression with three unlike terms. For example, $b^2 + 10b + 25$ is a *trinomial*. [page 443]

variable A quantity that can change or vary, or an unknown quantity. [page 4]

vector A line segment with an arrowhead used to describe translations. The length of the *vector* tells how far to translate and the arrowhead gives the direction. [page 313]

Español

sustitución Método para resolver un sistema de ecuaciones y que involucra el uso de las ecuaciones para escribir una expresión para una de las variables en términos de la otra variable y luego *sustituir* esa expresión en la otra ecuación. Por ejemplo: para resolver el sistema $y = 2x + 1$, $3x + y = 11$ podrías primero *sustituir* la y en la segunda ecuación con $2x + 1$.

sistema de ecuaciones Grupo de dos o más ecuaciones con las mismas variables.

término Parte de una expresión algebraica compuesta de números y/o variables que se multiplican entre sí. Por ejemplo: en la expresión $5x - 7x^2 + 2$, los términos son $5x$, $^-7x^2$ y 2.

transformación Una manera de crear una figura semejante o congruente a una figura original. Las reflexiones, las rotaciones, las traslaciones y las dilataciones son cuatro tipos de *transformaciones*.

traslación Una transformación dentro de un plano en que la figura se mueve una distancia específica en una dirección dada. Por ejemplo: la primera de las figuras que siguen se *trasladó* 1 pulgada a la derecha para obtener la segunda figura.

rezones triganametricas Las rezones de las medidas de dos lados de un triangula rectángulo.

trinomio Expresión con tres términos no semejantes. Por ejemplo: $b^2 + 10b + 25$ es un *trinomio*.

variable Cantidad que cambia o varía o cantidad desconocida.

vector Segmento de recta con punta de flecha que se usa para describir traslaciones. La longitud del *vector* indica la cantidad que hay que trasladar y la punta de flecha indica la dirección.

English	**Español**

x-intercept The x-coordinate of a point at which a graph crosses the x-axis. The *x-intercepts* of the graph of $f(x) = x^2 - 4x$ shown below are 0 and 4. [page 522]

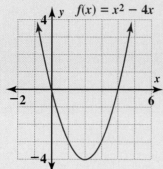

intersección x La coordenada x del punto donde la gráfica atraviesa el eje x. Las *intersecciones x* de la siguiente gráfica de $f(x) = x^2 - 4x$ son 0 y 4.

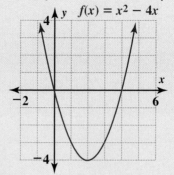

y-intercept The y-coordinate of a point at which a graph crosses the y-axis. The graph of a linear equation of the form $y = mx + b$, has y-intercept b. For example, the graph of $y = {}^-x - 2$, has y-intercept ${}^-2$. [page 31]

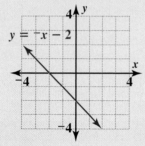

intersección y La coordenada y de un punto en el cual una gráfica atraviesa el eje y. La gráfica de una ecuación lineal de la forma $y = mx + b$, tiene intersección y de b. Por ejemplo: la gráfica de $y = {}^-x - 2$, tiene intersección y de ${}^-2$.

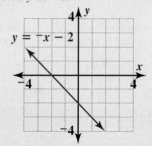

INDEX

PHOTO CREDITS